Totality

Totality

Eclipses of the Sun

THIRD EDITION

Mark Littmann
Fred Espenak
Ken Willcox

OXFORD
UNIVERSITY PRESS

Great Clarendon Street, Oxford OX2 6DP

Oxford University Press is a department of the University of Oxford.
It furthers the University's objective of excellence in research, scholarship,
and education by publishing worldwide in

Oxford New York

Auckland Cape Town Dar es Salaam Hong Kong Karachi
Kuala Lumpur Madrid Melbourne Mexico City Nairobi
New Delhi Shanghai Taipei Toronto

With offices in

Argentina Austria Brazil Chile Czech Republic France Greece
Guatemala Hungary Italy Japan Poland Portugal Singapore
South Korea Switzerland Thailand Turkey Ukraine Vietnam

Oxford is a registered trade mark of Oxford University Press
in the UK and in certain other countries

Published in the United States
by Oxford University Press Inc., New York

First published 1991
Second edition 1999
Third edition 2008

British Library Cataloguing in Publication Data

Data available

Library of Congress Cataloging in Publication Data

Data available

Typeset by Newgen Imaging Systems (P) Ltd., Chennai, India
Printed in Great Britain
on acid-free paper by
CPI Antony Rowe, Chippenham, Wiltshire

ISBN 978–0–19–953209–4

1 3 5 7 9 10 8 6 4 2

From Mark Littmann

To Peggy, Beth, and Owen, with love.

And in loving memory of Muriel Stein Littmann and Lewis Littmann, M.D., my parents.

From Fred Espenak

To Patricia Totten Espenak, my wife and best friend. We first met in the shadow of the Moon, half a world away and we've been chasing eclipses together ever since.

To Valerie Anne Delos-Reyes, my granddaughter who might see her first total eclipse from her own backyard in 2017.

And in memory of my parents Fred and Asie Espenak, and my sister Nancy J. Davies.

Contents

Foreword

Drama, travel, spectacle, science, biography—it's difficult to have it all, but total solar eclipses do. And the new edition of *Totality: Eclipses of the Sun* by Mark Littmann, Fred Espenak, and the late Ken Willcox indeed has it all, told with lots of interesting description for all types of reader.

Some chapters are lyrical, describing to readers why total eclipses are so spectacular to watch, and how they will get hooked into an eclipse addiction. The authors describe eclipses of the past and provide vignettes, personal accounts, from actual observers to add artistic verisimilitude.

Other chapters are historical, describing how from Stonehenge to the Maya to a seminal nineteenth-century expedition to India to the present, people have tried to keep track of eclipses and to predict where and when they will occur. The Indian eclipse of 1868 even led to the discovery of light that came from a hitherto unknown gas on the Sun, which was therefore named "helium"; it took decades before helium was found to also exist on Earth.

Still other chapters are biographical, describing the humans who figured out what the Sun is like by observing it at eclipses. These descriptions bring in many side matters, such as the early history of photography in the mid-nineteenth century, as insensitive daguerreotypes gave way to photographic plates sufficiently sensitive to provide proof that the corona seen around the Moon at eclipses was, in fact, part of the Sun.

In the middle of the book, the authors describe the science of the Sun and how energy gets from its core out through the corona that we see at eclipses and onward to us on Earth. They quote a number of contemporary eclipse scientists, myself included, to show a wide variety of observations that are made at eclipses. Some don't require any equipment at all—well, perhaps a watch to count time—while others take sophisticated cameras.

Most eclipse watchers these days are ecotourists, willing and able to travel the world to sit or stand in the glory of a total eclipse for its brief minutes. Of course, not only is the glory of the eclipse's pair of diamond rings, its pearly white corona, and its fearful darkening in the middle of the day part of their memories, but also the trips themselves, with their pleasures and hassles.

The authors tell us what's next. A total eclipse in 2008 crosses from Greenland over the north polar regions and down through Siberia, Mongolia, and northern China. A longer total eclipse in 2009 crosses from India through the Himalayas into China, where totality nearly as long as is ever possible will be visible especially near Shanghai, before an even longer totality goes out to sea. Further, the total eclipse of 2010 may not be the longest or the one with the highest chance of clear weather, but its visibility from Easter Island adds an extra fillip to the uniqueness of this eclipse.

My own first encounter with a total solar eclipse has led to the activity that has dominated nearly fifty years of my life. As an undergraduate, the members of my freshman seminar on astronomy were taken up in an airplane to view a total eclipse by our professor, Donald Menzel, a veteran solar and eclipse expert. The view was so dazzling that at least three of us became professional astronomers with interests in the Sun: I became an eclipse astronomer, one seminar mate became director of the National Solar Observatory, and another became a professional astronomer whose expertise backs up some television shows about astronomy that you may have seen. That seminar of 10 freshmen resulted in a remarkably high ratio of students continuing on in astronomy—and solar astronomy at that.

I have now seen 46 solar eclipses, and I can never choose among them when asked, as I often am, which was the best. When accused of liking the fact that the eclipses have taken me on or over all the Earth's continents, I can honestly point out that I saw my first eclipse, on that fateful airplane, within only a few miles of school, and thus it was clearly the science rather than the foreign travel that attracted me.

I've done my best to bring students along with me to see and study eclipses, and I am glad to have this opportunity to try to explain to you why we all find watching eclipses so appealing. Though my main interest is scientific, there is a wide range of categories into which eclipse observers fit. In the on-line Solar Eclipse Mailing List, a dozen or two messages a day circulate from people (almost all nonscientists) all over the world. Some people are planning future trips; others are figuring out how many special types of eclipses may occur over, say, the next few thousand years; others may be discussing which cameras are most useful or which computers can best control observing to leave the individuals free to look up at the sky rather than down at a viewfinder during totality; and others may discuss how to use today's personal computers, PC or Mac, to combine eclipse photographs and enhance detail, making them even more dazzling than previously. *Totality: Eclipses of the Sun* has something for each of these individuals, and I am sure that you will find that it has something for you.

One of the authors, Fred Espenak, is personally responsible for most of the eclipse map plotting that is used worldwide, and an ample selection of his maps is conveniently included in the book. We learn not only where to go in 2008, 2009, and 2010, but also where the next eclipse in the United States, 2017, will be visible.

Many eclipse tourists (I hate to call them eclipse chasers, since they get there first and wait for the eclipse to come overhead at more than a thousand of miles per hour, making it all but unchaseable) know years in advance where they will go on certain dates. *Totality: Eclipses of the Sun*, with not only its clear expositions but also its beautiful and well chosen set of photographs, may convert you to membership in this merry band. Pretty as the photographs are, no page or video screen can make it get a million times darker at midday, arousing some primal fear. You have to see and experience a total solar eclipse outdoors under an open sky to believe it.

Jay M. Pasachoff
Chair, International Astronomical Union Working Group on Eclipses
Field Memorial Professor of Astronomy
Williams College
Williamstown, Massachusetts

Acknowledgments

Enormous thanks to astronomers Charles Lindsey, Jay Pasachoff, and Larry Marschall for their generous advice and help through three editions of *Totality: Eclipses of the Sun*. And special thanks also to astronomer Joseph Hollweg for sharing his expertise with us through the second and third editions.

Totality is an unusual and richer work because noted authorities contributed vignettes (sidebars) for the book. We are very grateful to them:

Jay Anderson
Lucian V. Del Priore, M.D.
Drake Deming
Stephen J. Edberg
Patricia Totten Espenak
Alan D. Fiala
Carl Littmann
Laurence Marschall
Jay M. Pasachoff

There are 180 illustrations in this book. The magnificent eclipse photographs were graciously contributed by:

Greg Babcock
Michael F. Barrett
Cees Bassa
Fred Bruenjes
Greg Buchwald
Arne Danielsen
Kris Delcourte
Friedhelm Dorst
Miloslav Druckmüller
Alan Dyer
Patricia Totten Espenak
Daniel Fischer
Jacques Guertin
Johnny Horne
Odd Høydalsvik
Dave Kodama
Jeffrey Kuhn
Pete Lawrence
David Makepeace
Richard Nugent
Jay Pasachoff
Glenn Schneider
Robert Shambora
Geoff Sims
Olivier Staiger
Wolfgang Strickling
Tunç Tezel
Christian Viladrich
Alson Wong

The diagrams are primarily the work of solar astronomer Charles A. Lindsey, Will Fontanez and Tom Wallin of Cartographic Services at the University of Tennessee, and Fred Espenak.

Alan Clark, Drake Deming, David W. Dunham, Serge Koutchmy, Jeffrey Kuhn, Charles A. Lindsey, and Jay M. Pasachoff are some of the astronomers who continue to use eclipses to conduct research on the Sun. We are grateful to them for their advice and explanations for our chapter Modern Scientific Uses for Solar Eclipses.

A number of eclipse veterans—both professional and amateur astronomers—graciously offered their experiences and advice for our chapter on Observing a Total Eclipse. We thank them for their insight and eloquence:

Jay Anderson
John R. Beattie
Richard Berry
Dennis di Cicco
Stephen J. Edberg
Alan D. Fiala
Ruth S. Freitag
Joseph V. Hollweg
George Lovi
Frank Orrall
Jay M. Pasachoff
Leif J. Robinson
Virginia & Walter Roth
Roger W. Tuthill
Jack B. Zirker

Special additional thanks to John R. Beattie, who provided such a ringing moment-by-moment description of a total eclipse that it became the nucleus of our first chapter.

There are a host of other folks who helped us greatly in so many ways—astronomy, geology, archeoastronomy, history, mythology, translating, critiquing—in the course of three editions of *Totality*. Thanks so much:

Anthony F. Aveni
Eric Becklin
Kenneth Brecher
John Carper
Kevin Dieke
Patricia Totten Espenak
Andrew Fraknoi
Geoffrey K. Gay
Gerry Grimm
Robert S. Harrington
Karen Harvey
Don Hassler
Anne Hensley
Hugh Hudson
Kevin Krisciunas
David & Esther Littmann
Peggy Littmann
Eli Maor
Richard E. McCarron
John McNair
Larry November
Beatrice & Thomas Owens
Ann & Paul Rappoport
Gary Rottman
Sabatino Sofia
E. Myles Standish, Jr.

The index for both the second and third editions is the work of poet, writer, and master index maker Alexa Selph.

We are profoundly grateful to Iris Wiley, Executive Editor of the University of Hawaii Press, who first edited and published *Totality: Eclipses of the Sun* in 1991. We are indebted also to Joyce Berry,

Kim Torre-Tasso, Peter Grennan, and Michele LaForge at Oxford University Press for the handsome second edition of *Totality* they produced in 1999.

Our deep gratitude to Dr. Sonke Adlung, Oxford University Press's Senior Editor, Physical Sciences, for his interest in a new edition of *Totality* and for his enthusiasm, encouragement, and good ideas throughout the process. Our thanks too to the gracious and efficient Oxford University Press production team: Chloe Plummer, Kate Walker, Lynsey Livingston, and Lawrence Osborn.

Finally, Ken Willcox (1943–1999), we and so many others miss you—your good will, your infectious enthusiasm, and your knowledge. Your spirit is with us still and infuses the third edition of *Totality: Eclipses of the Sun*.

1
The Experience of Totality

> Some people see a partial eclipse and wonder why others talk so much about a total eclipse. Seeing a partial eclipse and saying that you have seen an eclipse is like standing outside an opera house and saying that you have seen the opera; in both cases, you have missed the main event.
>
> Jay M. Pasachoff (1983)

First contact. A tiny nick appears on the western side of the Sun.[1] The eye detects no difference in the amount of sunlight. Nothing but that nick portends anything out of the ordinary. But as the nick becomes a gouge in the face of the Sun, a sense of anticipation begins. This will be no ordinary day.

Still, things proceed leisurely for the first half hour or so, until the Sun is more than half covered. Now, gradually at first, then faster and faster, extraordinary things begin to happen. The sky is still bright, but the blue is a little duller. On the ground around you the light is beginning to dim. Over the next 10 to 15 minutes, the landscape takes on a steely gray metallic cast.

As the minutes pass, the pace quickens. With about a quarter hour left until totality, the western sky is now darker than the east, regardless of where the Sun is in the sky. The shadow of the Moon is approaching. Even if you have never seen a total eclipse of the Sun before, you know that something amazing is going to happen, is happening now—and that it is beyond normal human experience.

Less than fifteen minutes until totality. The Sun, a narrowing crescent, is still fiercely bright, but the blueness of the sky has deepened into blue-gray or violet. The darkness of the sky begins to close in around the Sun. The Sun does not fill the heavens with brightness anymore.

Five minutes to totality. The darkness in the west is very noticeable and gathering strength, a dark amorphous form rising upward and spreading out along the western horizon. It builds like a massive storm, but in utter silence, with no rumble of distant thunder. And now the darkness begins to float up above the horizon, revealing a yellow

Partial phases of the total solar eclipse of March 29, 2006 from Jalu, Libya. [Nikon D200 DSLR, Sigma 170–500 mm at 500 mm, f/11, 1/500 s, ISO 200, Thousand Oaks Type 3 solar filter. © 2006 Patricia Totten Espenak]

or orange twilight beneath. You are already seeing through the Moon's narrow shadow to the resurgent sunlight beyond.

The acceleration of events intensifies. The crescent Sun is now a blazing white sliver, like a welder's torch. The darkening sky continues to close in around the Sun, faster, engulfing it.

Minutes have become seconds. A ghostly round silhouette looms into view. It is the dark limb of the Moon, framed by a white opalescent glow that creates a halo around the darkened Sun. The corona, the most striking and unexpected of all the features of a total eclipse, is emerging. At one edge of the Moon the brilliant solar crescent remains. Together they appear as a celestial diamond ring.

Suddenly, the ends of the bare sliver of the Sun break into individual points of intense white light—Baily's beads—the last rays of sunlight passing through the deepest lunar valleys. The beads flicker, each lasting but an instant and vanishing as new ones form. And now there is just one left. It glows for a moment, then fades as if it were sucked into an abyss.

Totality.

Where the Sun once stood, there is a black disk in the sky, outlined by the soft pearly white glow of the corona, about the brightness of a full moon. Small but vibrant reddish features stand at the eastern rim of the Moon's disk, contrasting vividly with the white of the corona and the black where the Sun is hidden. These are the prominences, giant clouds of hot gas in the Sun's lower atmosphere. They are always a surprise, each unique in shape and size, different yesterday and tomorrow from what they are at this special moment.

You are standing in the shadow of the Moon.

It is dark enough to see Venus and Mercury and whichever of the brightest planets and stars happen to be close to the Sun's position and

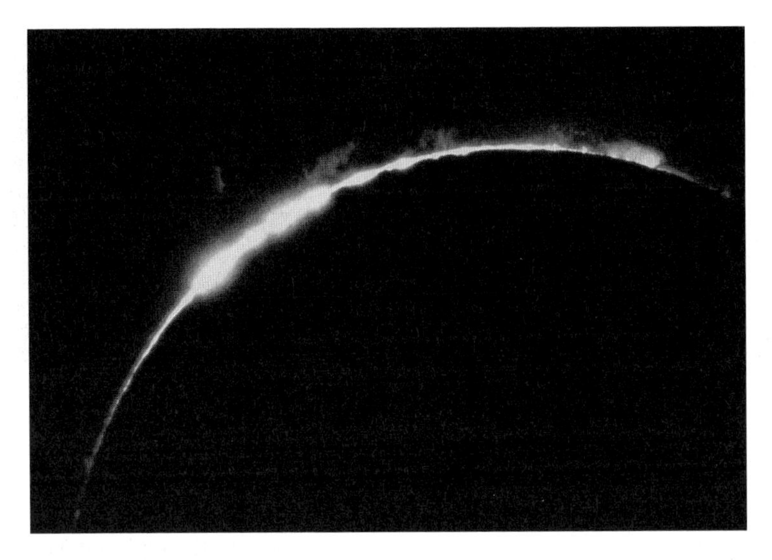

Baily's beads are seen amid a forest of red prominences during the total solar eclipse of August 11, 1999 from Lake Hazar, Turkey. [Pentax SLR, 94 mm Brandon refractor, f/30, 1/125 s, Ektachrome V100 film pushed to ISO 200. © 1999 Greg Babcock]

Diamond ring effect at the total solar eclipse of March 29, 2006 from Manavgat, Turkey. [Canon EOS 20D DSLR, Sigma 100–300 mm EX APO HSM at 300 mm with Sigma EX APO 2×, 1/1000 s, ISO 400. © 2006 Odd Høydalsvik]

2
The Great Celestial Cover-Up

> If God had consulted me before embarking upon creation, I would have recommended something simpler.
>
> Alfonso X, King of Castile (1252)

A total eclipse of the Sun is exciting and even profoundly moving.

But what causes a total solar eclipse? The Moon blocks the Sun from view. And that is all you absolutely need to know to enjoy a solar eclipse. So you can now skip to the next chapter.

If however you are reading this paragraph, you are right: there is more to tell—about dark shadows and oblong orbits and tilts and danger zones and amazing coincidences. Yet before you venture further, promise yourself one thing. If for any reason your eyes begin to glaze over, you will stop reading this chapter immediately and go right on to the next. You must not let celestial mechanics, or this explanation of it, stand in the way of your enjoyment of the wild, wacky, and wonderful things people have thought and done about solar eclipses.

Moon Plucking

How big is the Moon in the sky? What is its angular size?

Extend your arm upward and as far from your body as possible. Using your index finger and thumb, imagine that you are trying to pluck the Moon out of the sky ever so carefully, squeezing down until you are just barely touching the top and bottom of the Moon, trapping it between your fingers. How big is it? The size of a grape? A peach? An orange?

It is the size of a pea. (You can win bets at cocktail parties with this question.) The Moon has an angular size of only half a degree.

Now, how large is the Sun in the sky? Your friends will almost all immediately guess that it is bigger. Before they damage their eyes by trying the Moon pinch on the Sun, just remind them that a total eclipse is caused by the Moon completely covering the Sun, so the Sun must appear no bigger than a pea in the sky as well. It is the brightness of the

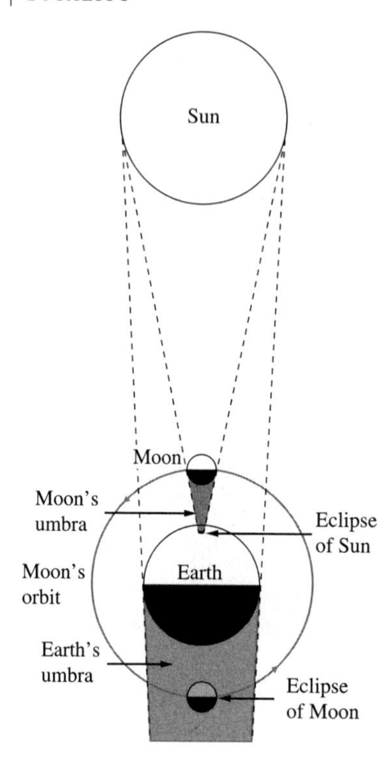

A total solar eclipse occurs when the Moon's umbra touches the Earth. A lunar eclipse occurs when the Moon passes into the Earth's shadow (umbra). The relative sizes and distances of the Sun, Moon, and Earth in this diagram are not to scale.

Moon and especially the Sun that deceives people into overestimating their angular size.

Now that you have collected on your bets and can lead a life of leisure, think about the remarkable coincidence that allows us to have total eclipses of the Sun. The Sun is 400 times the diameter of the Moon, yet it is about 400 times farther from the Earth, so the two appear almost exactly the same size in the sky. It is this geometry that provides us with the unique total eclipses seen on Earth when our Moon just barely covers the face of the Sun. If the Moon, 2,160 miles (3,476 kilometers) in diameter, were 169 miles (273 kilometers) smaller than it is, or if it were farther away so that it appeared smaller, people on Earth would never see a total eclipse.[1]

It is amazing that there are total eclipses of the Sun at all. As it is, total eclipses can just barely happen. The Sun is not always exactly the same angular size in the sky. The reason is that the Earth's orbit is not circular but elliptical, so the Earth's distance from the Sun varies. When the Earth is closest to the Sun (early January),[2] the Sun's disk is slightly larger in angular diameter, and it is harder for the Moon to cover the Sun to create a total eclipse.

The corona at the total solar eclipse of February 26, 1998, Oranjestad, Aruba. [Nikon FE SLR, Astrophysics 105 EDT refractor, fl = 620mm, AP 2× barlow, efl = 1300 mm, f/12, composite of 7 exposures: 1/125 s to 1/4 s, Ektachrome Elite II 100 slide film. © 1998 Fred Espenak]

An even more powerful factor is the Moon's elliptical orbit around the Earth. When the Moon is at its average distance from the Earth or farther, its disk is too small to occult the Sun completely. In the midst of such an eclipse, a circle of brilliant sunlight surrounds the Moon, giving the event a ringlike appearance; hence the name *annular* eclipse (from the Latin *annulus*, meaning ring).

Because the angular diameter of the Moon is smaller than the angular diameter of the Sun on the average, annular eclipses are more frequent than total eclipses.

But the Moon does not just dangle motionless in front of the Sun. It is in orbit around the Earth. It catches up with and passes the Sun's position in the sky about once a month (a period of time derived from this circuit of the Moon). The actual time for the Moon to complete this cycle is 29.53 days, and it is called a synodic month, after the Greek *synodos*, "meeting"—the meeting of the Sun and the Moon. Because the Moon gives off no light of its own and shines only by reflected sunlight, its orbit around the Earth changes its angle to the Sun and determines its phase. In 29.53 days, the Moon goes from new moon through full

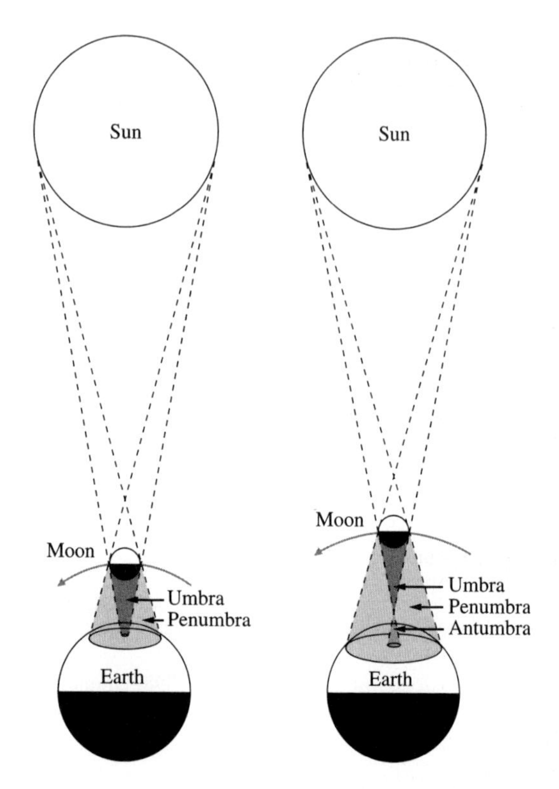

Configuration of a total solar eclipse *(left)* and an annular eclipse *(right)*. Within the Moon's umbra (dark converging cone), the entire surface of the Sun is blocked from view. In the penumbra (lighter diverging cone), a fraction of the sunlight is blocked, resulting in a partial eclipse. When the Moon's umbra ends in space *(right),* a total eclipse does not occur. Projecting the cone through the tip of the umbra onto the Earth's surface defines the region in which an annular eclipse is seen.

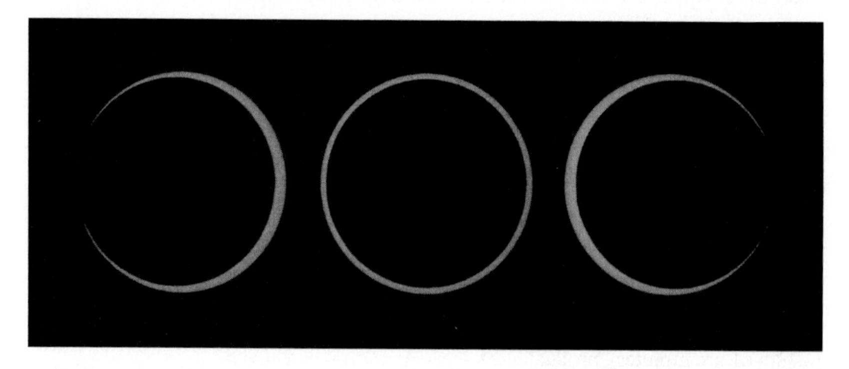

The beginning, middle, and end of annularity during the annular solar eclipse of October 3, 2005 from Spain. [Nikon D50 DSLR, 60 mm f/15 Unitron refractor, f/15, 1/125 s, ISO 200, Thousand Oaks Type II Filter, images combined in Photoshop. © 2005 Michael F. Barrett]

Diameters and Distances

	Sun	*Moon*	*Earth*
Diameter			
Miles	864,989	2,160	7,927
Kilometers	1,392,000	3,476	12,756
Mean distance from Earth			
Miles	92,960,200	238,870	—
Kilometers	149,598,000	384,400	—

Significance
The Moon's diameter is 1/400 of the Sun's.
The Moon's mean distance is 1/389 of the Sun's.
Thus the Moon and Sun are nearly the *same size* as seen from Earth.

The Shadow of the Moon

	Maximum	*Minimum*	*Mean*
Moon's distance from Earth (center to center)			
Miles	252,720	221,470	238,870
Kilometers	406,700	356,400	384,400
Length of Moon's shadow cone (umbra)			
Miles	236,050	228,200	232,120
Kilometers	379,870	367,230	373,540

Significance
On the *average* (in time), the Moon's shadow is too short to reach the Earth. Therefore, total solar eclipses occur *less* often than annular solar eclipses.

Angular Size of the Sun and Moon (as seen from Earth)

	Maximum	*Minimum*	*Mean*
Angular diameter of the Sun	32'31.9"	31'27.7"	31'59.3"
Angular diameter of the Moon	33'31.8"	29'23.0"	31'05.3"

Significance
The Moon's angular diameter can *exceed* the Sun's angular diameter by as much as 6.6% (2.1 arc minutes), producing a *total* eclipse of the Sun.

The Sun's angular diameter can *exceed* the Moon's angular diameter by as much as 10.7% (3.1 arc minutes), producing an *annular* eclipse of the Sun.

On the *average* (in time), the Moon's angular diameter is *smaller* than the Sun's angular diameter. Therefore, total solar eclipses occur *less* often than annular solar eclipses.

moon and back to new moon again. This period is called a lunar month, a *lunation*. Solar eclipses can take place only at new moon (dark-of-the-moon) and lunar eclipses may occur only at full moon.

So why don't we have an eclipse of the Sun every 29.53 days—every time the Moon passes the Sun's position? The reason is that the Moon's orbit around the Earth is tilted to the Earth's orbit around the Sun by about 5 degrees, so that the Moon usually passes above or below the Sun's position in the sky and cannot block the Sun from our view.

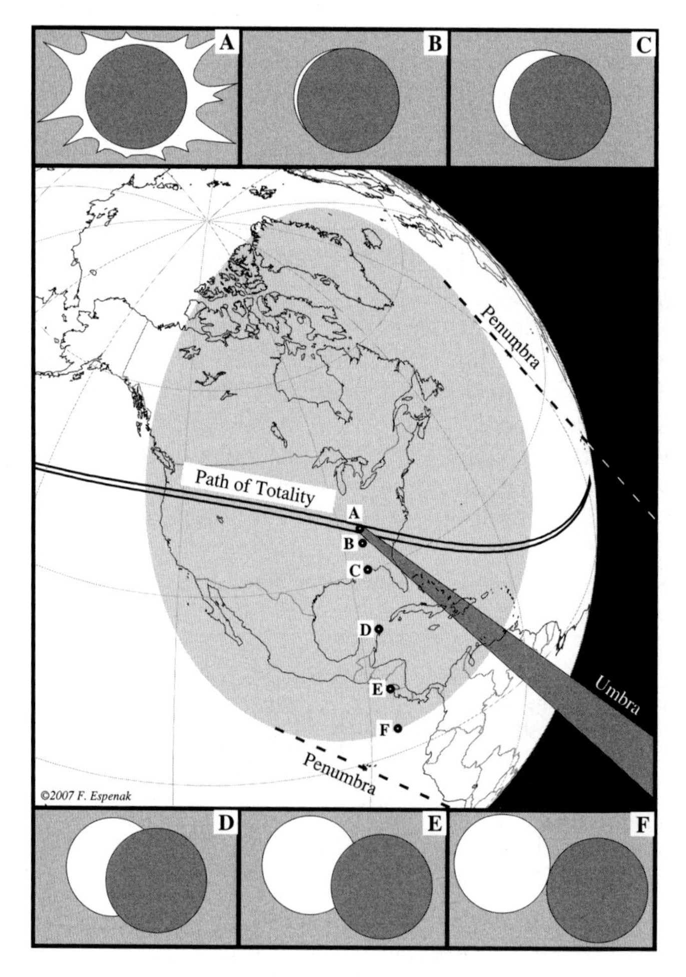

During a total eclipse of the Sun, the tip of the Moon's shadow touches the Earth and the Moon's orbital velocity carries the shadow rapidly eastward. Only along a narrow path is the eclipse total. Regions to the side of the path of totality experience varying degrees of partial eclipse. (This diagram illustrates the total eclipse of August 21, 2017.)

"Danger Zones"

The Moon's tilted orbit crosses the Earth's orbit at two places. Those intersections are called *nodes*. Node is from the Latin word meaning knot, in the sense of weaving, where two threads are tied together. The point at which the Moon crosses the plane of the Earth's orbit going northward is the ascending node. Going south, the Moon crosses the plane of the Earth's orbit at the descending node.

A solar eclipse can occur only when the Sun is near one of the nodes as the Moon passes.

If the Sun stood motionless in a part of the sky away from the nodes, there would be no eclipses, and you would not be agonizing over this. But the Earth is moving around the Sun, and, as it does so, the Sun appears to shift slowly eastward around the sky, through all the constellations of the zodiac, completing that journey in one year. In that yearly circuit, the Sun must cross the two nodes of the Moon. Think of it as a street intersection at which the Sun does not pause and runs the stop sign every time. It is an accident waiting to happen. When the Sun nears a node, there is the "danger" that the Moon will be coming and—crash!

No. The Moon is 400 times closer to the Earth than the Sun, so the worst—the best—that can happen is that the Moon will pass harmlessly but stunningly right in front of the Sun. The Sun's apparent pathway in the sky is called the *ecliptic* because it is only when the Moon is crossing the ecliptic that eclipses can happen. Thus twice a year, roughly, there is a "danger period," called an eclipse season, when the Sun is crossing the region of the nodes and an eclipse is possible.

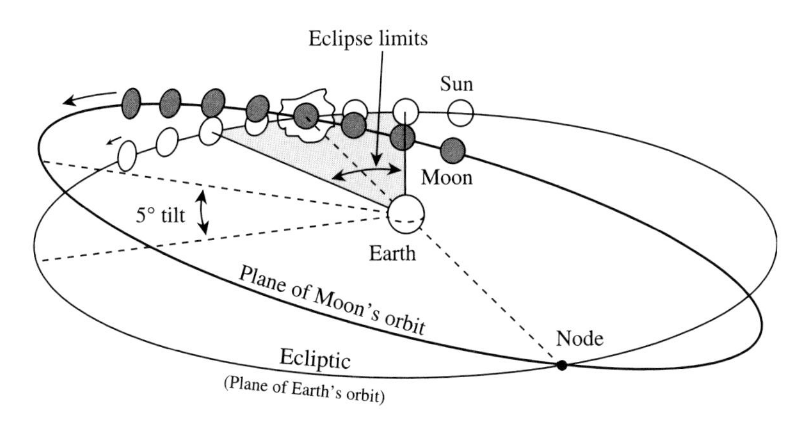

The paths of the Sun and Moon illustrate why eclipses occur only when the Sun is near the intersection (node) where the Moon crosses the ecliptic. The plane of the Moon's orbit is tilted approximately 5° to the ecliptic plane.

Eclipse Limits ("Danger Zones")

	Maximum[a]	*Minimum*[a]	*Mean*
A solar eclipse of some kind will occur at new moon if the Sun's angular distance from a node of the Moon is less than	18°31'	15°21'	16°56'
A central (total or annular) solar eclipse will occur at new moon if the Sun's angular distance from a node of the Moon is less than	11°50'	9°55'	10°52'

Sun's apparent eastward movement in the star field each day (due to the Earth orbiting the Sun): about 1°

Moon's synodic period (from new moon to new moon: the time the Moon takes to complete its eastward circuit of the star field and catch up with the Sun again): 29.53 days

Significance
The Sun cannot pass a node of the Moon without at least one solar eclipse occurring, and two are possible. If one occurs, it can be either a partial or a central eclipse (total or annular). If two occur, both will be partials, about one month apart.

The Sun can pass a node without a central solar eclipse (total or annular) occurring.

[a] Limits vary due to changes in the apparent angular size and speed of the Sun and Moon caused by the elliptical orbits of the Earth and Moon.

The Sun comes tootling up to the node traveling about 1 degree a day.[3] The hot-rod Moon, however, is racing around the sky at about 13 degrees a day. Now if the Sun and Moon were just dots in the heavens, they would have to meet precisely at a node for an eclipse to occur. But the disks of the Sun and Moon each take up about half a degree in the sky. And the Earth, almost 8,000 miles (12,800 kilometers) across, provides an extended viewing platform. Therefore, the Sun needs only to be *near* a node for the Moon to sideswipe it, briefly "denting" the top or bottom of the Sun's face. That will happen whenever the Sun is within 15⅓ degrees of a node.

An "eclipse alert" begins when the Sun enters the danger zone 15⅓ degrees west of one of the Moon's nodes and does not end until the Sun escapes 15⅓ degrees east of that node. The Sun must traverse 30⅔ degrees. Traveling at 1 degree a day, the Sun will be in the danger zone for about 31 days. But the Moon completes its circuit, going through all its phases, and catches up to the Sun every 29.53 days. It is not possible for the Sun to crawl through the danger zone before the Moon arrives. A solar eclipse *must* occur each time the Sun approaches a node and enters one of these danger zones, about every half year.

In fact, if the Moon nips the Sun at the beginning of a danger zone (properly called an *eclipse limit*), the Sun may still have 30 days of travel left within the zone. But the Moon takes only 29.53 days to orbit the

Earth and catch up with the Sun again. So it is possible for the Sun to be nipped by the Moon twice during a single node crossing, thereby creating two partial eclipses within a month of one another.

The closer the Sun is to the node when the Moon crosses, the more nearly the Moon will pass over the center of the Sun's face. In fact, if the Sun is within about 10 degrees of the node at the time of the Moon's crossing, a central eclipse will occur somewhere on Earth. Depending on the Moon's distance from the Earth and the Earth's distance from the Sun, this central eclipse will be total or annular.

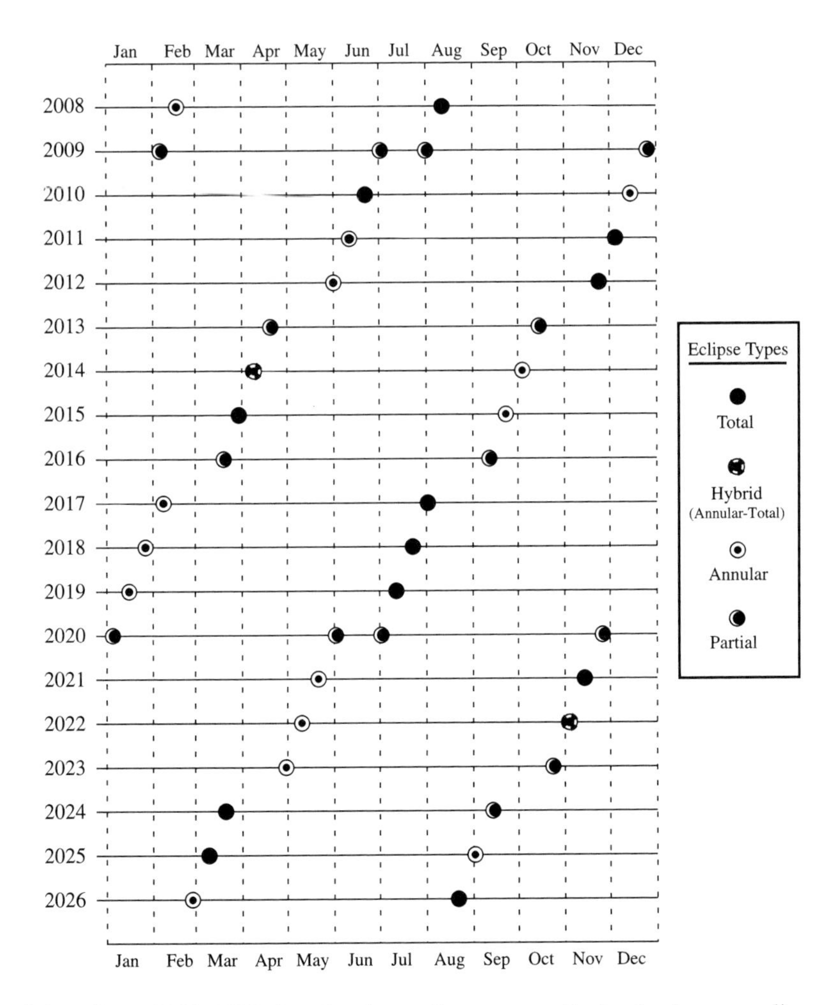

Solar eclipses 2008–2026 plotted to show eclipse seasons. Each calendar year, eclipses occur about 20 days earlier. Consequently, eclipse seasons shift to early months in the year.

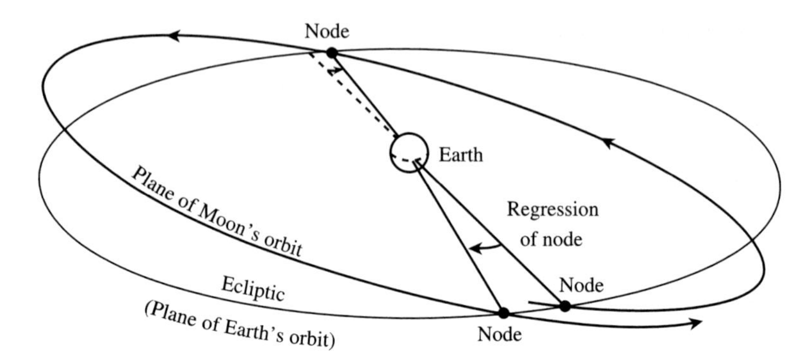

Each time the Moon completes an orbit around the Earth, it crosses the Earth's orbit at a point west of the previous node. Each year the nodes regress 19.4°, making a complete revolution in 18.61 years.

Solar Eclipses Outnumber Lunar Eclipses

There are more solar than lunar eclipses. This realization usually comes as a surprise because most people have seen a lunar eclipse, while relatively few have seen an eclipse of the Sun. The reason for this disparity is simple. When the Moon passes into the shadow of the Earth to create a lunar eclipse, the event is seen wherever the Moon is in view, which includes half the planet. Actually, a lunar eclipse is seen from more than half the planet because during the course of a lunar eclipse (up to 4 hours), the Earth rotates so that the Moon comes into view for additional areas.

In contrast, whenever the Moon passes in front of the Sun, the shadow it creates—a solar eclipse—is small and touches only a tiny portion of the surface of the Earth. On the average, your house will be visited by a total eclipse of the Sun only once in about 375 years.[a]

To be touched by the dark shadow (umbra) of the Moon is quite rare. But from either side of the path of a total eclipse, stretching northward and southward 2,000 miles (3,200 kilometers) and sometimes more, an observer sees the Sun partially eclipsed. Even so, this band of partial eclipse covers a much smaller fraction of the Earth's surface than a lunar eclipse. So more people have seen lunar eclipses than partial solar eclipses, and only a tiny fraction of people, about one in 10,000, have witnessed a total solar eclipse.

Yet, to the unaided eye, solar eclipses are substantially more frequent than lunar eclipses. Theodor von Oppolzer, in his monumental *Canon of Eclipses*, published shortly after his death in 1887, attempted to compute, with the help of a number of assistants, all eclipses of the Sun and Moon from 1208 B.C. to 2161 A.D. He cataloged 8,000 solar eclipses and 5,200 lunar eclipses. He found about three solar eclipses for every two lunar eclipses.

This ratio can be misleading, however. Oppolzer counted all solar eclipses, whether they were total (the Moon's *umbra* touches the Earth) or partial (only the Moon's *penumbra* touches the Earth). But for lunar eclipses, Oppolzer counted

Nodes on the March

There should be a solar eclipse or two every six months, whenever the Sun crosses one of the Moon's nodes. Actually, the Sun crosses the ascending node of the Moon, then the descending node, and returns to the ascending node in only 346.62 days—the *eclipse year.* Within this eclipse year there are two *eclipse seasons*, intervals of 30 to 37 days as the Sun approaches, crosses, and departs from a node. All eclipses will fall within eclipse seasons. There can be no eclipses outside this period of time. Because the Sun crosses a node about every 173 days, the eclipse seasons are centered about 173 days apart.

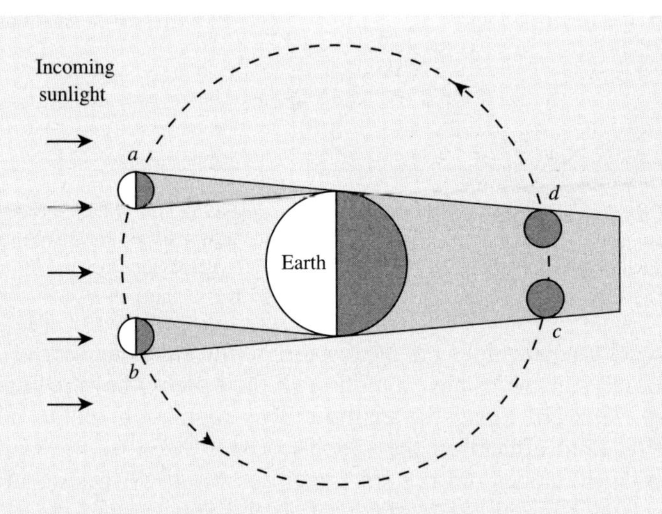

only those in which the Moon was totally or partially immersed in the Earth's *umbra.* He did not count *penumbral* lunar eclipses because they are virtually unnoticeable. If he had included penumbral lunar eclipses in his census so as to compare the number of all forms of solar and lunar eclipses, the ratio would have been close to even, with solar eclipses barely prevailing.

The reason why solar eclipses slightly outnumber lunar eclipses is most easily visualized if you imagine looking down on the Sun–Earth–Moon system and if you start by considering only total solar and total lunar eclipses.

The Moon will be totally eclipsed whenever it passes into the shadow of the Earth—between ***c*** and ***d*** on the diagram. At the Moon's average distance from Earth, that shadow is about 2.7 times the Moon's diameter. But there will be an eclipse of the Sun whenever the Moon passes between the Earth and Sun—between points *a* and *b.* The distance between *a* and *b* is longer than between *c* and *d*, so total solar eclipses must occur slightly more often.

---

[a]Jean Meeus: "The Frequency of Total and Annular Solar Eclipses at a Given Place," *Journal of the British Astronomical Association*, volume 92, April 1982, pages 124–126.

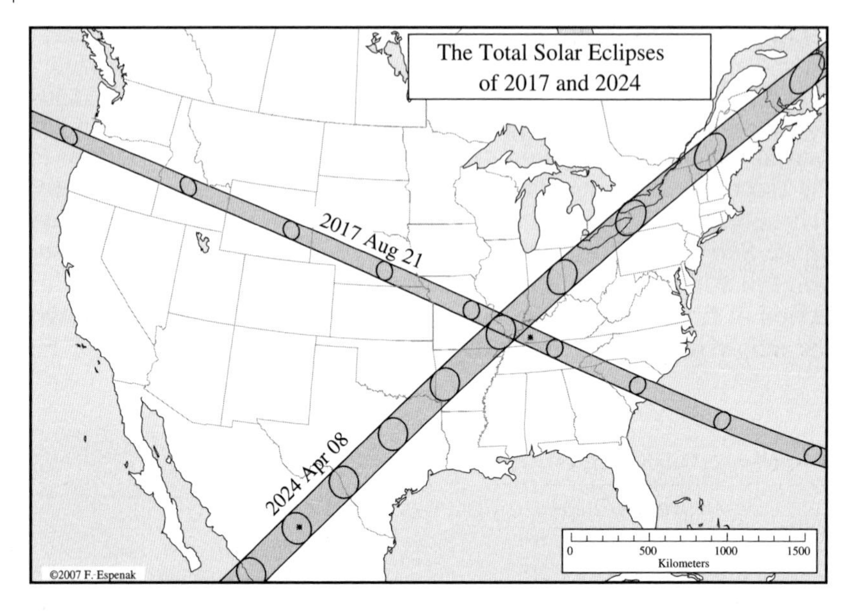

Paths of the total eclipses of 2017 and 2024. In a period of 6⅔ years, the people near the confluence of the Ohio and Mississippi Rivers will see two total solar eclipses. [Map and eclipse calculations by Fred Espenak]

The eclipse year does not correspond to the calendar year of 365.24 days because the nodes have a motion all their own. They are constantly shifting westward along the ecliptic. This regression of the nodes is caused by tidal effects on the Moon's orbit created by the Earth and the Sun. If the nodes did not shift, eclipses would always occur in the same calendar month year after year. If the Sun crossed the nodes in February and August one year to cause eclipses, the eclipses would continue to fall in February and August in succeeding years.

But the eclipse year is 346.62 days, 18.62 days shorter than a calendar year, so each ascending or descending node crossing by the Sun occurs 18.62 days earlier in the calendar year than the previous one of its kind. This migration of the eclipse seasons determines the number of eclipses that may occur each year. One solar eclipse must occur each eclipse season, so there have to be at least two solar eclipses each year (although both may be partial). But because of the width of the eclipse limits—up to 19 days on either side of the Sun's node crossing—and the slowness of the Sun's apparent motion, there can be two solar eclipses at each node passage (both partials). Thus occasionally there will be four solar eclipses in one calendar year.

There can actually be five. Because the eclipse year lasts 346.62 days, almost 19 days less than a calendar year, if a solar eclipse occurs before or on January 18 (or January 19 in a leap year), that eclipse year could conceivably bring two solar eclipses in January and two more around July. That eclipse year would then end in mid-December, and a new eclipse year would begin in time to provide one final solar eclipse before the end of December. At most, therefore, there can be five solar eclipses in a calendar year.

Heavenly Rhythm

Eclipses, then, are like fresh fruit—available only in season. Ancient peoples who kept written records, such as the Chaldeans from about

Frequency of Solar Eclipses

Solar eclipses by types (average over 5,000 years, –1999 to +3000):

Total:	26.7%
Annular:	33.2%
Annular/Total:	4.8%
Partial:	35.3%

Solar eclipses outnumber lunar eclipses almost 3 to 2 (excluding penumbral eclipses, which are seldom detectable visually)

Annular eclipses outnumber total eclipses about 5 to 4

Solar eclipses per century (average over 4,530 years): 238.9

Maximum number of solar eclipses per year: 5 (4 will be partial)

Minimum number of solar eclipses per year: 2 (both can be partial)

Maximum number of *total* solar eclipses per year: 2

Minimum number of *total* solar eclipses per year: 0

Maximum number of solar and lunar eclipses per year: 7 (4 solar and 3 lunar *or* 5 solar and 2 lunar)[a]

Minimum number of solar and lunar eclipses per year: 2 (2 solar and 0 lunar)[a]

Examples of years in which only 2 solar eclipses occur: 2008, 2009, 2010, 2112, 2013, 2014, 2015, 2016, 2017

Examples of years in which 4 solar eclipses occur: 2000, 2011, 2029, 2047

Examples of years in which 5 solar eclipses occur: 1805, 1935, 2206, 2709

Maximum diameter of the Moon's shadow cone (umbra) as it intercepts the Earth to cause a total eclipse: 170 miles (273 kilometers)

Maximum diameter of the Moon's "anti-umbra" as it intercepts the Earth to cause an annular eclipse: 232 miles (374 kilometers)

Source: Fred Espenak and Jean Meeus: *Five Millennium Canon of Solar Eclipses: –1999 to +3000* (Greenbelt, Maryland: NASA Goddard Space Flight Center, 2006; NASA Technical Publication 2006–214141). This publication can be downloaded from the web in PDF format at <http://eclipse.gsfc.nasa.gov/SEpubs/5MCSE.html>.

[a] Counts total and partial lunar eclipses but not penumbral lunar eclipses.

Total Eclipses—Duration of Totality

Longest duration (theoretical):	7 minutes 31 seconds
Longest duration in 10,000 years[a]:	7m 29s — July 16, 2186 A.D.
Longest duration by millennium:	
3000 B.C. to 2001 B.C.:	7m 21s — May 16, 2231 B.C.
2000 B.C. to 1001 B.C.:	7m 05s — July 3, 1443 B.C.
1000 B.C. to 1 B.C.:	7m 28s — June 15, 744 B.C.
1 A.D. to 1000 A.D.:	7m 24s — June 27, 363 A.D.
1001 A.D. to 2000 A.D.:	7m 20s — June 9, 1062 A.D.
2001 A.D. to 3000 A.D.:	7m 29s — July 16, 2186 A.D.
3001 A.D. to 4000 A.D.:	7m 18s — July 24, 3991 A.D.
4001 A.D. to 5000 A.D.:	7m 12s — August 4, 4009 A.D.
Longest duration of the 20th century[b]: (6 minutes or longer)	6m 29s — May 18, 1901
	6m 20s — September 9, 1904
	6m 51s — May 29, 1919
	7m 04s — June 8, 1937
	7m 08s — June 20, 1955
	7m 04s — June 30, 1973
	6m 53s — July 11, 1991
Longest duration of the 21st century[b]: (6 minutes or longer)	6m 39s — July 22, 2009
	6m 23s — August 2, 2027
	6m 06s — August 12, 2045
	6m 06s — May 22, 2096
Number of eclipses with 7 minutes or more of totality in 21st century:	0
Number of eclipses with 7 minutes or more of totality from July 1, 1098 A.D. to June 8, 1937 A.D. (839 years):	0

Note: All solar eclipses with long durations of totality have dates centered around July 4, the mean date for the Earth at aphelion (farthest from the Sun), when the Sun appears slightly smaller in size and is more easily covered by the Moon.

[a] Based on the 10,000-year period from 3000 B.C. to 7000 A.D.

[b] All the long total eclipses of the twentieth and twenty-first centuries are members of saros series 136 except for September 9, 1904 and May 22, 2096.

747 B.C. on, noticed after decades of observation that eclipses happen only at certain times of the year. These eclipse seasons are separated from one another by either five or six new moons.

From the expanse of their New Babylonian Empire in the Middle East, the Chaldeans could see only about half the lunar eclipses and only a small fraction of the solar eclipses, so, for them, the eclipse seasons were not periods during which one to three eclipses (1 or 2 solar; 0 or 1 lunar) would necessarily occur. Instead, the eclipse seasons were times of "danger" when an eclipse was possible. One of the two great

Annular Eclipses—Duration of Annularity

Longest duration (theoretical):	12 minutes 30 seconds
Longest duration in 10,000 years[a]:	12 m 24s — December 6, 150 A.D.
Longest duration by millennium:	
3000 B.C. to 2001 B.C.:	11m 02s — November 24, 2037 B.C.
2000 B.C. to 1001 B.C.:	12m 07s — December 12, 1656 B.C.
1000 B.C. to 1 B.C.:	12m 08s — December 22, 178 B.C.
1 A.D. to 1000 A.D.:	12m 24s — December 6, 150 A.D.
1001 A.D. to 2000 A.D.:	12m 09s — December 14, 1955 A.D.
2001 A.D. to 3000 A.D.:	11m 08s — January 15, 2010 A.D.
3001 A.D. to 4000 A.D.:	12m 09s — January 14, 3080 A.D.
4001 A.D. to 5000 A.D.:	11m 08s — January 20, 4885 A.D.
Longest duration of the 20th century[b]:	11m 01s — November 11, 1901
(9 minutes or longer)	11m 37s — November 22, 1919
	12m 00s — December 2, 1937
	12m 09s — December 14, 1955
	12m 03s — December 24, 1973
	11m 41s — January 4, 1992
Longest duration of the 21st century[b]:	11m 08s — January 15, 2010
(9 minutes or longer)	10m 27s — January 26, 2028
	9m 42s — February 5, 2046

Note: All solar eclipses with long durations of annularity have dates centered around January 3, the mean date for the Earth at perigee (closest to the Sun), when the Sun appears slightly larger in size and is harder for the Moon to cover.

Note: A long annularity means that the Moon is smaller than usual in apparent size, creating an annular eclipse that is farthest from being a total eclipse. Thus a long annularity will provide less landscape and sky darkening and less likelihood of a glimpse of Baily's beads.

[a] Based on the 10,000-year period from 3000 B.C. to 7000 A.D.

[b] All the long annular eclipses of the twentieth and twenty-first centuries are members of saros series 141.

celestial lights might be partially or totally darkened. The Chaldeans did not realize that the invisibility of an eclipse simply meant that it was occurring somewhere else on Earth.

As time passed and they accumulated more records, the Chaldeans and other ancient peoples recognized that a specific eclipse occurred a precise number of days after a previous eclipse and before a subsequent one. Eclipses had a long-term rhythm of their own.

The most famous and, perhaps, most useful of these eclipse rhythms was the *saros*, discovered by the Chaldeans and inscribed on clay tablets in their cuneiform writing.[4] The Chaldeans noticed that 6,585 days (18 years 11 days) after virtually every lunar eclipse, there was another very similar one. If the first was total, the next was almost always total.

And these eclipses, separated by 18 years, occurred in the same part of the sky, as if they were related to one another.

In a sense, they were. Imagine that a total solar eclipse occurs one day at the Moon's descending node. After 6,585 days, the Moon has completed 223 lunations (synodic periods) of 29.53 days each and returned to new moon at that same node. In that same period of 6,585 days, the Sun has endured 19 eclipse years of 346.62 days each and has returned to the descending node, forcing another solar eclipse to occur. And because 6,585 days is very close to an even 18 calendar years, this

Erratic Intervals Between Total Solar Eclipses

Location	*Dates*	*Interval*
An average town		average of 375 years
Unusually long intervals		
London	October 29, 878	
	May 3, 1715	837 years
Jerusalem	August 2, 1133	
	August 6, 2241	1,108 years
United States mainland	February 26, 1979	
	August 21, 2017	38 years
Unusually short intervals—past		
New Guinea (southern)	June 11, 1983	
	November 22, 1984	1½ years
Ukraine	March 14, 190 B.C.	
	July 17, 188 B.C.	
	October 19, 183 B.C.	3 in 7½ years
Sumatra	May 18, 1901	
	January 14, 1926	
	May 9, 1929	3 in 28 years
Spain	July 8, 1842	
	December 22, 1870	
	May 28, 1900	
	August 30, 1905	4 in 63 years
Jerusalem (region of)	March 1, 357 B.C.	
	July 4, 336 B.C.	
	April 2, 303 B.C.	3 in 54 years
Stonehenge, England	May 3, 1715	
	May 22, 1724	9 years
Yellowstone National Park, United States of America	July 29, 1878	
	January 1, 1889	10½ years
Mexico City	March 7, 1970	21 years
	July 11, 1991	

solar eclipse occurs at the same season of the year as its predecessor, and with the Sun very close to the same position in the zodiac that it occupied at the eclipse 18 years earlier. Even though 18 years have intervened, these two eclipses certainly seem to be relatives.

All the more so because another lunar cycle crucial to eclipses has a multiple that also adds up to 6,585 days. That cycle is the *anomalistic month*—the time it takes the Moon in its elliptical orbit around the Earth to go from *perigee* (closest to Earth) to *apogee* (farthest) and back to *perigee.*[5] When the Moon is near perigee, its angular size in the sky is just slightly larger, creating eclipses that are total rather than annular. The anomalistic month is 27.55 days long and 239 of these cycles add

Erratic Intervals Between Total Solar Eclipses

Location	*Dates*	*Interval*
Switzerland (almost all of)	May 12, 1706	
	May 22, 1724	18 years
Wichita, Kansas, U.S.A.	September 17, 1811	
	November 30, 1834	23 years
Turkey (northern)	August 11, 1999	
	March 29, 2006	6⅔ years
Lobito, Angola (just north of)	June 21, 2001	
	December 4, 2002	1½ years
Unusually short intervals—future		
Confluence of Ohio and	August 21, 2017	
Mississippi rivers, U.S.A.	April 8, 2024	6⅔ years
(southeast Missouri, southern		
Illinois, and western Kentucky)		
Australia	April 20, 2023	
	July 22, 2028	
	November 25, 2030	
	July 13, 2037	
	December 26, 2038	5 in 15 years
Florida panhandle, U.S.A.	August 12, 2045	
(Pensacola to Tallahassee)	March 30, 2052	6⅔ years
Paris	September 3, 2081	
	September 23, 2090	9 years
Bagé, Brazil	January 8, 2103	
(Brazil–Uruguay border)	February 8, 2111	
	September 15, 2118	3 in 15 years
New York City	May 1, 2079	
	October 26, 2144	65 years
Antwerp, Belgium	May 25, 2142	
	June 14, 2151	9 years

[7] In a sense, it has not vanished altogether. With each period of 6,585.32 days, the Sun's position with respect to the node continues to slip westward without experiencing or causing any eclipses until, after about 5,500 years, it encounters the eclipse limit of the opposite node, and that saros may be said to be reborn. George van den Bergh: *Periodicity and Variation of Solar (and Lunar) Eclipses* (Haarlem: H. D. Tjeenk Willink, 1955).

3
A Quest to Understand

> I look up. Incredible! It is the eye of God. A perfectly black disk, ringed with bright spiky streamers that stretch out in all directions.
>
> Jack B. Zirker (1984)

Ancient peoples around the world have left monuments and symbols of their reverence for the sky and the results of their efforts to record celestial motions. More than 2,500 years ago, some people could predict or at least warn of the possibility of eclipses, especially lunar eclipses, the fading of the Moon as it passed into the Earth's shadow. And the ability to predict eclipses may go back further still.

Stonehenge

The earliest and most famous of monuments that testify to a people's high level of astronomical knowledge is Stonehenge, near Salisbury in southern England. Awesome, haunting, strangely beautiful, Stonehenge is a permanent record of celestial knowledge in stone. Work was in progress on Stonehenge a generation before construction began on the first pyramids in Egypt. Stonehenge was already abandoned, a desolate mystery, when Moses led the Exodus from Egypt.

The familiar silhouette of archways, now mostly fallen, is the last of four major stages of Stonehenge development that ceased about 1500 B.C. The stones formed a circle of 30 linked archways, approximating the days in a lunar month.[1] Inside this Sarsen Circle was a horseshoe of five even larger freestanding archways, the trilithons, with uprights that weigh up to 50 tons. These massive, shaped boulders were dragged from a quarry 20 miles (32 kilometers) away to codify in stone the discoveries of an earlier people.

The most famous feature of Stonehenge is the line of sight from the center of the monument toward the northeast through an archway in the Sarsen Circle and over the Heel Stone, a 35-ton boulder set upright

Aerial view of Stonehenge. [English Heritage]

245 feet (75 meters) away. From this position, an observer sees the approximate point on the horizon where the Sun rises on the first day of summer, when it is farthest north of the equator and daytime lasts longest. With this alignment and others, the users of Stonehenge could time the beginning of summer and winter with high precision to create an accurate solar calendar, which could have been of great benefit to farmers.

Yet these massive stone archways, raised in the last phase of construction at Stonehenge, show little new in their orientations beyond what archaeologists have found in the original and essential Stonehenge, begun 1,300 years earlier.

The Essential Stonehenge

Stonehenge was begun about 2800 B.C. by a people who had no written language, no wheeled vehicles, no draft animals, and no metal tools. To dig holes in the ground, they used the antlers of deer.

The initial Stonehenge consisted of a circular embankment 350 feet (107 meters) in diameter, four marker stones set in a rectangle, some postholes, and the Heel Stone.[2] The Heel Stone was apparently the first of the great boulders brought to this site as construction commenced. But it may not have stood alone. A similar huge stone stood just to its left as seen from the center of Stonehenge.[3] In that ancient time, the

Viewed from the center of Stonchenge, the Sun rises just to the left of the remaining Heel Stone at the beginning of summer, the longest day of the year. [English Heritage]

Sun at the beginning of summer probably rose between the famed Heel Stone and its now-vanished companion, and the alignment with sunrise at the summer solstice was probably exact.

For someone standing at the center of Stonehenge, the embankment served to level the horizon of rolling hills. Within the embankment, four stones—the Station Stones—outlined a rectangle within the circle. The sides of this rectangle offered interesting lines of sight. The short side of the rectangle pointed toward the same spot on the horizon that the two Heel Stones framed, the position where the Sun rose farthest north of east, marking the commencement of summer. Facing in the opposite direction along the short side of the rectangle, an observer would see the place where the Sun set farthest south of west, signaling the beginning of winter.

In contrast, the long sides of the rectangle provided alignments for crucial rising and setting positions of the Moon. Looking southeast along the length of the rectangle, an observer was facing the point on the horizon where the summer full moon would rise farthest south. In the opposite direction, looking northwest, this early astronomer's gaze was led to the spot on the horizon where the winter full moon would set farthest north. These positions marked the north and south limits of the Moon's motion.

The structure of Stonehenge offers additional testimony to its builders' efforts to understand the motion of the Moon. Evidence of small holes near the remaining Heel Stone strongly suggests that the users of Stonehenge observed and marked the excursion of the Moon as much as 5° north and south of the Sun's limit.[4] This motion above and below the Sun's position is caused by the tilt of the Moon's orbit to the Earth's path around the Sun. Because of this tilt, the Moon does not pass directly in front of the Sun (a solar eclipse) or directly into the Earth's shadow (a lunar eclipse) each month.

Because the builders of Stonehenge had discovered and accurately recorded the range in the rising and setting positions of the Sun and Moon and had built a monument that marked these positions with precision, they may have been able to recognize when the Moon was on course to intercept the position of the Sun, to cause a solar eclipse. Perhaps they

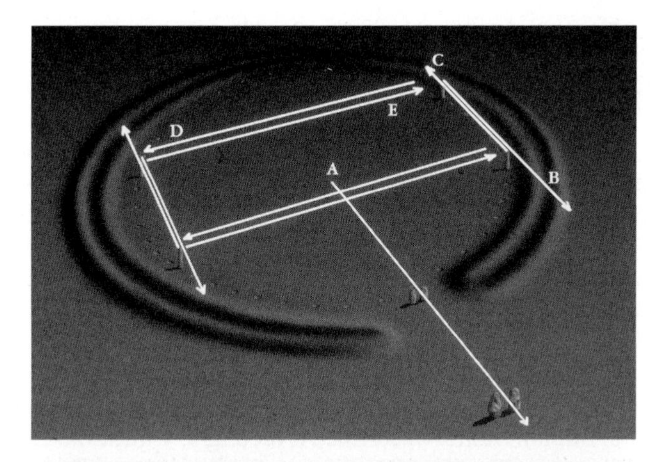

Stonehenge. *Top:* First phase of construction showing alignments A and B, northernmost sunrise: first day of summer; C, southernmost sunset: first day of winter; D, southernmost moonrise; E, northernmost moonset; *Bottom:* Final phase of construction. [© 1983 Hansen Planetarium]

could tell when the Moon was headed for a position directly opposite the Sun, which would carry it into the shadow of the Earth for a lunar eclipse. They almost certainly could not predict where or what kind of solar eclipse would be seen, but they might have been able to warn that on a particular day or night, an eclipse of the Sun or Moon was *possible.*

In the last phase of building at Stonehenge, two concentric circles of holes were dug just outside the Sarsen Circle—one with 30 holes and the other with 29. These circles reinforce the evidence that astronomers at Stonehenge were counting off the 29½-day cycle of lunar phases, from new moon to full moon and back to new moon again. Eclipses of the Sun can only take place at new moon; lunar eclipses can only occur at full moon. If indeed the lunar phasing cycle was watched carefully, perhaps some ancient genius noticed a periodicity in eclipses as well. With a knowledge of that period, that early astronomer could have converted a mere warning of a possible eclipse into a prediction of a likely eclipse, especially for lunar eclipses, which are visible over half the Earth.[5]

The builders of Stonehenge left no written records of their objectives or results, so we must judge from the monument and its alignments what they knew. Whatever that was, they thought it so worth celebrating that the rulers and apparently the common people were willing to devote vast amounts of time, physical effort, and ingenuity to raising a lasting monument of great size, precision, and beauty.

China

A frequently recounted Chinese story says that Hsi and Ho, the court astronomers, got drunk and neglected their duties so that they failed to predict (or react to) an eclipse of the Sun. For this, the emperor had them executed. So much for negligent astronomers.

If this story were an account of an actual event, the dynasty mentioned would place the eclipse somewhere between 2159 and 1948 B.C., making it by far the oldest solar eclipse recorded in history. But all serious attempts to identify one particular eclipse as the source of this story have been abandoned as scholars have recognized that the episode is mythological.

In ancient Chinese literature, Hsi-Ho is not two persons but a single mythological being who is sometimes the mother of the Sun and at other times the chariot driver for the Sun. Later, in the *Shu Ching* (Historical Classic), parts of which may date from as early as the seventh or sixth century B.C., this single character is split, not into two, but into six. In the *Shu Ching* story, the legendary Chinese emperor Yao commissions the eldest of the Hsi and Ho brothers "to calculate and delineate the sun, moon, the stars, and the zodiacal markers; and so to deliver respectfully the seasons to the people."[6] In further orders, he sends a younger Hsi brother to the east and another to the south; he orders a younger Ho

The Hsi and Ho brothers receive their orders from Emperor Yao to organize the calendar.

brother to the west and another to the north. Each is responsible for a portion of the rhythms of the days and seasons, to turn the Sun back at the solstices and to keep it moving at the equinoxes.

These mythological magicians are always charged with the prevention of eclipses, hence the story that appears later in the *Shu Ching* about the emperor's anger with his servants for failing to *prevent* an eclipse, not just predict or respond ceremonially to it. The story appears in a chapter that is a exhortation by the Prince of Yin, commander of the armies, to government officials to fulfill their duties to the administration, thereby making the emperor "entirely intelligent." If anyone neglects this requirement, "the country has regular punishments for you."

> *Now here are Hsi and Ho. They have entirely subverted their virtue, and are sunk and lost in wine. They have violated the*

duties of their office, and left their posts. They have been the first to allow the regulations of heaven to get into disorder, putting far from them their proper business. On the first day of the last month of autumn, the sun and moon did not meet harmoniously in Fang. The blind musicians beat their drums; the inferior officers and common people bustled and ran about. Hsi and Ho, however, as if they were mere personators of the dead in their offices, heard nothing and knew nothing;—so stupidly went they astray from their duty in the matter of the heavenly appearances, and rendering themselves liable to the death appointed by the former kings. The statutes of government say, "When they anticipate the time, let them be put to death without mercy; when they are behind the time, let them be put to death without mercy."[7]

We never hear whether Hsi and Ho were ever tracked down and executed.

The story of Hsi and Ho as drunken astronomers was a myth. But the myth did come true in a sense about 33 centuries later. Chinese history records that in 1202 A.D., for the second time in four years, the chief court astronomer made an eclipse forecast that was not as accurate as predictions from people with no official scientific credentials or status. "The astronomical officials were found guilty of negligence and severely punished."[8]

The earliest Chinese word for eclipse, *shih*, means "to eat" and referred to the gradual disappearance of the Sun or Moon as if it were eaten by a celestial dragon.[9] The Chinese were early in recording eclipses but late in recognizing their cause. Not until the third or fourth century A.D. did they understand solar and lunar eclipses well enough to be able to predict them accurately.

The Maya

In the New World, there were ancient people who, like the Chaldeans and the Chinese, used writing to record eclipses and from these records detected a rhythm by which they could predict them or at least warn of their likelihood. Those people were the Maya and we know of their achievement through one of their books—one of only four that survived the Spanish conquest and its zealous destruction of the religious beliefs of the native peoples.

All that we know of Maya accomplishments in recognizing the patterns of eclipses comes from the Dresden Codex, written in hieroglyphs and pictures in color paints on processed tree bark with pages that open and shut in accordion folds. The book dates from the eleventh century A.D. and is probably a copy of an older work.

We can only wonder what was lost when the conquering Spaniards destroyed by the thousands the books of the Maya and other Mesoamerican peoples. What remains is impressive enough. The Maya realized that discernible eclipses occur at intervals of five or six lunar months. Five or six full moons after a lunar eclipse, there was the *possibility* of another lunar eclipse. Five or six new moons after a solar eclipse, another solar eclipse was *possible*.

The Maya had discovered in practical, observable terms the approximate length of the eclipse year, 346.62 days, and the eclipse half year of 173.31 days. The interval for one complete set of lunar phases is 29.53 days. Six lunations amount to approximately 177.18 days, close enough to the eclipse half year (173.31 days) so that there is the "danger" of an eclipse at every sixth new or full moon, but not a certainty. After

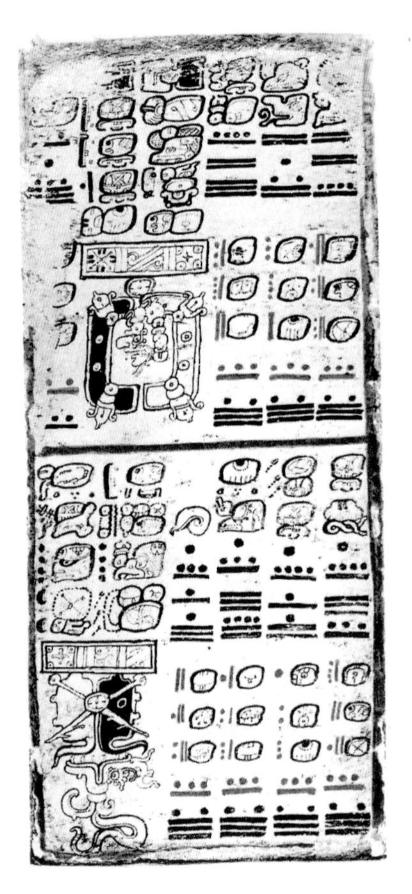

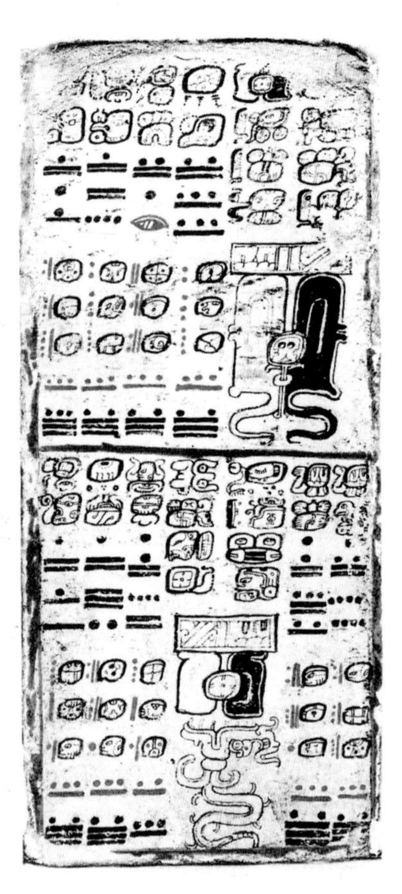

Portion of the Maya solar eclipse prediction tables from the Dresden Codex: at the bottom are the day counts that lead up to a solar eclipse, indicated, *bottom right*, by a serpent swallowing a symbol for the Sun. [American Philosophical Society]

another six lunar months, the passing days have amounted to 354.36, nearly 8 days *too long* to coincide with the Sun's passage by the Moon's node. An eclipse is less likely. As the error mounts, the need increases to substitute a five-lunar-month cycle into the prediction system rather than the standard six-lunar-month count.

Some great genius must have noticed after recording a sizeable number of eclipses that major eclipses were occurring only at intervals of 177 days (6 lunar months) or 148 days (5 lunar months). Using the date of an observed solar or lunar eclipse, it would then have been possible to predict the likelihood of another eclipse, even though in some cases an eclipse would not occur and in others it would not be visible from Mesoamerica.

In the Dresden Codex there are eight pages with a variety of pictures representing an eclipse. Each depiction is different, but most show the glyph for the Sun against a background half white and half black. In two of the pictures, the Sun and background are being swallowed by a serpent. Leading up to each picture is a sequence of numbers: a series of 177s ending with a 148. Each sequence adds up to the number of days in well-known three- to five-year eclipse cycles. At the end of each burst of numbers stands the giant, haunting symbol of an eclipse.

Astronomer-anthropologist Anthony F. Aveni notes that "The reduction of a complex cosmic cycle to a pair of numbers was a feat equivalent to those of Newton or Einstein and for its time must have represented a great triumph over the forces of nature."[10]

From the Maya, we have the numbers that demonstrate one of the greatest of their many discoveries about the rhythms of the sky, but we have no account of the emotion the astronomer-priests or the common folk felt when they observed an eclipse. Perhaps the closest we can come is a passage in the Florentine Codex of the Aztecs, who inherited and used the Mesoamerican calendar but apparently knew little of the astronomy discovered by the Maya a thousand years and more before.

> *When the people see this, they then raise a tumult. And a great fear taketh them, and then the women weep aloud. And the men cry out, [at the same time] striking their mouths with [the palms of] their hands. And everywhere great shouts and cries and howls were raised....And they said: "If the sun becometh completely eclipsed, nevermore will he give light; eternal darkness will fall, and the demons will come down. They will come to eat us!"*[11]

Notes and References

Epigraph: Jack B. Zirker: *Total Eclipses of the Sun* (New York: Van Nostrand Reinhold, 1984), page vi.

[1] One of these 30 uprights is only half the diameter of all the others, as if to suggest 29½, the length of time in days that it takes the Moon to complete a cycle of phases.
[2] There were also 56 chalk-filled Aubrey Holes just inside the embankment, but their function is still not generally agreed upon.
[3] The hole for this stone was discovered in 1979 under the shoulder of the road that passes close to the Heel Stone. The original stone is gone. Michael W. Pitts: "Stones, Pits and Stonehenge," *Nature*, volume 290, March 5, 1981, pages 46–47.
[4] The limits of the Moon's motion north and south of the ecliptic can differ by as much as 10 minutes of arc from the mean inclination of 5 degrees 8 minutes because of gravitational perturbations on the Moon caused by the Sun and the equatorial bulge of the Earth.
[5] How Stonehenge may have been used as a computer to predict eclipses is proposed and explained in Gerald S. Hawkins (with John B. White): *Stonehenge Decoded* (Garden City, N.Y.: Doubleday, 1965); Gerald S. Hawkins: *Beyond Stonehenge* (New York: Harper & Row, 1973); and Fred Hoyle: *On Stonehenge* (San Francisco: W. H. Freeman, 1977).
[6] James Legge, editor and translator: *The Chinese Classics*, volume 3, *The Shoo King* [Shu Ching] (Hong Kong: Hong Kong University Press, 1960), part 1, chapter 2, paragraphs 3–8 (pages 18–22). Legge romanizes Hsi's name as He. In some recountings, Hsi's name appears as Hi. The *Shu Ching* is one of five books in the *Wu Ching* (Five Classics), the sourcebooks of the Confucian tradition.
[7] James Legge, editor and translator: *The Chinese Classics*, volume 3, *The Shoo King* (Hong Kong: Hong Kong University Press, 1960), part 3, book 4, chapter 2, paragraph 4 (pages 165–166; notes continue to page 168).
[8] Joseph Needham and Wang Ling: *Science and Civilisation in China*, volume 3, *Mathematics and the Sciences of the Heavens and the Earth* (Cambridge: At the University Press, 1959), page 422. For the record, there were no total solar eclipses visible in China through this four-year period, and no partials of any consequence. Perhaps, if there is any historical foundation to this story, the court astronomer predicted that a major eclipse would occur and it did not, or perhaps the problem concerned lunar eclipses.
[9] Joseph Needham and Wang Ling: *Science and Civilisation in China*, volume 3, *Mathematics and the Sciences of the Heavens and the Earth* (Cambridge: At the University Press, 1959), page 409. Most of this information is also available in Colin A. Ronan's abridgment of Needham's work, *The Shorter Science and Civilisation in China*, volume 2 (Cambridge: Cambridge University Press, 1981).
[10] Anthony F. Aveni: *Skywatchers of Ancient Mexico* (Austin: University of Texas Press, 1980), page 181.
[11] Bernardino de Sahagún: *Florentine Codex; General History of the Things of New Spain*, book 7, *The Sun, Moon, and Stars, and the Binding of the Years*, translated from the Aztec by Arthur J. O. Anderson and Charles E. Dibble (Santa Fe, New Mexico: School of American Research; Salt Lake City: University of Utah, 1953), pages 36 and 38.

4
Eclipses in Mythology

> [T]here was at the same time something in its singular and wonderful appearance that was appalling: and I can readily imagine that uncivilised nations may occasionally have become alarmed and terrified at such an object...
>
> Francis Baily (1842)

Two great lights brighten the heavens. Life depends on them. The disappearance of one threatens the order of the universe and life itself. Through the ages, most cultures responded to eclipses of the Sun and Moon with folklore to explain and endure the eerie events.

Solar eclipse mythology might be divided into several themes, and each of these themes is found scattered throughout the world:

- A celestial being (usually a monster) attempts to destroy the Sun.
- The Sun fights with its lover the Moon.
- The Sun and Moon make love and discreetly hide themselves in darkness.
- The Sun god grows angry, sad, sick, or neglectful.[1]

Within these myths is a great truth. The harmony and well-being of Earth is dependent on the Sun and Moon. Abstract science cannot convey this profound realization as powerfully as a myth in which the celestial bodies come to life.

The Sun for Lunch

Most often in mythology a solar eclipse is considered to be a battle between the Sun and the spirits of darkness. The fate of Earth and its inhabitants hangs in the balance. With so much at stake, the people enduring an eclipse were anxious to help the Sun in this struggle if they could.

In the mythology of the Norse tribes, Loki, an evil enchanter, is put in chains by the gods. In revenge, he creates giants in the shape of wolves. The mightiest is Mânagarmer (Moon-Wolf, also called Hati), who causes a lunar eclipse by swallowing the Moon. Sköll, another of these wolf-like giants, follows the Sun, always seeking a chance to devour it. Old French and German expressions and incantations echo this belief: "God protect the Moon from wolves."

In India, Rāhu is one of the Asurases, demons who are the elder brothers of the gods. The Asurases and the gods fight over the possession of Lakṣmī, goddess of wealth and beauty, and the possession of ambrosia. The Asurases had captured the ambrosia and Rāhu was drinking it when the god Nārāyana caught up with him, threw his discus, and sliced Rāhu in half. That is why Rāhu has a head but no body. The head flew off into the sky where it attacks the Sun and Moon, swallowing them to cause an eclipse. The severed lower body of Rāhu is Ketu and has become the constellations.

Buddhism placed the Buddha in a position to correct the celestial terrorism of Rāhu. In the midst of an eclipse, the Sun or Moon cries out to the Buddha for help. "Rāhu," says the Buddha, "let go of the [Sun], for the Buddhas pity the world." Rāhu departs in terror, fearing that, if he harms the Sun or the Moon, the Buddha will cause his head to shatter into seven parts.[2]

The Indian tale of Rāhu spread eastward into China and northward into Mongolia and eastern Siberia, where the monster's name became Arakho or Alkha. The Buryat people, living east of Lake Baikal, told of Alkha, a monster who continually pursued and swallowed the Sun and Moon. Finally, the gods were exasperated by the repeated darkening of the world and cut Alkha in two. His lower body fell to Earth, but the upper portion lives on and continues to haunt the sky. This is why the Sun and Moon still disappear from time to time: Alkha swallows them. But they soon reappear because Alkha's body cannot retain them. When eclipses are in progress, say the Buryats, the Sun and Moon pray for help, and the people respond by screaming and by throwing stones and shooting weapons into the sky to scare away the monster.

Another Buryat myth says that the eclipse-maker is Arakho, a beast who formerly lived on Earth. In those days long ago, the people were quite hairy. Arakho roamed the Earth eating the hair off their bodies until the people had become the nearly hairless creatures they are today. This annoyed the gods, who chopped Arakho in two. Thus Arakho no longer grazes on human hair. Instead, the upper portion of his body, which is still alive, eats the Sun and Moon, causing eclipses.

Rāhu appears again in Indonesia and Polynesia as Kala Rau, all head and no body, who eats the Sun, burns his tongue, and spits the Sun out.

Ancient Egypt produced a tale of the black pig, the evil god Set in disguise, who leaped into the eye of Horus, the Sun god. The story becomes confused before we learn how the eye was healed, but it may have been the work of Thout, the Moon, who regulates such disturbances as eclipses and is also the healer of eyes.[3]

Egyptian emblem of the winged Sun at the top of the Gateway of Ptolemy at Karnak. [William Tyler Olcott: *Sun Lore of All Ages*]

In northwestern France (Ille-et-Vilaine), the people recognized that eclipses occur when the Moon blocks the Sun from view. For them, the Moon's aggression took a different form. If the Moon were to cover the Sun entirely, it would stick in that position so that the Sun would never shine again.[4]

From around the world come stories of many different monsters intent on devouring the Sun and Moon. In China, it was a heavenly dog who causes eclipses by eating the Sun. In South America, the Mataguaya Indians of the pampas saw eclipses as a great bird with wings outspread, assailing the Sun or Moon. In Armenian mythology, eclipses were the work of dragons who sought to swallow the Sun and Moon. By contrast, another Armenian myth says that a sorcerer can stop the Sun or Moon

Drawing of the corona by Samuel P. Langley from the top of Pike's Peak, July 29, 1878. The elongated corona resembles the Egyptian emblem of the winged Sun. [Mabel Loomis Todd: *Total Eclipses of the Sun*]

in their courses, deprive them of light, and even force them down from the skies. Despite the Moon's size, once it has been brought down to Earth, the sorcerer can milk it like a cow.[5]

On rare occasions in mythology, the Sun and Moon are not totally innocent victims. In a variant Hindu myth, the Sun and Moon once borrowed money from a member of the savage Ḍom tribe and failed to pay it back. In retribution, the Ḍom occasionally devours the two heavenly bodies.[6]

The Original Black Holes

In two instances, the hungry monster who swallows the Sun or Moon becomes quite scientifically sophisticated in character. A western Armenian myth, said to be borrowed from the Persians, tells of two dark bodies, the children of a primeval ox. These dark bodies orbit the Earth closer than the Sun and Moon. Occasionally they pass in front of the Sun or Moon and thereby cause an eclipse.[7]

Still more remarkable is a Hindu myth which speaks of the Navagrahas, "the nine seizers." These nine "planets" that wander through the star field include the usual seven familiar to the Greeks—Sun, Moon, Mercury, Venus, Mars, Jupiter, and Saturn—plus Rāhu and Ketu, "regarded as the ascending and descending nodes" of the Moon,

Modern digital cameras allow for hand-held snapshots of totality such as this one during the 2006 eclipse in Libya. [Nikon D100 DSLR, 300 mm lens, 1/30 s, ISO 400, no tripod. © 2006 Greg Buchwald]

the shifting points in the sky where the Moon crosses the apparent path of the Sun.[8] Thus, quite correctly, the Sun would be at risk of an eclipse whenever it passed by Rāhu or Ketu.

Buddhism carried Rāhu and Ketu from India to China in the first century A.D., where they became Lo-Hou and Chi-Tu. They were imagined as two invisible planets positioned at the nodes in the Moon's path: Lo-Hou at the ascending node and Chi-Tu at the descending node. These "dark stars" were numbered among the planets and were considered to be the cause of eclipses.[9]

Love, Marriage, and Domestic Violence

A Germanic myth explained eclipses differently. The male Moon married the female Sun. But the cold Moon could not satisfy the passion of his fiery bride. He wanted to go to sleep instead. The Sun and Moon made a bet: whoever awoke first would rule the day. The Moon promptly fell asleep, but the Sun, still irritated, awoke at 2 a.m. and lit up the world. The day was hers; the Moon received the night. The Sun swore she would never spend the night with the Moon again, but she was soon sorry. And the Moon was irresistibly drawn to his bride. When the two come together, there is a solar eclipse, but only briefly. The Sun and

Moon begin to reproach one another and fall to quarreling. Soon they go their separate ways, the Sun blood-red with anger.

It was not always a fight between the Sun and Moon that caused an eclipse. Sometimes it was love, and modesty. The Tlingit Indians of the Pacific coast in northern Canada explained a solar eclipse as the Moon-wife's visit to her husband. Across the continent, in southeastern Canada, the Algonquin Indians also envisioned the Sun and Moon as loving husband and wife. If the Sun is eclipsed, it is because he has taken his child into his arms.[10]

For the Tahitians, the Sun and Moon were lovers whose union creates an eclipse. In that darkness, they lose their way and create the stars in order to light their return.

An Angry Sun

Sometimes, as in a folktale from eastern Transylvania, it is the perversion of mankind that brings on an eclipse. The Sun shudders, turns away in disgust, and covers herself with darkness. Stinking fogs gather. Ghosts appear. Dogs bark strangely and owls scream. Poisonous dews fall from the skies, a danger to man and beast. Neither humankind nor animals should consume water or eat fresh fruit or vegetables. Such beliefs persisted into the nineteenth century. This poisonous dew that supposedly accompanied eclipses could be the source of an outbreak of the plague or other epidemics. If people had to leave their homes, they wrapped a towel around their mouths and noses to strain out the noxious vapors. Clothes caught drying outdoors during a solar eclipse were considered to be infected.

The Germans were not alone in their belief that a solar eclipse brought a dangerous form of precipitation. Eskimos in southwestern Alaska believed that an unclean essence descended to Earth during an eclipse. If it settled on utensils, it would produce sickness. Therefore, when an eclipse began, every Eskimo woman turned all her pots, buckets, and dishes upside down.[11]

Yet it was not always an angry Sun that brought darkness to the Earth. When U.S. Coast Survey scientist George Davidson observed the eclipse of August 7, 1869 from Kohklux, Alaska, he found that the Indians there attributed the eclipse to an illness of the Sun. He had alerted them to the impending eclipse, but they doubted it. Halfway through the partial phase, the Indians and their chief quit work and hid in their houses: "[T]hey looked upon me as the cause of the Sun's being 'very sick and going to bed.' They were thoroughly alarmed, and overwhelmed with an indefinable dread."[12]

Sometimes in mythology, an eclipse is not a monster devouring the Sun, not a sickness of the Sun, not a fight between the Sun and Moon, not even the result of the always abundant sins of mankind. Sometimes an eclipse is what in sports would be called an unforced error. The Bella Coola Indians of the Pacific coast in Canada had a myth that began with a remarkable observational description of the Sun's apparent annual pathway though the sky. The trail of the Sun, they said, is a bridge whose width is the distance between the summer and winter solstices, the northernmost and southernmost positions of the Sun. In the summer, the Sun walks on the right side of this bridge; in the winter, he walks on the left. The solstices are where the Sun sits down. Accompanying the Sun on his journey are three guardians who dance about him. Sometimes the Sun simply drops his torch, and thus an eclipse occurs.[13]

Warding Off the Evil

Corruption and death are a frequent theme of eclipse myths. Evil spirits descend to Earth or emerge from underground during eclipses.

On June 16, 1406, says an enlightened and bemused French chronicler, "between 6 and 7 a.m., there was a truly wonderful eclipse of the Sun which lasted nearly half an hour. It was a great shame to see the people withdrawing to the churches and believing that the world was bound to end. However the event took place, and afterward the astronomers gathered and announced that the occurrence was very strange and portended great evil."[14] (Half an hour is too long for totality, but the right length for conspicuously reduced light surrounding totality.)

For Hindus, the place to be during an eclipse was in the water, especially in the purifying current of the Ganges.[15] This Hindu practice of immersing oneself in water was known to the French philosopher and popularizer of science Bernard Le Bovier de Fontenelle, as recorded in his *Entretiens sur la pluralité des mondes* (Conversations on the Plurality of Worlds) in 1686.

> *All over India, they believe that when the sun and moon are eclipsed, the cause is a certain dragon with very black claws which tries to seize those two bodies, wherefore at such times the rivers are seen covered with human heads, the people immersing themselves up to the neck, which they regard as a most devout position, and implore the sun and moon to defend themselves well against the dragon.*[16]

A Participatory Event

From around the world come reports of people trying to assist the Sun and Moon against the peril of eclipse. Screaming, crying, and shouting are supposed to encourage the Sun and Moon to escape the clutches of the evil spirit. Historians recorded that Germans watching a lunar eclipse in the Middle Ages chanted in unison, "Win, Moon."

The Sun and Moon were the supreme gods of the Indians of Colombia. When these gods were threatened by an eclipse, they seized their weapons and made warlike sounds on their musical instruments. They also shouted to the gods, promising to mend their ways and work hard. To prove it, they watered their corn and worked furiously with their tools during the eclipse.[17]

People frequently augmented their voices with the clanging of metal pots, pans, and knives. The Chippewa Indians in the northeastern United States and southeastern Canada went even further. Seeing the Sun's light being extinguished, though not by a monster, they shot flaming arrows into the sky, hoping to rekindle the Sun.[18] The Sencis of eastern Peru did the same, but to scare off a savage beast attacking the Sun.

Ethnographers descended on the Kalina tribe in Suriname to collect their folklore and watch their behavior as the total eclipse of June 30, 1973 neared. In Kalina mythology, the Sun and Moon are brothers. They usually get along well together, but occasionally they have sudden and ferocious quarrels that endanger mankind. At such times, it is important to separate the combatants by making a maximum of noise: banging on tools, hollow objects, and instruments. The fading of the Sun or Moon means that one has been knocked unconscious. When this happens, the tribesmen yell, "Wake up, Papa!" Papa, here, is a term of respect, not an indication that the Kalina consider themselves descended from the Sun or Moon. After the 1973 eclipse, invisible because of clouds but noticeable because of darkening, the old women (the pot makers) rounded up the children and used branches to spread white clay all over them, somewhat too vigorously for the children's enjoyment. Then they smeared the women and finally the men from head to foot with white clay. The white clay was the blood of the injured Moon that had dripped onto the ground. It was necessary to wash oneself with the Moon's blood to restore purity in man and whiteness to the Moon. After an hour, the tribesmen washed themselves off in the river.[19]

In ancient Mexico and Central America, the most important god was represented as a plumed serpent. For the Maya, he was Kukulcán; for the Aztecs, Quetzalcóatl. At an eclipse of the Moon and most especially during an eclipse of the Sun, a special snake was killed and eaten.[20]

In prehistoric times, screams, cries, banging noises, and prayer may not have been deemed adequate to ward off eclipses or their effects. In many places around the world, human sacrifice was performed at the appearance of unexpected and confusing sights, such as an eclipse or a comet. Yet few eclipse myths refer to human sacrifice, suggesting that this practice had largely been abandoned before most eclipse myths were preserved. An exception was in Mexico and Central America, where the

Spanish invaders saw the Aztecs and their neighbors carry out human sacrifice in the early sixteenth century. For the Aztecs, almost any natural or political event was commemorated with a sacrifice. On the occasion of an eclipse, the Sun was in need of help from people, just as he had constant help from the dog Xolotl (Sho-LOT-uhl). Xolotl was the god of human monstrosities (including twins), so it was humpbacks and dwarfs who were sacrificed to the Sun to help him prevail.[21]

Solar eclipses boded ill for everyone, it seems, except prospectors. In Bohemia, people believed that a solar eclipse would help them find gold.

Notes and References

Epigraph: Francis Baily: "Some Remarks on the Total Eclipse of the Sun, on July 8th, 1842," *Memoirs of the Royal Astronomical Society*, volume 15, 1846, page 6.

[1] Viktor Stegemann: "Finsternisse" in Hanns Bächtold-Stäubli, editor: *Handwörterbuch des Deutschen Aberglaubens*, Bände 2 (Berlin: W. de Gruyter, 1930), columns 1509–1526. Germanic eclipse lore described in this chapter comes from this article unless otherwise noted.

[2] Arthur Berriedale Keith: *Indian Mythology*, volume 6 of *The Mythology of All Races* (Boston: Marshall Jones, 1917), pages 151, 192.

[3] Wilhelm Max Müller: *Egyptian Mythology*, volume 12 of *The Mythology of All Races* (Boston: Marshall Jones, 1918), pages 33, 90, 124–125.

[4] Paul Yves Sébillot: *Le folk-lore de France*, tome 1, *Le ciel et la terre* (Paris: Librairie orientale & américaine, 1904), page 40.

[5] John C. Ferguson: *Chinese Mythology*, volume 8 of *The Mythology of All Races* (Boston: Marshall Jones, 1928), page 84. Hartley Burr Alexander: *Latin-American Mythology*, volume 11 of *The Mythology of All Races* (Boston: Marshall Jones, 1920), page 319. Mardiros H. Ananikian: *Armenian Mythology*, volume 7 of *The Mythology of All Races* (Boston: Marshall Jones, 1925), page 48.

[6] Arthur Berriedale Keith: *Indian Mythology*, volume 6 of *The Mythology of All Races* (Boston: Marshall Jones, 1917), pages 232–233.

[7] Mardiros H. Ananikian: *Armenian Mythology*, volume 7 of *The Mythology of All Races* (Boston: Marshall Jones, 1925), page 48.

[8] Arthur Berriedale Keith: *Indian Mythology*, volume 6 of *The Mythology of All Races* (Boston: Marshall Jones, 1917), page 234.

[9] Joseph Needham and Wang Ling: *Science and Civilisation in China*, volume 3, *Mathematics and the Sciences of the Heavens and the Earth* (Cambridge: At the University Press, 1959), page 228.

[10] Hartley Burr Alexander: *North American Mythology*, volume 10 of *The Mythology of All Races* (Boston: Marshall Jones, 1916), pages 25, 277.

[11] James George Frazer: *Balder the Beautiful*, volume 1; *The Golden Bough*, volume 10 (London: Macmillan, 1930), page 162.

[12] Mabel Loomis Todd: *Total Eclipses of the Sun*, revised edition (Boston: Little, Brown, 1900), page 131.

[13] Hartley Burr Alexander: *North American Mythology*, volume 10 of *The Mythology of All Races* (Boston: Marshall Jones, 1916), page 255.
[14] Paul Yves Sébillot: *Le folk-lore de France*, tome 1, *Le ciel et la terre* (Paris: Librairie orientale & américaine, 1904), page 52. The chronicler was Jean Juvénal des Ursins.
[15] Arthur Berriedale Keith: *Indian Mythology*, volume 6 of *The Mythology of All Races* (Boston: Marshall Jones, 1917), page 234.
[16] A free translation from the "Second Soir." Compare Bernard Le Bovier de Fontenelle: *A Plurality of Worlds*, translated by John Glanvill (England: Nonesuch Press, 1929), with the excerpt in François Arago: *Popular Astronomy*, volume 2, translated by W. H. Smyth and Robert Grant (London: Longman, Brown, Green, Longmans, and Roberts, 1858), page 349.
[17] Hartley Burr Alexander: *Latin-American Mythology*, volume 11 of *The Mythology of All Races* (Boston: Marshall Jones, 1920), pages 277–278.
[18] James George Frazer: *The Magic Art*, volume 1; *The Golden Bough*, volume 1 (London: Macmillan, 1926), page 311.
[19] Patrick Menget: "30 juin 1973: station de Surinam," *Soleil est mort; l'éclipse totale de soleil du 30 juin 1973* (Nanterre, France: Laboratoire d'ethnologie et de sociologie comparative, 1979), pages 119–142.
[20] Hartley Burr Alexander: *Latin-American Mythology*, volume 11 of *The Mythology of All Races* (Boston: Marshall Jones, 1920), page 135.
[21] Hartley Burr Alexander: *Latin-American Mythology*, volume 11 of *The Mythology of All Races* (Boston: Marshall Jones, 1920), page 82.

5
The Strange Behavior of Man and Beast

> The Sun...
> In dim eclipse disastrous twilight sheds
> On half the nations, and with fear of change
> Perplexes monarchs.
>
> John Milton (1667)

The Human Response

One of the most dramatic responses in history to a total solar eclipse is presented by Herodotus, the first Greek historian, writing around 430 B.C.

> *[W]ar broke out between the Lydians and the Medes [major powers in Asia Minor], and continued for five years, with various success. In the course of it the Medes gained many victories over the Lydians, and the Lydians also gained many victories over the Medes.... As, however, the balance had not inclined in favour of either nation, another combat took place in the sixth year, in the course of which, just as the battle was growing warm, day was on a sudden changed into night. This event had been foretold by Thales, the Milesian, who forewarned the Ionians of it, fixing for it the very year in which it actually took place. The Medes and Lydians, when they observed the change, ceased fighting, and were alike anxious to have terms of peace agreed upon.*

A treaty was quickly made and sealed by the marriage of the daughter of the Lydian king to the son of the Median king.[1] This story indicates the awe that ancient people felt when confronted with a total eclipse of the Sun.

Modern astronomers, armed with the dates of the kings described in the account and a knowledge of the dates and paths of ancient eclipses, have generally settled upon May 28, 585 B.C. as the eclipse to which the story refers, if Herodotus, given to fanciful embellishment, can be trusted about an event that occurred a century before he was born.[2]

In his account, Herodotus credits Thales with predicting this eclipse. If so, Thales would have been the first person *known* to have calculated a future solar eclipse. Cuneiform writing on clay tablets from the Chaldean (or New Babylonian) Empire dating about two centuries later shows recognition of an 18-year-11-day rhythm in eclipses—the saros. Perhaps Thales borrowed this eclipse rhythm from the Babylonians as it was being developed. However, such a rhythm predicts not just the year but the month and precise day of the eclipse. Yet Herodotus seems amazed that Thales could be accurate to "the very year in which it actually took place." Was Herodotus so surprised that Thales could predict an eclipse accurate to the day that he simply could not believe that degree of precision and used the more conservative "year" instead? That would be out of character for the flamboyant Herodotus. Yet predicting a solar eclipse accurate to a year is not much of a trick since there are a minimum of two solar eclipses a year. The problem is to predict a total eclipse for a particular location on Earth. Could Thales have accomplished this? It is doubtful.

The saros period is actually 18 years 11⅓ days, so the Earth has spun through an extra eight hours, so each subsequent eclipse falls about one-third of the way around the world westward from the one before it. Thus successive eclipses in a saros series are almost never visible from the same site.

The eclipse in the same saros series that preceded 585 B.C. occurred on May 18, 603 B.C., with an early morning path from the northern portion of the Red Sea to the northern tip of the Persian Gulf, about 600 miles (1,000 kilometers) distant from the end of the path of the May 28, 585 B.C. eclipse. Thales could have heard reports of the 603 B.C. eclipse and used it to calculate the date for the 585 B.C. eclipse. But the saros projection would not have told him where the eclipse would be visible. Thales, then, first of the great Greek philosophers, could have warned of the *possibility* of a solar eclipse, but he could not predict from the saros period that it would be visible in Asia Minor. And there is no evidence that he had the celestial knowledge or the mathematics to calculate it from orbital considerations.

Of course the key to appreciating the story of the solar eclipse that stopped a war is the realization that people long ago were stunned by a total eclipse of the Sun and incredulous that someone could predict such an event. Quite often in ancient history, eclipses are reported to have played a decisive role in the turn of events.

Herodotus tells of another turning point in world history that he says hinged on a solar eclipse. Xerxes and his Persian army were about to march from Sardis to Abydos on their advance toward Greece.

> *At the moment of departure, the sun suddenly quitted his seat in the heavens, and disappeared, though there were no clouds in sight,*

During a battle between the Lydians and the Medes on May 28, 585 B.C., a total eclipse of the Sun occurred. It scared the soldiers so badly that they stopped fighting and signed a treaty. [Mabel Loomis Todd: *Total Eclipses of the Sun*]

> *but the sky was clear and serene. Day was thus turned into night; whereupon Xerxes, who saw and remarked the prodigy, was seized with alarm, and sending at once for the Magians, inquired of them the meaning of the portent. They replied—"God is foreshadowing to the Greeks the destruction of their cities; for the sun foretells for them, and the moon for us." So Xerxes, thus instructed, proceeded on his way with great gladness of heart.*[3]

To disaster! He reached and burned Athens, but his navy was destroyed by the Greeks and his forces had to withdraw. Twice more Xerxes invaded Greece, but each time his armies were crushed. After his last defeat, his nobles assassinated him.

Xerxes' first march against Greece actually occurred in 480 B.C., but the only major eclipse visible in the region near that date was the total eclipse of February 17, 478 B.C. Thus the story tells us less about observational astronomy in that era than about the power exercised by eclipses over the minds of men and the effectiveness of their use to heighten the drama of a story.

One final story illustrates the advance of the Greeks from superstitious dread of eclipses to an understanding of what causes them. On August 3, 430 B.C., Pericles and his fleet of 150 warships were about to sail for a raid upon their enemies.

> *But at the very moment when the ships were fully manned and Pericles had gone onboard his own trireme, an eclipse of the sun took place, darkness descended and everyone was seized with panic, since they regarded this as a tremendous portent. When Pericles saw that his helmsman was frightened and quite at a loss what to do, he held up his cloak in front of the man's eyes and asked him whether he found this alarming or thought it a terrible omen. When he replied that he did not, Pericles asked, "What is the difference, then, between this and the eclipse, except that the eclipse has been caused by something bigger than my cloak?" This is the story, at any rate, which is told in the schools of philosophy.*[4]

The eclipse was a large partial at Athens and annular about 600 miles (1,000 kilometers) to the northeast. This eclipse had also been recorded by Thucydides, without the didactic story, but exhibiting an increased awareness of the cause of eclipses: "The same summer, at the beginning of a new lunar month, the only time by the way at which it appears possible, the sun was eclipsed after noon. After it had assumed the form of a crescent and some stars had come out, it returned to its natural shape."[5]

By 1654, Paris was a center of enlightenment, but on August 12, "at the mere announcement of a total eclipse, a multitude of the inhabitants of Paris hid themselves in deep cellars."[6]

No wonder then that the sight of a total eclipse on July 29, 1878 had a powerful effect on Native Americans near Fort Sill, Indian Territory (now Oklahoma). A non-Indian described it this way:

> *It was the grandest sight I ever beheld, but it frightened the Indians badly. Some of them threw themselves upon their knees and invoked the Divine blessing; others flung themselves flat on the ground, face downward; others cried and yelled in frantic excitement and terror. Finally one old fellow stepped from the door of his lodge, pistol in hand, and, fixing his eyes on the darkened Sun, mumbled a few unintelligible words and raising his arm took direct aim at the luminary, fired off his pistol, and after throwing his arms about his head in a series of extraordinary gesticulations retreated to his*

The Shawnee Prophet Uses an Eclipse

Tenkskwatawa, the Shawnee Prophet (1775?–1837?), was an important Indian religious leader in Ohio and Indiana in the early nineteenth century. He saw great danger for his people as they increasingly adopted the customs of the European settlers, especially alcohol. He urged them to return to traditional Indian ways and to unite into a single Indian nation under the leadership of his brother Tecumseh to resist the encroachment of white men with their fraudulent treaties.

General William Henry Harrison, later president of the United States, was at that time the governor of Indiana Territory, where the Shawnee Prophet was successfully recruiting converts to his Indian religious revival. Seeking to undermine the credibility of the Shawnee Prophet as a shaman, Harrison urged Indians to demand proof from the Prophet that he could perform miracles. Thinking in biblical terms, Harrison asked if Tenkskwatawa could "cause the Sun to stand still, the Moon to alter its course, the rivers to cease to flow, or the dead to rise from their graves."

The followers of the Shawnee Prophet did not need such displays, but Tenkskwatawa was a canny politician. He proclaimed that on June 16, 1806, he would blot the Sun from the sky as a sign of his divine powers. Whether he knew of this total eclipse from a British agent or from an almanac is uncertain, but a great many Indians gathered at the Shawnee Prophet's camp as the appointed day dawned clear.

At the proper moment, the Prophet, in full ceremonial regalia, pointed his finger at the Sun, and the eclipse began. When the Prophet called out to the Good Father of the Universe to remove his hand from the face of the Sun, the light gradually returned to the Earth. Response to the Prophet's performance was overwhelming and his fame spread rapidly and widely. Harrison's condescension had backfired, to his embarrassment.

But the westward migration of European settlers was unstoppable. In the Battle of Tippecanoe in 1811, Harrison destroyed the Shawnee Prophet's religious center, killing many Indians, and breaking the power of Tenkskwatawa.[a]

---

[a] See especially Laurence A. Marschall: "A Tale of Two Eclipses," *Sky & Telescope*, volume 57, February 1979, pages 116–118.

> *own quarters. As it happened, that very instant was the conclusion of totality. The Indians beheld the glorious orb of day once more peep forth, and it was unanimously voted that the timely discharge of that pistol was the only thing that drove away the shadow and saved them from... the entire extinction of the Sun.*[7]

The Animal Response

Noting their own primal response to the daytime darkening of the Sun, people through the ages have been fascinated by the reaction of animals to a total eclipse. Reports go back more than 750 years. In describing the eclipse of June 3, 1239, Ristoro d'Arezzo wrote: "[W]e saw the whole body of the Sun covered step by step... and it became night... and all the animals and birds were terrified; and the wild beasts could easily be caught... because they were bewildered."[8]

As a university student in Portugal, astronomer Christoph Clavius saw the total eclipse of August 21, 1560: "[S]tars appeared in the sky, and (miraculous to behold) the birds fell down from the sky to the ground in terror of such horrid darkness."[9]

In 1706 at Montpellier in southern France, observers reported that "bats flitted about as at the beginning of night. Fowls and pigeons ran precipitately to their roosts." In 1715, the French astronomer Jacques Eugène d'Allonville, Chevalier de Louville, traveled to London for the eclipse and observed that at totality "horses that were laboring or employed on the high roads lay down. They refused to advance."[10]

By 1842, some people were even conducting behavioral experiments on their pets. "An inhabitant of Perpignan [France] purposely kept his dog without food from the evening of the 7th of July. The next morning, at the instant when the total eclipse was going to take place, he threw a piece of bread to the poor animal, which had begun to devour it, when the sun's last rays disappeared. Instantly the dog let the bread fall; nor did he take it up again for two minutes, that is, until the total obscuration had ceased; and then he ate it with great avidity."[11]

William J. S. Lockyer, son of the pioneering solar spectroscopist, traveled to Tonga for an eclipse in 1911. The weather conditions were miserable and the insects numerous and very hungry. He and his colleagues caught only a brief view of the corona through thin clouds and the scientific results were meager. The only members of his team with good results were those studying animal behavior. The horses did not seem to notice the darkening, but fowl ran home to roost and pigs lay down. Flowers closed. But most memorable of all were the insects, which had been completely silent until the moment of totality and then sang as if it were night. "The noise," recalled Lockyer, "was most impressive, and will remain in my memory as a marked feature of that occasion."[12]

Notes and References

Epigraph: John Milton: *Paradise Lost*, book 1, lines 594 and 597–599. See John Milton: *The Complete Poems* (New York: Crown Publishers, 1936), page 24.

[1] Herodotus: *The History*, volume 1, translated by George Rawlinson; Everyman's Library, volume 405 (London: J. M. Dent, 1910), book 1, chapter 74, pages 36–37.

[2] Robert R. Newton lists three annular eclipses seen in the region during a 50-year period which he feels are equally likely to have given rise to this story, although an annular eclipse is not nearly as spectacular as one that is total. *Ancient Astronomical Observations and the Accelerations of the Earth and Moon* (Baltimore: Johns Hopkins Press, 1970), pages 94–97.

[3] Herodotus: *The History*, volume 2, translated by George Rawlinson; Everyman's Library, volume 406 (London: J. M. Dent, 1910), book 7, chapter 37, page 136.

[4] Plutarch: *The Rise and Fall of Athens: Nine Greek Lives*, translated by Ian Scott-Kilvert (Baltimore: Penguin Books, 1960), pages 201–202. Ironically, Pericles' raid was disastrous for the Athenian forces; they fell victim to the plague. Pericles was fined and temporarily stripped of power.

[5] Thucydides: *History of the Peloponnesian War*, translated by Richard Crawley (New York: E. P. Dutton, 1910), book 2, paragraph 28. Stars would not have been visible at Athens. Perhaps Thucydides heard reports from where the eclipse was annular and incorporated them into his account.

[6] From the chapter "Second soir" in Bernard Le Bovier de Fontenelle: *Entretiens sur la pluralité des mondes* (Paris: Chez la veuve C. Blageart, 1686), cited in François Arago: *Popular Astronomy*, volume 2, translated by W. H. Smyth and Robert Grant (London: Longman, Brown, Green, Longmans, and Roberts, 1858), pages 359–360.

[7] Mabel Loomis Todd: *Total Eclipses of the Sun*, revised edition (Boston: Little, Brown, 1900), pages 141–142.

[8] F. Richard Stephenson and David H. Clark: *Applications of Early Astronomical Records*, Monographs on Astronomical Subjects, number 4 (New York: Oxford University Press, 1978), page 9.

[9] F. Richard Stephenson and David H. Clark: *Applications of Early Astronomical Records*, Monographs on Astronomical Subjects, number 4 (New York: Oxford University Press, 1978), page 14.

[10] François Arago: *Popular Astronomy*, volume 2, translated by W. H. Smyth and Robert Grant (London: Longman, Brown, Green, Longmans, and Roberts, 1858), page 359.

[11] François Arago: *Popular Astronomy*, volume 2, translated by W. H. Smyth and Robert Grant (London: Longman, Brown, Green, Longmans, and Roberts, 1858), page 362.

[12] William J. S. Lockyer: "The Total Eclipse of the Sun, April 1911, as Observed at Vavau, Tonga Islands," in Bernard Lovell, editor: *Astronomy*, volume 2, The Royal Institution Library of Science (Barking, Essex: Elsevier Publishing, 1970), pages 190–191.

illusion, since the Sun is gaseous throughout. We are really seeing the layer of the Sun in which the density and ionization of atoms is so great that the gas becomes opaque. This region that provides the Sun with the appearance of a surface is called the *photosphere* ("light sphere"). It is only about 200 miles (300 kilometers) deep. It is there that we notice sunspots, areas of magnetic disturbance on the Sun the size of Earth or Jupiter or even larger, appearing and disappearing and riding along in the photosphere with the Sun's rotation. The sunspots increase and decrease in number over a period of about 11 years.

If we examine the photosphere more closely, we see that it has a mottled appearance, created by rising columns of hot, bright gas surrounded by darker haloes where the gas has cooled and is descending, to be heated again. The photosphere is boiling. These *granulations* of upwelling and downfalling gases are typically 500 miles (800 kilometers) in diameter and last for five or ten minutes. The gases in the granulations, the topmost layer of the convection zone, carry with them magnetic fields from deep within the Sun. As these gases tumble up and down, the magnetic lines of force twist and snap, and this magnetic turbulence governs the behavior of gases in the photosphere and in the solar atmosphere above it.

Above the photosphere is the *chromosphere* ("color sphere"), aptly named for its vibrant reddish color. Seen with a telescope, the rim of the

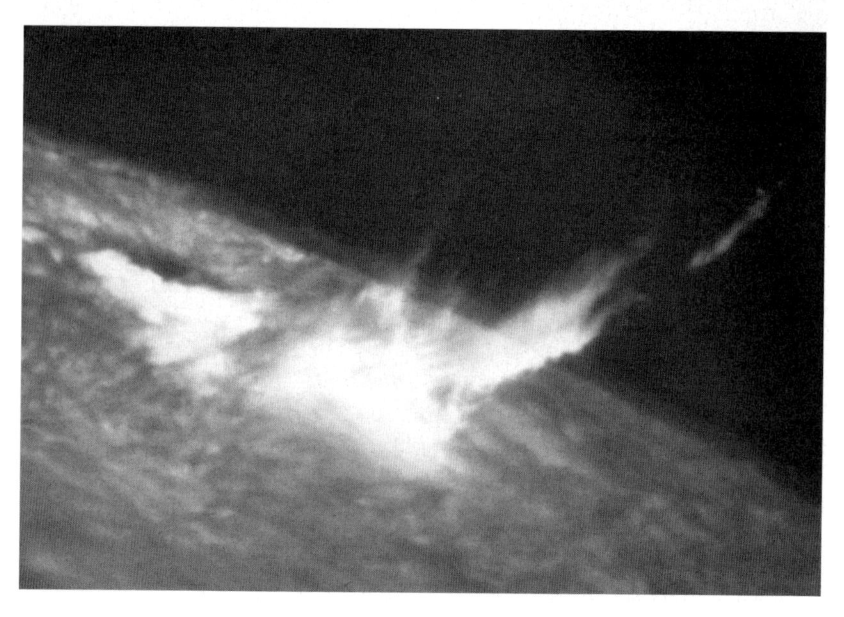

Solar flares are violent explosions in the Sun's atmosphere that release huge amounts of energy and subatomic particles. [National Optical Astronomy Observatories (NOAO)]

Sun looks like a fire-ocean.[3] Its lower levels are cooler than the white-hot photosphere, with temperatures of about 7,200 °F (4,000 °C). Its upper layers, however, are much hotter than the photosphere—about 18,000 °F (10,000 °C). This temperature causes hydrogen atoms to emit a red wavelength that is primarily responsible for the chromosphere's fiery color.[4] The chromosphere is a thin atmospheric layer, only about 1,600 miles (2,500 kilometers) thick, although there are no sharp boundaries above or below.

Seen under high magnification along the rim of the Sun, the chromosphere is not a smooth layer of gas. It looks like an erratic forest of thin, topless trees. These spiked features that compose the chromosphere are known as *spicules*. They are less than 400 miles (700 kilometers) in diameter but may tower thousands of miles high, reaching out of the chromosphere and into the corona. These spicules may be not only columns of rising gas but also the outlines of magnetic flux tubes that transfer magnetic fields from the photosphere to the corona.

It is in the photosphere and chromosphere that prominences and flares are rooted and stretch upward into the corona. They too are

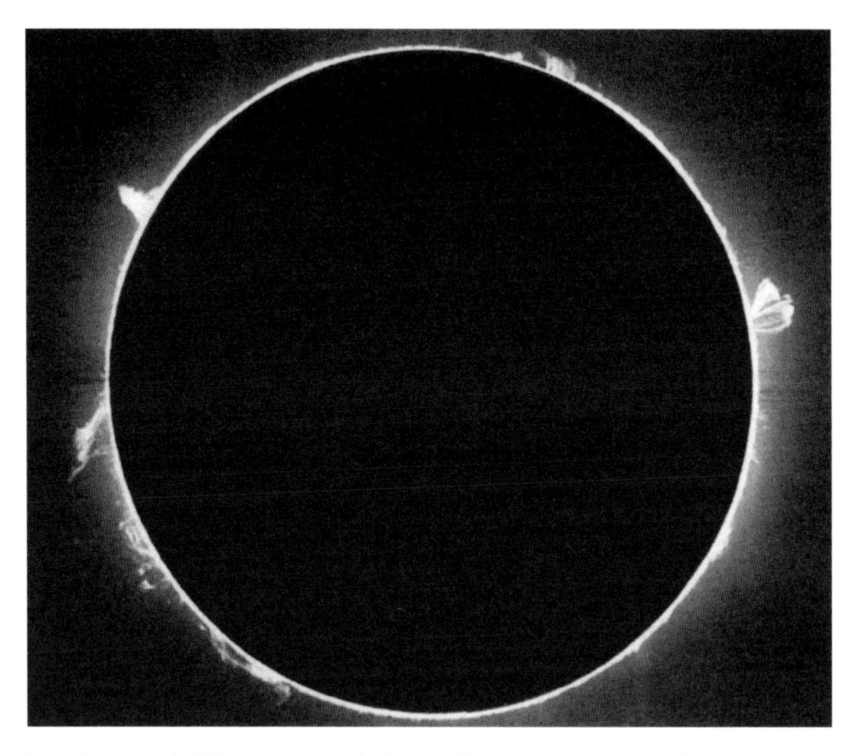

Prominences of different sizes and shapes silhouetted against the limb of the Sun, December 9, 1929. [Carnegie Observatories]

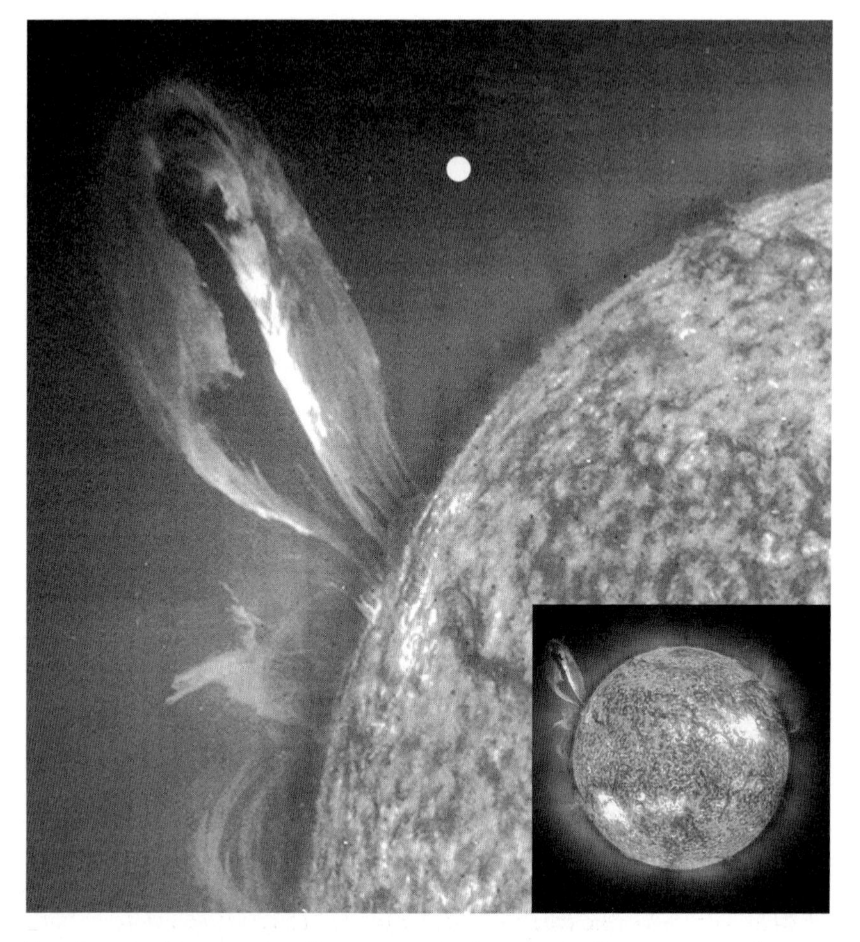

Eruptive prominences can affect communications, navigation systems, and power grids, while producing auroras in the night skies. This eruptive prominence on July 24, 1999 extends 35 Earths out from the Sun. The small white disk is the size of Earth for comparison. [ESA/NASA SOHO EIT]

transient features of the Sun that vary with the rhythm of the sunspots below them. *Prominences* are the same temperature as the upper chromosphere and glow with the same red color of excited hydrogen. They are condensed clouds of solar gas, but much cooler and denser than the surrounding corona. The prominences are bent and twisted by local magnetic fields. Magnetic forces keep the gases in the prominences from all falling back to the surface, which would take only about 15 minutes if gravity were the only force acting. Most often these clouds, the prominences, do slowly rain material back toward the surface, but occasionally they erupt outward.

Flares are much stronger eruptions, also triggered by the magnetic activity of the Sun, that launch great torrents of mass and energy from the Sun at millions of miles an hour.

The chromosphere is also the birthplace of *solar tornados*, almost the size of the Earth, whirling at speeds up to 300,000 miles per hour (500,000 kilometers per hour), spiraling up and through the corona, probably contributing fast-moving atomic fragments to the solar wind of particles flowing spaceward from the Sun.[5] How these solar tornados form and function is not certain, but, as with virtually every feature seen on the Sun's surface and in its atmosphere, the explanation almost certainly involves twisted magnetic fields.

In the upper reaches of the chromosphere and extending outward into the corona is a realm called the *transition region*, so named because the temperature there suddenly climbs to over 1.8 million °F (1 million °C).

Why should the temperature of the *corona* ("crown"), the outer atmosphere of the Sun, be so high? It is much hotter than the photosphere and chromosphere, yet those layers are closer to the core where the Sun generates its energy. The answer again lies in the Sun's magnetic fields. The Sun not only behaves like a giant bar magnet (like the Earth with its magnetic poles), but the Sun is also pocked by many local magnetic regions with intensities greater than the Sun's polar magnetism. These surface sites of magnetic activity, the largest marked by sunspots, are induced by varying magnetic fields in the Sun's interior. These magnetic fields are borne to the surface by the columns of hot gases seen as granulations in the photosphere. The random motion of the rising and falling gases contort the lines of magnetic force. These magnetic fields stretch into the chromosphere and corona as loops and arches, as if they were invisible electric wires attached to terminals in the photosphere. The twisted and coiled magnetic fields induce electrical currents in the corona, which heat the gases there. But explaining exactly how this heating is accomplished in the near vacuum of the corona remains a challenge to solar astronomers.[6]

The gases rising from the surface of the Sun are all so hot that their atoms are missing some or all of their electrons. This plasma (ionized gas) expands as it rises from the surface of the Sun. As it expands, its density declines and these charged gases are more easily warped by the Sun's magnetic fields. The patterns of this magnetic control can be seen in the loops and arches of the prominences that extend high into the corona. Here and there a magnetic loop in the corona is stressed so greatly that it snaps and perhaps 10 billion tons of million-degree plasma is slung into space as a huge expanding bubble traveling as fast as 4.5 million miles per hour (2,000 kilometers per second). Such outpourings are called *coronal mass ejections*.[7] Coronal mass ejections, eruptive prominences, and flares are probably different parts of the

Coronal loops are fountains of multimillion-degree electrified gas in the atmosphere of the Sun that are 300 times hotter than the Sun's visible surface. The shape and structure of the coronal loops map out the magnetic field in the lower corona. [NASA Transition Region and Coronal Explorer (TRACE) spacecraft]

same phenomenon. They involve the expulsion of hot gases and twisted ropes of magnetic fields from the Sun, they often occur together, and all are more frequent at sunspot maximum.[8]

Coronal mass ejections enhance and create shock waves in the normal solar wind with charged particles (primarily protons and electrons) that are escaping the Sun's gravitational hold at speeds of 900,000 miles per hour (400 kilometers per second). Coronal mass ejections also carry with them some of the coronal magnetic field, and distort the existing field in the solar wind. When the altered magnetic fields and these especially energetic subatomic particles strike the Earth, they create colorful displays of the northern and southern lights—the aurora. But solar windstorms can also damage electronic equipment on satellites in Earth orbit; disrupt telephone, radio, and television transmission; and overload electric power lines, causing blackouts (as one did for 9 hours to 6 million Canadians in Quebec on March 13, 1989). Scientists now think that coronal mass ejections, rather than flares, are the principal cause of the aurora and other geomagnetic events.

A coronal mass ejection blasts immense bubbles of hot plasma into the solar system. This January 8, 2002 image shows a widely spreading coronal mass ejection (*right half of picture*) hurling more than a billion tons of electrons and protons into space at millions of kilometers per hour. [ESA/NASA SOHO LASCO]

In visible light the corona shows graceful, delicate streamers and brushlike features.[9] However, in X-ray pictures, which better capture the activity of high-temperature gases, the corona is a riot of everchanging loops, plumes, eruptions, and contrasting light and dark regions. These dark regions of lesser activity are called *coronal holes*. It is primarily through these coronal holes, where magnetic fields are weaker, that the solar wind escapes into space. Slower and more variable solar wind flows from coronal streamers.

The coronal holes and streamers seen during a total eclipse mark the visible departure from the Sun of the solar wind. The Sun loses about 10 million tons of material a year in the solar wind. But such a loss is

A Sun-grazing comet too near the Sun for its survival, December 23, 1996. [ESA/NASA SOHO LASCO C2 coronagraph, courtesy of Steele Hill]

insignificant compared to the Sun's total mass and the rate at which the Sun is converting mass into energy at its core.

The solar wind blows outward in all directions. The Earth is a small target 93 million miles (150 million kilometers) away, so only two out of every billion particles in the solar wind will reach the Earth, to cause mischief here. But they are enough. Scientists are studying "space weather" with the hope of predicting when the Earth will be struck by a stormy blast of enhanced solar wind.

To the eye, the white flame brushes of the corona can extend outward from the Sun 2 million miles (3 million kilometers) or more until they are so tenuous that they are no longer visible. But the corona is still measurable out to the Earth and beyond, in the form of the solar wind. The Earth orbits the Sun within the Sun's rarefied outer atmosphere.

We do not normally see the chromosphere or the corona. They are concealed from our view by the overwhelming glare of the photosphere, half a million times brighter than the corona. To study the Sun, early scientists had only its surface to rely on—just the photosphere and sunspots. Progress was slow.

We do not see the chromosphere or the corona unless something blocks the glare of the photosphere so that the faint atmosphere of the Sun is revealed. In the nineteenth century, astronomers discovered that the Moon was their scientific collaborator, obscuring the Sun's surface from time to time so that they might see a part of the Sun that had never been studied before.

In less than a century, total solar eclipses, and the scientific instruments and theories they helped to stimulate, revealed the composition of the Sun, the structure of the Sun's interior, and the wonder of how the Sun shines.

Notes and References

Epigraph: Amos 8:9, *New American Standard Bible.*

[1] Solar physicists Joseph Hollweg, Charles Lindsey, and Jay Pasachoff calculated or reviewed the statistics and information in this chapter. Personal communication, October 20–November 3, 1998. Lindsey calculates that 4.26 million metric tons per second (4.69 million English tons per second) of mass are converted to energy in the core of the Sun.

[2] The figure of 14.8 trillion years is based on the Sun converting *all* its mass to energy, which, of course, it cannot do. Only 0.7% of the hydrogen mass is converted into energy; the rest becomes helium. If only the total amount of hydrogen available is considered, the Sun's life span falls to about 77 billion years. Of course, only the hydrogen close to the center of the Sun is under sufficient pressure so that the temperatures are great enough for fusion to occur. Only about 10% of the Sun's hydrogen will undergo fusion, reducing the Sun's lifetime to 8–10 billion years.

To generate and sustain a hydrogen-to-helium fusion reaction, the core of a star must have a temperature of at least 18 million °F (10 million °C). To have enough gravity to generate sufficient pressure to obtain this high a core temperature, a star must have at least 8 percent the mass of the Sun.

[3] The expressions "fire-ocean" to describe the chromosphere and "flame-brushes" to describe features in the corona were used by Agnes M. Clerke: *A Popular History of Astronomy during the Nineteenth Century*, 4th edition (London: A. and C. Black, 1902), pages 68, 175.

[4] Spectroscopically, the red of the chromosphere is produced by the hydrogen-alpha line.

[5] Solar tornados were discovered in 1998 by David Pike and Helen Mason using the Solar and Heliospheric Observatory (SOHO), a collaboration of the European Space Agency (ESA) and NASA (United States).

[6] The temperature of the corona can be misleading. The Sun's magnetic fields cause the electrically charged atoms of the corona to move at great speeds (high temperature), but the density of these ions is so low that the corona has relatively little heat (energy in a given volume). The corona is so rarefied that if you had a box there 100 miles (160 kilometers) on each side (1 million cubic miles; 4.1 million cubic kilometers), you would entrap less than a pound (0.4 kilogram) of matter. The corona is a good vacuum by laboratory standards on Earth.

Solar physicist Joe Hollweg points out that it is also hard to explain the heating of the chromosphere, which requires as much energy as the corona. Personal communication, August 3, 1998.

[7] Coronal mass ejections were discovered using NASA's Orbiting Solar Observatory 7 between 1971 and 1973 and confirmed using NASA's Solar Maximum Mission satellite in 1980 and after 1984, following repair in orbit by Space Shuttle astronauts. In the 1990s, a new generation of solar spacecraft studied corona mass ejections: Yohkoh (Japan), SOHO (*So*lar and *H*eliospheric *O*bservatory—ESA and NASA), and TRACE (*T*ransition *R*egion *a*nd *C*oronal *E*xplorer—NASA). In the 2000s, a new generation of solar observatories continues the quest: Hinode (Japan) and STEREO (*S*olar-*Te*rrestrial *Re*lations *O*bservatory—NASA).

[8] At sunspot minimum, about one coronal mass ejection a week is observed. Near sunspot maximum, two or three coronal mass ejections are observed each day on the average.

[9] Coronal structures trace out the magnetic field lines, like iron filings trace out the field around a bar magnet.

7
The First Eclipse Expeditions

> I did not expect, from any of the accounts of preceding eclipses that I had read, to witness so magnificent an exhibition as that which took place.
>
> Francis Baily (1842)

An Unlikely Beginning

Francis Baily, the man who might be said to have founded the field of solar physics, received only an elementary education, was not trained in science, and did not get around to astronomy until the age of 37. Like his father, a banker, he entered the commercial world as an apprentice when he was 14. But adventure called. When his seven years of apprenticeship expired, he sailed for the New World and spent the next two years, 1796–1797, exploring unsettled parts of North America, narrowly escaping from a shipwreck, flatboating down the Ohio and Mississippi Rivers from Pittsburgh to New Orleans, and then hiking nearly 2,000 miles back to New York through territory inhabited mostly by Indians. He liked the United States so well that he planned to marry and become a citizen, but he finally abandoned those plans and returned home in 1798.

Back in England, he began efforts to mount an expedition to explore the Niger River in Africa. He could not raise enough money, however, so he became a stockbroker. To dedication and enthusiasm he quickly added a reputation for intelligence and integrity, and he made a fortune. He exposed stock-exchange fraud and helped clean it up. He published a succession of explanations of life insurance methods and comparisons of insurance companies, which became wildly popular. He also published a chart of world history that was equally popular, confirming the nickname given to him in his apprentice days: the Philosopher of Newbury (his birthplace).

His first astronomical paper (1811) tried to identify the solar eclipse allegedly predicted by Thales. In 1818, he called attention to an annular

Francis Baily. [Royal Astronomical Society]

eclipse of the Sun coming in 1820, and he observed it from south-eastern England. That same year, he became one of the founders of the Astronomical Society of London, later the Royal Astronomical Society.

In 1825, he retired from the stock market to devote all his time to his new profession. He was 51 years old. His revisions of a series of old star catalogs were considered so valuable that the Royal Astronomical Society twice awarded him its Gold Medal and four times elected him president. Although he was not renowned as an observer, he had an abiding fascination with eclipses, a good eye for detail, and the ability to express what he saw.

Thus it was in 1836 that a few words from Francis Baily sparked the immediate, intense, and unending study of the physical properties of the Sun that had been generally ignored or discounted until then. He traveled to an annular eclipse of the Sun in southern Scotland and watched on May 15, 1836 as mountains at the Moon's limb occulted the face of the Sun but allowed sunlight to pour through the valleys between them so that the ring of sunlight around the rim of the Moon was broken up into "a row of lucid points, like a string of bright beads."[1] With those

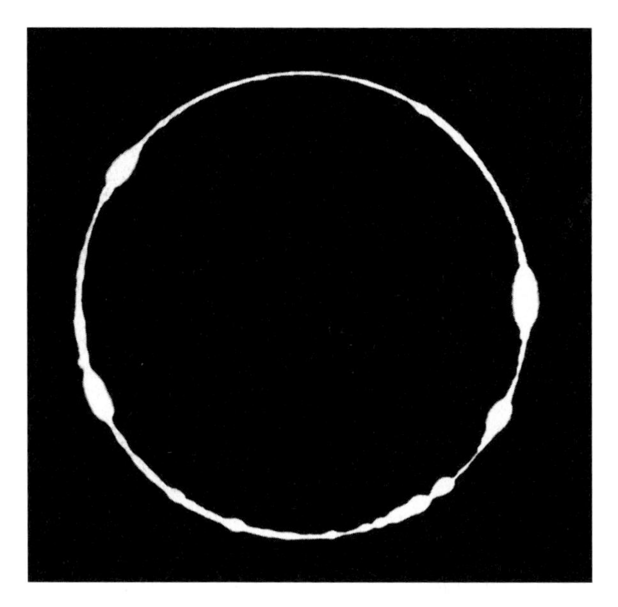

Baily's beads during the annular eclipse of April 28, 1930. [Lick Observatory]

words, Baily generated fervor for solar physics and founded the industry of eclipse chasing.

The Surprise of Totality

At the next accessible eclipse, July 8, 1842, a high percentage of the astronomers of Europe migrated to southern France and northern Italy to see "Baily's beads." Baily, now 68 years old, went too.

This was not an annular eclipse, as Baily had seen twice before. It was total. No European astronomer then alive had ever seen a total eclipse.

Baily set up his telescope at an open window in a building at the university in Pavia, Italy: "[A]ll I wanted was to be left *alone* during the whole time of the eclipse, being fully persuaded that nothing is so injurious to the making of accurate observations as the intrusion of unnecessary company."[2]

Again Baily's beads were visible—to Baily at least. George Airy, England's Astronomer Royal, observing from Turin, Italy, did not see them. Baily was just jotting down the time of appearance and duration of the beads

> *. . . when I was astounded by a tremendous burst of applause from the streets below, and at the same moment was electrified at the*

> *sight of one of the most brilliant and splendid phenomena that can well be imagined. For, at that instant the dark body of the moon was suddenly surrounded with a corona, or kind of bright glory, similar in shape and relative magnitude to that which painters draw round the heads of saints, and which by the French is designated an auréole.*

Baily was not the first to use the word *corona* to designate the glowing outer atmosphere of the Sun visible during a total eclipse, but his striking description of it caught everyone's attention as it never had before and forever united the word with the phenomenon.

> *[W]hen the total obscuration took place, which was instantaneous, there was an universal shout from every observer... I had indeed anticipated the appearance of a luminous circle round the moon during the time of total obscurity: but I did not expect, from any of the accounts of preceding eclipses that I had read, to witness so magnificent an exhibition as that which took place.... It riveted my attention so effectually that I quite lost sight of the string of beads, which however were not completely closed when this phenomenon first appeared.*

There was so much to see, said Baily, that in future eclipses, each observer should be assigned a single observing task.

> *Splendid and astonishing, however, as this remarkable phenomenon really was, and although it could not fail to call forth the admiration and applause of every beholder, yet I must confess that*

Drawings of corona and prominences at different eclipses. *Left*: July 8, 1842; *right*: July 28, 1851. [François Arago: *Popular Astronomy*, edited and translated by W. H. Smyth and Robert Grant]

> *there was at the same time something in its singular and wonderful appearance that was appalling: and I can readily imagine that uncivilised nations may occasionally have become alarmed and terrified at such an object, more especially in times when the true cause of the occurrence may have been but faintly understood, and the phenomenon itself wholly unexpected.*

It was the last eclipse Francis Baily was to see. Two years later he died. Of him, historian Agnes M. Clerke wrote: "He was gentle as well as just; he loved and sought truth; he inspired in an equal degree respect and affection.... Few men have left behind them so enviable a reputation."[3]

Before Baily

Never again would a total eclipse over an inhabited land mass go unattended by professional and amateur astronomers, even when the observers had to travel to remote sites halfway around the world. The corona, Baily's beads, prominences, shadow bands—all had been seen before, many times, throughout the world. But now they commanded attention and explanation.

The first written record of the corona may be Chinese characters inscribed on oracle bones from about 1307 B.C. that say "three flames ate up the Sun, and a great star was visible."[4] Or were these flames prominences instead? The first unequivocal description of the corona comes from a chronicler more than 2,000 years later, observing from Constantinople the eclipse of December 22, 968 A.D.

The great astronomer Johannes Kepler did not see a total eclipse himself, but from the reports he read, he concluded that the corona must be material around the Sun and not the Moon. Giacomo Filippo Maraldi, an Italian-born French astronomer, provided evidence that the corona is part of the Sun because the Moon traverses the corona during a solar eclipse; the corona does not move with the Moon but stays fixed around the Sun.

The Spanish astronomer José Joaquin de Ferrer, well ahead of his time, traveled to the New World to observe total eclipses in Cuba in 1803 and Kinderhook, New York in 1806. He was probably the first to use the word "corona" to describe the glow of the outer atmosphere of the Sun seen during a total eclipse.

> *The disk [of the Moon] had round it a ring of illuminated atmosphere, which was of a pearl colour... From the extremity of the ring, many luminous rays were projected to more than 3 degrees distance.—The lunar disk was ill defined, very dark, forming a contrast with the luminous corona...*[5]

Ferrer quite correctly attributed the corona to the Sun. If this glow belonged to the Moon, he calculated, the lunar atmosphere would extend upward 348 miles (560 kilometers)—50 times more extensive than the atmosphere of Earth. Thus, he concluded, the corona "must without any doubt belong to the Sun." Baily, too, at the 1842 eclipse, attributed the corona to the Sun.

The first sure report of solar prominences came from Julius Firmicus Maternus in Sicily, who noticed them during the annular eclipse of July 17, 334. Edmond Halley, the great English astronomer, saw them clearly as bright red protrusions during the total eclipse of May 3, 1715.

Baily had pointed the way. But what was the significance of the corona and the prominences? New techniques—photography and spectroscopy—were emerging with the capability to record and explore these features of the Sun. In photography and spectroscopy, observational astronomy gained its two most powerful tools to augment the telescope.

The stage was set for the eclipse of July 28, 1851. The astronomical world gathered along its path of totality through Scandinavia and Russia. They solved one mystery and uncovered another.

The Debut of Photography

The first successful photograph of the Sun in total eclipse was a daguerreotype taken on July 28, 1851 by a professional photographer

First photograph of the Sun in total eclipse, July 28, 1851. Only the last name of the photographer is known: Berkowski. [Courtesy of Dorrit Hoffleit]

named Berkowski, assigned to the task by August Ludwig Busch, director of the observatory in Königsberg, Prussia.[6] The inner corona and prominences are clearly visible. Yet, for now, photography was just a curiosity, a promising experiment. Hard-core science was still carried out by visual observations made through telescopes.

At the 1851 eclipse, two teams of astronomers provided proof of what most astronomers suspected: that prominences were part of the Sun. Robert Grant and William Swan from the United Kingdom and Karl Ludwig von Littrow from Austria documented how the eastward motion of the Moon across the face of the Sun covered prominences along the east rim of the Sun as totality began and then uncovered prominences along the west rim as totality ended, demonstrating that the prominences belonged to the Sun and that the Moon was only passing in front of them.

Often in science, observations that lead to the solution of one problem simultaneously reveal a fascinating new problem. As George B. Airy, England's Astronomer Royal, was observing this eclipse, he noticed a jagged edge to the solar atmosphere just above the edge of the Moon. He called it the *sierra*, thinking that he might be looking at mountains on the Sun.

Drawing by Lilian Martin-Leake from a telescopic view of the chromosphere and the corona, May 28, 1900, showing red spicules in the chromosphere that George Airy had thought were mountains. [Annie S. D. Maunder and E. Walter Maunder: *The Heavens and Their Story*]

Airy thus became the first to call attention to the chromosphere, the lowest level of the transparent solar atmosphere immediately above the opaque photosphere that creates the appearance of a surface for the Sun. The jaggedness that Airy observed was actually innumerable small jets of rising gas called spicules.

By 1860, daguerreotypes were obsolete, superseded by the faster wet plate (collodion) photographic process. Before photography, astronomers could only describe or sketch what they saw. The early photographic emulsions were not as sensitive to detail as desired, but they were more objective than the human eye.

On July 18, 1860, a total eclipse was visible from Europe and found astronomers waiting with improved cameras. Prominences remained a matter of high priority and they encountered the resourcefulness of British astronomer Warren De La Rue and Italian astronomer and Jesuit priest Angelo Secchi.

De La Rue's well-to-do family provided him with a solid education, after which he entered his father's printing business. There he demonstrated a great affinity for machinery. He could make any instrument or device run better. He was also a fine draftsman, and it was this talent that lured him into astronomy. He could produce drawings of the planets, Moon, and Sun that were better than those of astronomers. But no sooner had he produced his first excellent drawings than he found out about a new invention called photography. He never saw a mechanical device that did not fascinate him, so he was soon making improvements

Angelo Secchi. [Mabel Loomis Todd: *Total Eclipses of the Sun*]

in cameras, inventing specialized cameras for solar photography, and photographing the Moon and Sun stereographically so that lunar features appeared in relief and sunspots revealed themselves as depressions, not mountains, in the photosphere.[7]

Observing the 1860 eclipse 250 miles (400 kilometers) away from De La Rue was Angelo Secchi. He too had a remarkable aptitude for inventing and coaxing instruments. Secchi was from a poor family and received his education through the Catholic Church in the demanding Jesuit tradition. From the beginning, he showed brilliance in mathematics and astronomy. When the Jesuits were expelled from Italy by a liberal, anticlerical government in 1848, Secchi spent a year in the United States as assistant to the director of the Georgetown University observatory. When the ban against Jesuits was lifted in 1849, he returned to Italy to become director of the Pontifical (now Vatican) Observatory of the Collegio Romano (or Gregorian University). He transformed it into a modern, well-equipped center for research in the new field of astrophysics. Secchi was one of the pioneers in applying spectroscopy to astronomy. He surveyed more than 4,000 stars and realized that stellar spectra could all be grouped into a handful of classifications.

De La Rue and Secchi ambushed the 1860 eclipse with improved cameras using the wet-plate process, which greatly reduced exposure time and thereby increased the clarity with which objects in motion could be seen. They captured the prominences and compared them. The prominences looked the same from the photographers' widely separated sites, so Secchi and De La Rue could conclude that they were indeed part of the Sun. If they had been features on the Moon, so much closer than the Sun, the difference in viewing angles (parallax) from Secchi's and De La Rue's separate sites would have given them a different appearance.

The Debut of Spectroscopy

On August 18, 1868, a great eclipse touched down near the Red Sea and swept across India and Malaysia. Once again, the international scientific community had assembled.

From the United Kingdom to the path of the eclipse had come, among others, James Francis Tennant and John Herschel (son of John F. W. Herschel, grandson of William Herschel, both renowned astronomers). Norman Pogson, born in England, represented India and the observatory he directed there. The French delegation included Georges Rayet and Jules Janssen. Each of them carried a new weapon just added to the scientific arsenal for prying secrets from the Sun during an eclipse. That tool was a spectroscope. It would prove as indispensable to eclipse studies of the structure of the Sun as it was rapidly demonstrating itself to be in other realms of astronomy and in all other physical and biological sciences. By passing the light of the corona or the prominences through a prism, it could be broken down into a spectrum of lines and colors. From this spectrum, a scientist could identify the chemical elements present and even the temperature and density of its source.

As the eclipse sped along its course, spectroscopes pointed upward toward the prominences. They showed that the prominences emitted bright lines, and most of these were quickly identified with hydrogen. More and more it seemed that the Sun must be composed primarily of gas and that hydrogen must be a major constituent.[8] The spectroscopists, properly pleased with their results, packed their equipment and headed home.

All but one. His name was Jules Janssen and the brightness of the prominences and the strength of their spectral lines had given him an idea. He wanted to look for them again when the eclipse was over, when the Moon was not blocking the intense glare of the Sun from view. Might it be possible for him to see the prominences and their spectrum in broad daylight?

The weather was cloudy the rest of that day. He would have to wait until tomorrow.

That Extra Step

Pierre Jules César Janssen was 12 years old when Baily called attention to the beads visible during the annular eclipse of 1836. A childhood accident had left Janssen lame and he never attended elementary or high school. His family was cultured, but his father was a struggling musician, so Jules had to go to work at a young age.

While employed at a bank, he earned his college degree in 1849. He then went on to gain a certificate as a science teacher and served as a substitute teacher at a high school. In 1857, he traveled to Peru as part of a government team to determine the position of the magnetic equator. There he became severely ill with dysentery and was sent home. At age 33, he seemed destined for a quiet life in teaching, if he could get a job. He became the tutor for a wealthy family in central France. At their steel mills, he noticed that the eye could watch molten metal without fatigue or injury, while the skin had to be protected from the heat. He wrote a careful study of how the eye protects itself against heat radiation, which earned him his doctorate in 1860.

The glow of molten metal had led him to spectroscopic analysis, which he then applied to the Sun and, in 1859, identified several Earth elements present in the Sun. He moved to Paris in 1862 to dedicate himself to solar physics and scientific instrument-making. His work had already made him a leader in solar spectroscopy. He used the changing

Jules Janssen (1824–1907)

Two years after his breakthrough at the 1868 eclipse, Jules Janssen again planned to apply spectral analysis to a solar eclipse, this one on December 22, 1870 in Algeria. When it came time for departure, however, the Franco-Prussian War was in progress and Paris was under siege. Colleagues in Britain had obtained from the Prussian prime minister safe passage for Janssen from Paris, but Janssen wouldn't accept favors from his country's enemies. He had a different plan: "France should not abdicate and renounce taking part in the observation of this important phenomenon.... [A]n observer would be able, at an opportune moment, to head toward Algeria by the aerial route..."[a]

Although he had never been in a balloon before, on December 2, with a sailor as an assistant and himself as pilot, he ascended from Paris and headed west. Despite violent winds, he landed safely near the Atlantic coast. He reached Algeria in time—only to have the eclipse clouded out. From that experience, however, Janssen designed an aeronautical compass and ground speed indicator and prophesied methods of air travel that would "take continents, seas, and oceans in their stride." In 1898 and subsequently, he used balloons to study meteor showers from above the clouds, pioneering high altitude astronomy and foreseeing the advantages of space observations.

Janssen was also a leader in astrophotography. "[T]he photographic plate is the retina of the scientist," he wrote. It was for spectroscopy, however, that Janssen was most renowned. His methods opened up the Sun's atmosphere to continuous study. The French government tried to find him an observatory position, but the director of the Paris Observatory did not want him. So Janssen was allowed to pick a site near Paris for a new observatory dedicated to astrophysics. He chose Meudon and directed the observatory from its founding in 1876 until his death in 1907.

"There are very few difficulties that cannot be surmounted by a firm will and a sufficiently thorough preparation," he wrote. But he was too modest in his self-assessment. To everything he investigated, he brought imagination and insight. Jules Janssen was one of the most creative scientists of any era.

[a]Quotations are from the sketch of Janssen by Jacques R. Lévy in the *Dictionary of Scientific Biography*.

spectrum of the Sun in its daily journey from horizon to horizon to separate spectral lines caused by the Earth's atmosphere from those originating in the Sun and to demonstrate the composition and density of the Earth's atmosphere through which the sunlight passed. Janssen then applied spectroscopy to the other planets and, in 1867, discovered water in the atmosphere of Mars.

He traveled to Guntur, India for the total eclipse of August 18, 1868 to use his spectroscope on solar prominences. The great contemporary English spectroscopist Norman Lockyer spoke ringingly of that pivotal moment: "Janssen—a spectroscopist second to none—...was so struck with the brightness of the prominences rendered visible by the eclipse that, as the sun lit up the scene, and the prominences disappeared, he exclaimed, '*Je reverrai ces lignes la!*' "[9] [I will see those lines again!] The next morning he succeeded. He had found a way to study the atmosphere of the Sun without waiting for a total eclipse, traveling halfway around the world to see it, and hoping for good weather at the critical moment.

For two weeks Janssen continued to map solar prominences by this technique and continued to perfect it on his circuitous way home, with a stop in the Himalayas to observe at high altitude. He proved that prominences change considerably from one day to the next.

A standard spectroscope breaks down the light of a glowing object into the characteristic colors of its spectrum. Janssen modified the spectroscope by blocking unwanted colors so that the observer could view an object in the light of one spectral line at a time. He had invented the spectrohelioscope. The Sun could now be analyzed in detail on a daily basis.

A month after the eclipse, on his way home to France, Janssen wrote up his findings and sent them to the Academy of Sciences in Paris. His paper arrived a few minutes after one from England that reported precisely the same discovery.

Jules Janssen. [Mary Lea Shane Archives of the Lick Observatory]

Coincidence

Joseph Norman Lockyer came from a well-to-do family with scientific interests. He received a classical education, traveled in Europe, and then entered civil service. So wide were his interests that he wrote on everything from the construction dates and astronomical purposes of Egyptian pyramids and temples to Tennyson to the rules of golf. "The more one has to do, the more one does," was his motto.[10]

When Gustav Kirchhoff and Robert Bunsen showed in 1859 how spectroscopy could be used to determine the chemical composition of objects in space, Lockyer saw the discovery as a key to what had seemed the locked door of the universe. Had not French philosopher Auguste Comte confidently asserted only 24 years earlier that never, by any means, would we be able to study the chemical composition of celestial bodies, and every notion of the true mean temperature of the stars would always be concealed from us.[11] It was clear that Comte was wrong. Lockyer bought a spectroscope, attached it to his 6¼-inch (16-centimeter) refracting telescope, and began his observations.

Although he had never seen a solar eclipse, it occurred to him that, since prominences were probably clouds of hot gas, he should be able to use a spectroscope to analyze prominences without waiting for an eclipse. This idea struck Lockyer two years before the 1868 eclipse that inspired Janssen to the same realization. Lockyer tried the experiment in 1867 but found his spectroscope inadequate to the task. So he ordered a new spectroscope to his specifications. Because of construction delays, however, it did not arrive until October 16, 1868, two months after the

Medallion created by the Academy of Sciences in Paris to honor Janssen and Lockyer for their independent discovery of how to observe prominences without waiting for an eclipse. The front of the medal shows the heads of the two scientists. The reverse shows the Sun god Apollo pointing to prominences on the Sun.

eclipse that Janssen saw in India. Lockyer rapidly and excitedly calibrated his new instrument and on October 20, 1868, he trained it on the rim of the Sun and recorded bright lines typical of hot gases under very little pressure. He wrote up his findings and sent them to the Academy of Sciences in Paris for presentation by his friend Warren De La Rue.

Just minutes before De La Rue was to speak, Janssen's letter arrived and both papers were read at the same session of the Academy to great acclaim for both scientists. A special medal was struck to honor them. It showed the heads of Janssen and Lockyer side by side.[12]

A New Element

Lockyer continued to examine the spectrum of the gases at the rim of the Sun. He recognized that the lower atmosphere of the Sun, what Airy had called the sierra, was decidedly reddish in color, so he named it the *chromosphere*, and it has been known by that name ever since.

J. Norman Lockyer (1836–1920)

In 1869, the year after he independently showed how the atmosphere of the Sun could be analyzed without the benefit of an eclipse and discovered the element helium, Norman Lockyer founded the scientific journal *Nature*. He edited it for 50 years, until just before his death, keeping it alive through many crises.

While the French government was establishing a special observatory for Janssen, the British government likewise recognized the importance of Lockyer's contribution and set about creating a solar physics observatory for him. For its opening in 1875, Lockyer collected old and modern instruments and placed them on display. The display became permanent and grew. Lockyer had founded London's world famous Science Museum.

Lockyer was not shy in interpreting his findings to form startling theories. Often he was wrong, but always he provided useful data and often there was a nucleus of truth in his grand speculations. He thought that all atoms shared certain spectral lines and were therefore made of smaller common constituents. He was wrong about the spectra, but right about the composition of atoms.

He offered dates for the construction of ancient Egyptian temples based on their alignments with the rising and setting positions of the Sun and certain stars. His dates were wrong, but he was right that many of the temples had astronomical orientations and his work helped to establish the field of archeoastronomy.

When he died, a colleague wrote of him: "Lockyer's mind had the restless character of those to whom every difficulty is a fresh inspiration. His enthusiasm never failed him, despite repeated disappointments and opposition."[a]

[a] Alfred Fowler: "Sir Norman Lockyer, K.C.B., 1836–1920," *Proceedings of the Royal Society of London*, series A, volume 104, 1923, pages i–xiv.

J. Norman Lockyer in 1895, the year helium, the element he discovered on the Sun, was finally found to exist on Earth as well.

Lockyer was not done yet. In examining the spectrum of the prominences, he noticed a yellow line that he could not identify. It did not seem to belong to any element known on Earth. So he announced the existence of a new element and proposed the name *helium* for it, because it had been found in the Sun—*helios* in Greek. Most scientists rejected the idea of a new element, suggesting that this line was produced by a known element under unusual physical conditions. But Lockyer clung tenaciously to his interpretation. Finally, in 1895, William Ramsay found trapped in radioactive rocks on Earth an unknown gas that exhibited the mysterious spectral line that Lockyer had discovered on the Sun. Helium was an element. Lockyer had been right. In 1869, just after he discovered helium, Lockyer had urged: "[L]et us... go on quietly deciphering one by one the letters of this strange hieroglyphic language which the spectroscope has revealed to us—a language written in fire on that grand orb which to us earth-dwellers is the fountain of light and heat, and even of life itself."[13]

The total eclipse of 1868 had raised the strong possibility of the existence of a new element—and, for the first time, one discovered not on Earth but in the heavens. One year later, on August 7, 1869, the United States lay on the path of a total eclipse. Two American astronomers, Charles A. Young and William Harkness, working separately, observed the event with spectroscopes. Each noticed a green line in the spectrum

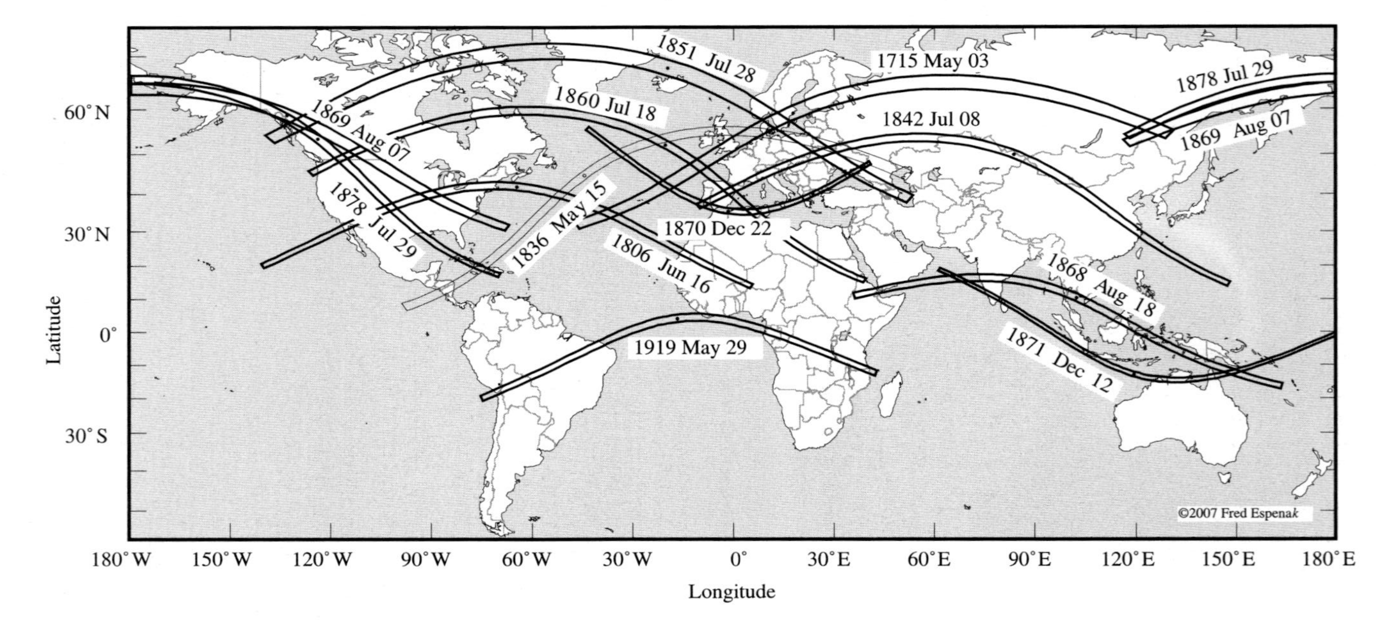

Paths of totality for the solar eclipses of 1715, 1806, 1836, 1842, 1851, 1860, 1868, 1869, 1870, 1871, 1878, and 1919. [Map and eclipse calculations by Fred Espenak]

of the corona that defied identification with known elements. This suspected new element was called *coronium*.

In 1895, Lockyer's helium was identified in rocks on Earth, but coronium remained a spectral presence seen only on the Sun during total eclipses. As time passed, more elements were discovered on Earth until the periodic table of stable chemical elements was nearly complete. There was no room left for coronium to be an element. What could it be?

Might it be an already known element under such unusual conditions that it emitted a spectrum never before seen in a laboratory? Walter Grotrian of Germany in 1939 pointed the way and Bengt Edlén of Sweden in 1941 identified the green line of coronium as the element iron with 13 electrons missing—a "gravely mutilated state."[14] To ionize iron so greatly, the temperature of the corona had to be about 2 million °F (1 million °C) and its density had to be less than a laboratory vacuum. Because the conditions necessary for the production of such lines cannot be achieved in a laboratory, they are known as forbidden lines.

The Reversing Layer

Astronomy was a family tradition for Charles Augustus Young. His maternal grandfather and his father had been professors of astronomy at Dartmouth College. Charles entered Dartmouth at age 14 and four years later graduated first in his class. He immediately began teaching—classics!—at an elite prep school, and commenced studies at a seminary to become a missionary. But in 1856 he changed his plans and became professor of astronomy at Western Reserve College, with a break in his duties to serve in the Civil War. He returned to Dartmouth in 1866 at the age of 32 as professor of astronomy, holding the same chair as his father and grandfather before him. There he pioneered in spectroscopy, especially applied to the Sun.

At the eclipse of December 22, 1870, which he observed at Jerez, Spain, Young noticed that the dark lines in the Sun's spectrum become bright lines for a few seconds at the beginning and end of totality. He had discovered the *reversing layer*, the lowest 600 miles (1,000 kilometers) of the chromosphere, which is cooler than the photosphere and thus absorbs radiation of specific wavelengths, producing the ordinary dark-line spectrum of the Sun. However, when the Moon blocks the photosphere from view and the reversing layer of the chromosphere can be seen momentarily before it too is eclipsed, the bright-line spectrum of its glowing gases briefly flashes into view. Here at last was the layer responsible for the dark-line spectrum of the Sun seen on ordinary days. A long-missing piece in the puzzle of the structure, composition, and density of the solar atmosphere was fitted into place.

Charles A. Young. [Mary Lea Shane Archives of the Lick Observatory]

In 1877, Young was lured away from Dartmouth by the offer of more equipment and research time at the College of New Jersey, now Princeton University. Not only was he a great researcher, but a revered teacher and acclaimed writer. His textbooks were the standard of his day.

The Legacy of Eclipses

Throughout the final three decades of the nineteenth century, Janssen, Lockyer, and Young led expeditions to the major total eclipses and, weather permitting, always contributed useful data and often new discoveries.

Many scientists had noticed that the corona changes its appearance from one eclipse to the next. But it was Jules Janssen who first spotted a pattern to those variations. He compared the coronas of the 1871 and 1878 eclipses and concluded that the shape of the corona varies according to the sunspot cycle. In 1871, the Sun was near sunspot maximum and the corona was round. In 1878, near sunspot minimum, the corona was more concentrated at the Sun's equator.

As the twentieth century began, solar eclipses were still the principal means of gathering information about the workings of the Sun. Every total eclipse over land was attended by scientists willing to travel great distances, endure hostile climates, and risk complete failure because of clouds for a few minutes' view of the corona—vital for the systematic study of the Sun launched more than half a century earlier by Francis Baily in his report on the annular eclipse of 1836.

One could see the Sun best when it was obscured.

What Eclipses at Jupiter Taught Us

by Carl Littmann

As the moons in the solar system revolve around their planets, they too create and undergo periodic solar and lunar eclipses. These events allowed the Danish astronomer Ole Römer in 1676 to prove, contrary to prevailing opinion, that light travels at a finite speed. He even succeeded in making the first good estimate of the speed of light. Such luminaries as Aristotle, Kepler, and Descartes had been certain that the speed of light was infinite. Gian Domenico Cassini, director of the Paris Observatory where Römer made his discovery, refused to believe the results, and Römer's triumph was not fully appreciated for half a century.

In observations of Io, innermost of Jupiter's four large moons, Römer noticed discrepancies between the observed times of its disappearance into the shadow of the planet and the calculated times for these events. He correctly explained that these discrepancies were due to the travel time of light between Jupiter and

(Left) Solar eclipse on Jupiter. The black dot left of center is the shadow cast by Io. Io, slightly larger than Earth's Moon, is visible over Jupiter's clouds in the upper right center. [NASA Hubble Space Telescope—WFPC 2]

(Right) Solar eclipse on Saturn caused by its ring system and a moon. At the bottom are moons Tethys and Dione. Tethys' shadow on Saturn can be seen at the upper right, just below the rings. [NASA/Jet Propulsion Laboratory *Voyager 1*]

Saturn's shadow eclipses its rings [NASA/Jet Propulsion Laboratory—*Cassini* mission]

the Earth. When the Earth is approaching Jupiter, the interval between satellite eclipses is shorter because the distance light must travel is decreasing. When the Earth is moving farther from Jupiter, the interval between eclipses lengthens because the distance light must travel is increasing. Römer determined that light from Io took about 22 minutes longer to reach the Earth when the Earth was farthest from Jupiter than when it was closest. Thus light required about 22 minutes to cross the orbit of Earth. The diameter of the Earth's orbit was not known at the time. Modern measurements show that light actually requires about $16\frac{2}{3}$ minutes to make this journey.

When Römer returned to Denmark, the king gave him a succession of appointments, including master of the mint, chief judge of Copenhagen, chief tax assessor (everyone said he was fair!), mayor and police chief of Copenhagen, senator, and head of the state council of the realm—all this and more while he served as director of the Copenhagen observatory and Astronomer Royal of Denmark. He discharged all his duties with distinction.

Carl Littmann is a physicist and historian.

Notes and References

Epigraph: Francis Baily: "Some Remarks on the Total Eclipse of the Sun, on July 8th, 1842," *Memoirs of the Royal Astronomical Society*, volume 15, 1846, page 4.

[1] Francis Baily: "On a Remarkable Phenomenon That Occurs in Total and Annular Eclipses of the Sun," *Memoirs of the Royal Astronomical Society*, volume 10, 1838, pages 1–42. Baily, searching back through the records, realized that Edmond Halley in 1715 and many other observers had seen and reported this bead-like apparition before him. Among the

Bolivia, November 12, 1966 | Mexico, March 7, 1970 | India, February 16, 1980

Siberia, July 31, 1981 | Java, June 11, 1983 | Philippines, March 18, 1988

The shape of the corona changes with sunspot activity. Near sunspot minimum (1965, 1976, 1987), the corona is elongated at the Sun's equator. Near maximum (1969, 1980, 1990), the corona is more symmetrical. These photographs were taken with a radial density filter developed by Gordon Newkirk that captures faint detail in the outer corona without overexposing the inner corona—similar to what the eye sees. The corona never appears the same twice. [High Altitude Observatory/National Center for Atmospheric Research]

previous observers of the beads, he named José Joaquin de Ferrer, who also described the corona and first called it by that name. Perhaps it was Ferrer's account and his use of the word corona that came to Baily's mind when he saw the eclipse of 1842.

[2] This and the following Baily quotations are from Francis Baily: "Some Remarks on the Total Eclipse of the Sun, on July 8th, 1842," *Memoirs of the Royal Astronomical Society*, volume 15, 1846, pages 1–8.

[3] Agnes M. Clerke: "Baily, Francis," *The Dictionary of National Biography*, volume 1 (London: Oxford University Press, 1921), page 903.

[4] Joseph Needham and Wang Ling: *Science and Civilisation in China*, volume 3, *Mathematics and the Sciences of the Heavens and the Earth* (Cambridge: At the University Press, 1959), page 423.

[5] José Joaquin de Ferrer: "Observations of the Eclipse of the Sun, June 16th, 1806, Made at Kinderhook, in the State of New-York," *Transactions of the American Philosophical Society*, volume 6, 1809, pages 264–275.

[6] Dorrit Hoffleit: *Some Firsts in Astronomical Photography* (Cambridge, Massachusetts: Harvard College Observatory, 1950). The first successful photograph of the uneclipsed Sun, also a daguerreotype, was achieved by the French scientists Hippolyte Fizeau and Léon Foucault in 1845.

[7] De La Rue made that discovery in 1861. Earlier evidence for sunspots as depressions had come from observations by Scottish astronomer Alexander Wilson in 1774. He noted that the geometry of sunspots seemed to change as they were seen from different angles as the Sun's rotation carried them across the solar disk and toward the limb.

[8] The Sun as a sphere of hot gas had been proposed independently by Angelo Secchi and John Frederick William Herschel (William's son) in 1864.

[9] J. Norman Lockyer: "On Recent Discoveries in Solar Physics Made by Means of the Spectroscope," in Bernard Lovell, editor: *Astronomy*, volume 1, The Royal Institution Library of Science (Barking, Essex: Elsevier Publishing, 1970) page 90.

[10] Alfred Fowler: "Sir Norman Lockyer, K.C.B., 1836–1920," *Proceedings of the Royal Society of London*, series A, volume 104, December 1, 1923, pages i–xiv.

[11] Auguste Comte: *The Essential Comte, Selected from Cours de philosophie positive*, translated by Margaret Clarke (London: Croom Helm, 1974), pages 74, 76.

[12] A. J. Meadows: *Science and Controversy: A Biography of Sir Norman Lockyer* (London: Macmillan, 1972), page 53.

[13] J. Norman Lockyer: "On Recent Discoveries in Solar Physics Made by Means of the Spectroscope," in Bernard Lovell, editor: *Astronomy*, volume 1, The Royal Institution Library of Science (Barking, Essex: Elsevier Publishing, 1970), pages 101–102.

[14] "A gravely mutilated state" is the picturesque description found in Gabrielle Camille Flammarion and André Danjon, editors: *The Flammarion Book of Astronomy*, translated by Annabel and Bernard Pagel (New York: Simon and Schuster, 1964), page 227.

8
The Eclipse That Made Einstein Famous

Oh leave the Wise our measures to collate.
One thing at least is certain, LIGHT has WEIGHT
One thing is certain, and the rest debate—
Light-rays, when near the Sun, DO NOT GO STRAIGHT.

Arthur S. Eddington (1920)

Of all the lessons that scientists learned from eclipses, the most profound and momentous was the confirmation of Einstein's general theory of relativity by the eclipse of May 29, 1919.

The Birth of Relativity

In 1905, an obscure Swiss patent examiner third class named Albert Einstein published three articles in the same issue of the leading German scientific journal *Annalen der Physik* that utterly changed the course of physics. One proved the existence of atoms. The second laid the cornerstone for quantum mechanics. The third was a revolutionary view of space and time known as the special theory of relativity. It required only high-school algebra, yet its implications were so profound that this theory baffled many of the leading scientists of the day.

In the decade that followed the publication of the special theory of relativity, Einstein labored mightily to expand his concept to accelerated systems. In 1907, he formulated his principle of equivalence: there is no way for a participant to distinguish between a gravitational system and an accelerated system. For an observer in a closed compartment, does a ball fall to the floor by gravity because the compartment is resting on a planet or does the ball fall because the compartment is accelerating toward the ball? In both cases, the ball falls to the floor. In both cases, the observer feels weight. It is impossible to tell whether that weight is from gravity or acceleration.

Albert Einstein in 1922. [Burndy Library & AIP Emilio Segrè Visual Archives]

Einstein's principle of equivalence. A ball falls in a compartment. Does it fall by gravity because the compartment is resting on a planet or does it fall because the compartment is accelerating upward? Einstein realized that an observer in the compartment could not distinguish whether gravity or acceleration caused the ball to fall.

Einstein then realized from his principle of equivalence that relativity required gravity to bend light rays, much as it bends the paths of particles. In a closed accelerating compartment, a light on one wall is aimed directly at the opposite wall, across the line of motion. The beam travels at a finite speed, so in the time it takes to traverse the compartment, the opposite wall has moved upward. For an observer in the compartment, the light beam has struck the opposite wall below where it was aimed. The observer concludes that light has been bent.

What about that same experiment performed in a compartment at rest on a planet? According to Einstein's principle of equivalence, there can be no difference between phenomena measured in the two compartments. Therefore, for the observer in the gravitational environment, light must be bent by gravity. But the effect is very small. It takes a lot of mass to bend light enough to be measured. In 1911, Einstein realized that this peculiar idea might be tested during a total solar eclipse.

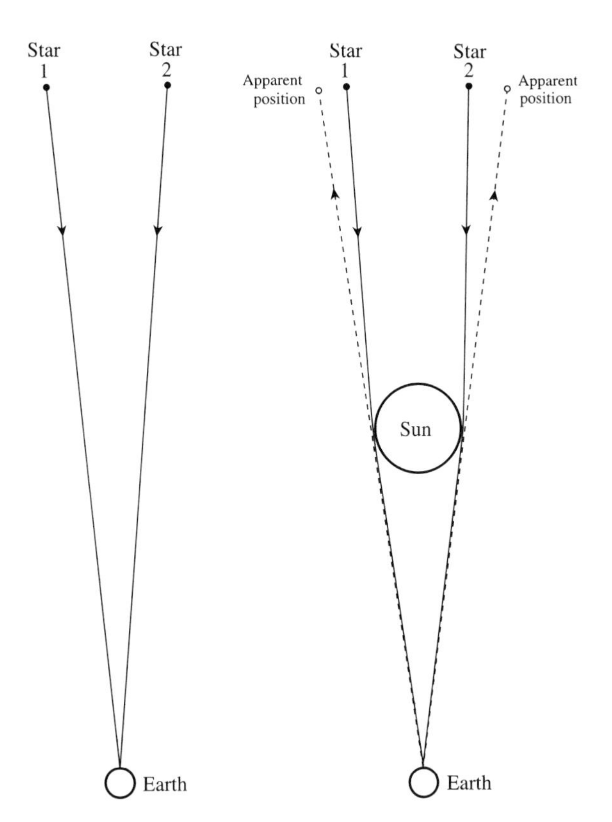

In a perfect vacuum with no gravity, light from distant stars travels in a straight line *(left)*. The gravitational field of the Sun bends light *(right)*, making the stars appear slightly farther apart than they actually are.

next passage across the face of the Sun was due, the planet never kept the appointment. It did not exist. But it was not until Einstein formulated the general theory of relativity that the 43 seconds per century anomaly could be explained.

Einstein was more than just pleased when he realized that his theory could account for this discrepancy in the motion of Mercury. "I was beside myself with ecstasy," he wrote.[8] This explanation of a perplexing problem gave general relativity high credibility. But the power of the general theory would be even more evident if it could predict something never before contemplated or detected.

Einstein offered two such predictions: that starlight passing close to the Sun would be bent, and that light leaving a massive object would have its wavelengths extended so that the light would be redder. This *gravitational redshift* was so small an effect that it could not be detected in the Sun with the equipment available at the time, so this proof of general relativity had to wait many years. It was finally detected in 1959 by Robert V. Pound and Glen A. Rebka, Jr., using the recently discovered Mössbauer effect in which the gamma rays emitted by atomic nuclei serve as the most precise of clocks. One of these atomic clocks was placed in the basement of a building and moved up and down so that its depth in the Earth's gravitational field varied minutely. The deeper in the basement the clock was, the longer the wavelength of its radiation. It had taken 45 years, but the gravitational redshift predicted by Einstein had at last been confirmed.[9]

In contrast, the gravitational deflection of starlight predicted by Einstein could be tested at most total eclipses of the Sun. Between 1911 and 1915, Einstein revised his calculation, using his new general theory of relativity. He found the deflection to be twice the initially assigned value. Starlight passing near the Sun would be bent 1.75 arc seconds. (Einstein and the world were fortunate that his initial prediction was not tested before it was revised; otherwise his later figure, although rigorously honest, might have seemed to be a manipulation to make the numbers come out right. There would have been far less drama in the confirmation of relativity.)

In 1916 Einstein published his complete general theory of relativity. But the world hardly noticed. For two years, nations had been locked in World War I. Feelings against Germans ran strong in France and Great Britain, just as the Germans hated the French and British. Einstein sent his paper to a friend, scientist Willem de Sitter in the Netherlands. De Sitter, in turn, forwarded the paper to Arthur Eddington in England. At the age of 34, Eddington was already famous for his pioneering work in how a star emits energy. Eddington instantly recognized the significance of Einstein's discoveries and was deeply impressed by its intellectual beauty. He shared the paper with other scientists in Britain.

The 1919 Test

Frank Dyson, the Astronomer Royal of England, began planning for a British solar eclipse expedition in 1919 to test relativity. It was the perfect eclipse for the purpose because the Sun would be standing in front of the Hyades, a nearby star cluster, so there would be a number of stars around the eclipsed Sun bright enough for a telescope to see. But the timing could hardly have been worse. In 1917 Britain was in the midst of a terrible war whose issue was still very much in doubt. Yet somehow Dyson managed to persuade the government to fund the expedition, despite the fact that its purpose was to test and probably confirm the theory of a scientist who lived in Germany, the leader of the hostile powers.

Meanwhile, American astronomers had an earlier opportunity to verify or disprove relativity. In the final months of World War I, a total eclipse passed diagonally across the United States from the state of Washington to Florida. On June 8, 1918, a Lick Observatory team led by William Wallace Campbell and Heber D. Curtis, observing from Goldendale, Washington, pointed their instruments skyward to render a verdict on relativity. The weather was mostly cloudy, but the Sun broke through for three minutes during totality. Measuring and interpreting the plates had to wait several months, however, until comparison

The total solar eclipse of May 29, 1919 featured an enormous prominence. Arthur Eddington's observation of this eclipse from Principe provided the first observational confirmation of Einstein's general theory of relativity. [Royal Astronomical Society]

pictures could be taken of the same region of the sky at night, without the Sun in the way. By this time, Curtis was in Washington, D.C. working on military technology for the government as the war came to a close. Curtis returned to Lick in May 1919 and the results were announced in June. The star images were not as pointlike as desired. Curtis could detect no deflection of starlight. By this time, word was spreading about the findings of the British expedition, and the Lick paper was never published.[10]

Four months after the armistice, two British scientific teams were poised for departure. Andrew C. D. Crommelin and Charles R. Davidson, heading one party, were to set sail for Sobral, about 50 miles (80 kilometers) inland in northeastern Brazil. Eddington, Edwin T. Cottingham, and their team were headed for Principe, a Portuguese island about 120 miles (200 kilometers) off the west coast of Africa in the Gulf of Guinea. On the final day before sailing, the four team leaders met with Dyson for a final briefing. Eddington was extremely enthusiastic and confident that Einstein was right. A deflection of 1.75 arc seconds would confirm relativity. Half that amount—a deflection of starlight by 0.87 arc seconds—would reconfirm Newtonian physics.[11]

"What will it mean," asked Cottingham, "if we get double the Einstein deflection?" "Then," said Dyson, "Eddington will go mad and you will have to come home alone!"

Arthur S. Eddington. [AIP Emilio Segrè Visual Archives, gift of S. Chandrasekhar]

On Principe, worrisome weather conditions greeted Eddington and Cottingham. Every day was cloudy. However, May was the beginning of the dry season and no rain fell—until the morning of the eclipse. The fateful day, May 29, 1919, dawned overcast, and heavy rain poured down. The thunderstorm moved on about noon, but the cloud cover remained. The Sun finally peeped through 18 minutes before the eclipse became total but continued to play peekaboo with the clouds. Said Eddington: "I did not see the eclipse, being too busy changing plates, except for one glance to make sure it had begun and another half-way through to see how much cloud there was."

Eddington had much cause for worry. He was not interested in prominences or the corona. He needed to see clearly the region around the Sun. With great care, the plates were developed one at a time, only two a night. Eddington spent all day measuring the plates. Clouds had interfered with the view, but five stars were visible on two plates. When he had finished measuring the first usable plate, Eddington turned to his colleague and said, "Cottingham, you won't have to go home alone."

English astronomers photographed the total solar eclipse of 1919 from two sites to test Einstein's general theory of relativity. Andrew Crommelin and Charles Davidson took this photograph from Sobral, Brazil. The white bars mark the location of stars recorded during totality. Careful measurements of their positions showed shifts because the stars' light, passing close to the Sun, was bent by the Sun's gravity, just as Einstein predicted. [Royal Society of London]

Eddington measured the displacement in the stars' positions, extrapolated to the limb of the Sun, to be between 1.44 to 1.94 arc seconds, for a mean value of 1.61 ± 0.30 arc seconds. The deflection agreed closely with Einstein's prediction. In later years, Eddington referred to this occasion as the greatest moment of his life.[12]

The expedition to Sobral had equally threatening weather but there was a clear view of totality except for some thin fleeting clouds in the middle of the event. Crommelin and Davidson stayed in Brazil until July to take reference pictures of the star field without the presence of the Sun and then brought all their photographic plates back to Britain before measuring them. They found that their largest telescope had failed because heat caused a slight change in focus, which spoiled the pinpoint images of the stars. But the other instrument had worked well, and its plates also supported Einstein's prediction. Their mean value was 1.98 ± 0.12 arc seconds.

It was now September and no news about eclipse results had been published. Einstein was curious and inquired of friends. Hendrik Antoon Lorentz, the Dutch physicist, used his British contacts to gather news and telegraphed Einstein: "Eddington found star displacement at rim of Sun..."[13] He also announced the favorable results at a scientific meeting in Amsterdam on October 25, with Einstein in attendance. But no reporters were present, and no word of the discovery was published.

Albert Einstein and Arthur S. Eddington in Eddington's garden, 1930. [Royal Greenwich Observatory]

Finally, on November 6, 1919, the Royal Society and the Royal Astronomical Society held a joint meeting to hear the results of the eclipse expeditions. The hall was crowded with observers, aware that an age was ending. From the back wall, a large portrait of Newton looked down on the proceedings. Joseph John Thomson, president of the Royal Society and discoverer of the electron, chaired the meeting, praising Einstein's work as "one of the highest achievements of human thought." He called upon Frank Dyson to summarize the results and introduce the reports of the eclipse team leaders. The Astronomer Royal concluded his presentation by saying: "After careful study of the plates I am prepared to say that there can be no doubt that they confirm Einstein's prediction. A very definite result has been obtained that light is deflected in accordance with Einstein's law of gravitation."[14]

Einstein awoke in Berlin on the morning of November 7, 1919 to find himself world famous. Hordes of reporters and photographers converged on his house. He genuinely did not like this attention, but he found a way to turn the disturbance to the benefit of others. He told the reporters about the starving children in Vienna. If they wanted to take his picture, they first had to make a contribution to help those children. Suddenly Einstein was a celebrity.

More Tests

Expeditions from many nations set off to retest relativity by measuring the deflection of starlight during the eclipse of September 21, 1922. The shadow passed across Somalia, the Indian Ocean, and Australia. Most of the teams had bad luck; the weather failed them. But the successful American delegation, a Lick Observatory team again headed by Campbell, observed the eclipse from Wallal on the northwest coast of Australia. After the Lick Observatory disappointments in 1914 and 1918, this expedition was an all-out effort, with the best of equipment and lots of rehearsal. The results had to wait, though, until comparison photographs had been taken and until Campbell and Robert J. Trumpler had measured the plates with utmost care. In April 1923, seven months after the eclipse, Campbell announced: "The agreement with Einstein's prediction from the theory of relativity ... is as close as the most ardent proponent of that theory could hope for."[15] They measured the displacement at 1.72 ± 0.11 arc seconds.

When Campbell was asked for his personal reaction to this new confirmation of relativity, he replied: "I hoped it would not be true."[16] Campbell was not the only one reluctant to accept such a new and difficult concept. There were a number of scientists who did not share Eddington's delight in Einstein's victory.

Makeshift mount for the principal telescope used by the Lick Observatory team to test relativity at the 1922 eclipse in Wallal, Australia. [Mary Lea Shane Archives of the Lick Observatory]

Women's work at the 1922 Lick Observatory eclipse expedition to Wallal, Australia. Note the mosquito netting. [Mary Lea Shane Archives of the Lick Observatory]

Solar eclipse photographed by the Lick Observatory expedition at Wallal, Australia, September 21, 1922. [Mary Lea Shane Archives of the Lick Observatory]

be the most distant objects in the universe—so distant that they were essentially fixed markers. Their angular distance from one another provided a new coordinate system against which the positions of all other objects could be referred. Because some of the quasars lay near the ecliptic, the Sun's apparent path through the heavens in the course of a year, the Sun's gravity would annually deflect the quasars' radiation, making them seem to shift slightly in position. Only this time, the radiation would be radio waves rather than visible light. And no eclipse was necessary because radio telescopes do not require darkness. In 1974 and 1975, Edward B. Fomalont and Richard A. Sramek used a 22-mile (35-kilometer) separation between radio telescopes to measure the deflection of light at the limb of the Sun as 1.761 ± 0.016 arc seconds, a result that not only confirmed relativity, but also favored Einsteinian relativity over slightly variant relativity formulations by other scientists.[18]

Solar eclipses may no longer be the most accurate way of determining the relativistic deflection of starlight, but when the general theory of relativity needed proof from a phenomenon predicted but never before observed, it was solar eclipses that provided the first and most dramatic demonstration of Einstein's masterpiece and brought relativity and Einstein to the attention of the entire world.

Notes and References

Epigraph: Arthur S. Eddington as cited in Allie Vibert Douglas: *The Life of Arthur Stanley Eddington* (London: T. Nelson, 1956), page 44.

[1] The angular displacement of a star is inversely proportional to the angular distance of that star from the Sun's center.

[2] In 1911, Einstein had asked Freundlich to investigate another aspect of his emerging general theory of relativity: the motion of Mercury. Freundlich reviewed the anomaly in Mercury's motion, recognized since the work of Le Verrier in 1859, and reconfirmed that Mercury was indeed deviating slightly from the law of gravity as formulated by Newton. Freundlich's results, which matched Einstein's relativistic recalculation for the precession of Mercury's orbit, were published in 1913 over the objections of his superiors.

[3] To make certain that the telescope optics and the camera system introduced no unknown deflection in the positions of stars, it was important to have for comparison with the eclipse plate a picture of that same region taken with the same equipment when the Sun present was not present. No such plates were available.

[4] Charles Dillon Perrine, an American who was directing the Argentine National Observatory, prepared to test the bending of starlight during the October 10, 1912 eclipse in Brazil, but was rained out.

[5] Ronald W. Clark: *Einstein, the Life and Times* (London: Hodder and Stroughton, 1973), page 176.

[6] Freundlich's team was not alone in the Crimea. An American team from the Lick Observatory also journeyed to Russia to test relativity, but they were clouded out. In 1918, Freundlich left the Royal Observatory to work full time with Einstein at the Kaiser Wilhelm Institute. In 1920, he was appointed observer and then chief observer and professor of astrophysics at the newly created Einstein Institute at the Astrophysical Observatory, Potsdam. Freundlich continued to be plagued by miserable luck on his eclipse expeditions to measure the deflection of starlight. He returned empty-handed in 1922 and 1926 because of bad weather. He finally got to see an eclipse in Sumatra in 1929, although he obtained a deflection (now known to be erroneous) considerably greater than Einstein predicted. When Hitler came to power, Freundlich left Germany and eventually settled in Scotland, where he changed his name to Finlay-Freundlich, based on his mother's maiden name, Finlayson. He built the first Schmidt-Cassegrain telescope, the prototype for almost all large photographic survey telescopes today.

[7] Banesh Hoffmann with the collaboration of Helen Dukas: *Albert Einstein, Creator and Rebel* (New York: Viking Press, 1972), page 116.

[8] Banesh Hoffmann with the collaboration of Helen Dukas: *Albert Einstein, Creator and Rebel* (New York: Viking Press, 1972), page 125.

[9] Robert V. Pound and Glen A. Rebka, Jr.: "Resonant Absorption of the 14.4-kev Gamma Ray from 0.10-microsecond Fe^{57}," *Physical Review*

Letters, volume 3, December 15, 1959, pages 554–556. The gravitational redshift was detected previously, based on Einstein's suggestion, in light emitted from extremely dense white dwarf stars, but to nowhere near the same degree of certainty.

[10] The Lick team was working with mediocre equipment and improvised mounts. Their excellent regular equipment had stood ready under cloudy skies in Russia in 1914, but it was too cumbersome to transport home after the outbreak of World War I. After the war, it was delayed in shipment home and missed the eclipse of 1918.

[11] The "Newtonian prediction" is calculated on the basis that light energy has a mass equivalent (Einstein's $E = mc^2$). That mass is then treated as ordinary matter using Newton's equations for gravity.

[12] Allie Vibert Douglas: *The Life of Arthur Stanley Eddington* (London: T. Nelson, 1956), pages 40–41.

[13] Ronald W. Clark: *Einstein, the Life and Times* (London: Hodder and Stroughton, 1973), page 226.

[14] Frank W. Dyson, Andrew C. D. Crommelin, and Arthur S. Eddington: "Joint Eclipse Meeting of the Royal Society and the Royal Astronomical Society," *The Observatory*, volume 42, November 1919, pages 389–398. The article includes dissenting comments from scientists in the audience. The article based on eclipse results was Frank W. Dyson, Arthur S. Eddington, and Charles R. Davidson: "A Determination of the Deflection of Light by the Sun's Gravitational Field, From Observations Made at the Total Eclipse of May 29, 1919," *Philosophical Transactions of the Royal Society of London*, series A, volume 220, 1920, pages 291–333. A summary of the eclipse results had appeared the week after the joint meeting: Andrew C. D. Crommelin: "Results of the Total Solar Eclipse of May 29 and the Relativity Theory," *Nature*, volume 104, November 13, 1919, pages 280–281. Eddington had yet another reason to be proud: "By standing foremost in testing, and ultimately verifying, the 'enemy' theory, our national observatory kept alive the finest traditions of science; and the lesson is perhaps still needed in the world today" (Clark: *Einstein*, page 284). British physicist Robert Lawson made a similar observation: "The fact that a theory formulated by a German has been confirmed by observations on the part of Englishmen has brought the possibility of cooperation between these two scientifically minded nations much closer. Quite apart from the great scientific value of his brilliant theory, Einstein has done mankind an incalculable service" (Clark: *Einstein*, page 297). The world soon forgot this lesson.

[15] Jeffrey Crelinsten: "William Wallace Campbell and the 'Einstein Problem': an Observational Astronomer Confronts the Theory of Relativity," *Historical Studies in the Physical Sciences*, volume 14, part 1, 1983, pages 1–91. Clarence Augustus Chant of Canada also measured a deflection favoring Einstein's theory.

[16] Allie Vibert Douglas: *The Life of Arthur Stanley Eddington* (London: T. Nelson, 1956), page 44.

[17] Jack B. Zirker: *Total Eclipses of the Sun* (New York: Van Nostrand Reinhold, 1984), pages 177–178.

[18] Edward B. Fomalont and Richard A. Sramek: "A Confirmation of Einstein's General Theory of Relativity by Measuring the Bending of Microwave Radiation in the Gravitational Field of the Sun," *Astrophysical Journal*, volume 199, August 1, 1975, pages 749–755. The result reported in their article, 1.775 ± 0.019 arc seconds, was later revised to 1.761 ± 0.016 arc seconds. (Personal communication, March 1990.) They used three 85-foot (26-meter) antennas and a distant 45-foot (14-meter) antenna, instruments of the National Radio Astronomy Observatory at Green Bank, West Virginia. The idea of using radio waves and radio telescopes to test the bending of light was first proposed by Irwin I. Shapiro: "New Method for the Detection of Light Deflection by Solar Gravity," *Science*, volume 157, August 18, 1967, pages 806–807.

9
Modern Scientific Uses for Eclipses

I have a little shadow that goes in and out with me,
And what can be the use of him is more than I can see.

Robert Louis Stevenson (1885)

When scientists first turned their attention to solar eclipses, they put them to use to clock the motions of the Moon around the Earth and the Earth around the Sun. By trying to predict the exact time and location of the path of a solar eclipse, astronomers could take note of their errors and refine their knowledge of the orbits of the Earth and Moon. This work was pioneered by Edmond Halley in 1715 for a total eclipse crossing southern England.

More than a century later, in the early days of astrophysics, a second use for total eclipses emerged. The eclipse of 1842, carving a path across southern France and northern Italy, gave European scientists a front-row seat to see for themselves the occasionally reported effects surrounding totality: the corona, prominences, and chromosphere. They were awed and realized that the Sun, by covering its face, was revealing physical aspects of itself that were not visible at any other time. By studying the atmosphere of the Sun during total eclipses and the visible surface of the Sun at all other times, scientists began to perceive how hot the interior of the Sun must be.

Then, unexpectedly, early in the twentieth century, there arose a third great scientific use for solar eclipses: to confirm or deny the peculiar new theory of gravity and the structure of the cosmos offered by Albert Einstein. The first affirmative answer came from an eclipse in 1919, with data from subsequent eclipses over the remainder of the century adding to the certainty.

Meanwhile, the original scientific uses for total eclipses waned. The U.S. Naval Observatory no longer refines the orbits of the Moon or Earth from eclipse timings. Equipment and techniques developed during the twentieth century allow astronomers to study the prominences of the Sun independent of eclipses. In 1930, Bernard Lyot of France invented the coronagraph, a telescope with a special optical system to create an artificial eclipse so that the brighter regions of the corona can be studied without waiting for a precious few moments of eclipse totality in some remote corner of the world. And the relativistic bending of starlight can now be checked more precisely by using radio waves, which can be received during broad daylight, without waiting for an eclipse.

Have eclipses been used to exhaustion by scientists and then abandoned to the care of aesthetes and amateur astronomers?

Dogging the Sun's Diameter

David Dunham is not about to consign eclipses to the historical archives of scientific relics. He, Joan Dunham, Alan Fiala, and other members of the International Occultation Timing Association (IOTA), journey to eclipses around the world for the express purpose of almost missing them—that is, missing what almost everyone else is most anxious to see: the longest possible duration of totality. You will never find Dunham and his colleagues along the central line. Instead, they and any friends and local people who wish to join them position themselves just inside the northern and southern limits of the eclipse path so that they witness just a few moments of totality.

From their positions at the edges of totality, the top or bottom of the Moon just briefly covers the full face of the Sun. The corona appears and then fades away as the Sun reemerges. Here Dunham finds precisely the information he seeks. He and his co-workers are trying to measure minute changes in the diameter of the Sun, and have enlisted the Moon for their service. The size and distance of the Moon are well known. The distance of the Sun is known to great accuracy. Therefore, by measuring the size of the Moon's shadow, they can derive the size of the body responsible for that shadow: the Sun.

By stretching a team of observers perpendicular to the expected edge of the eclipse path, typically from 0.5 mile (0.8 kilometer) outside to 2.0 miles (3.2 kilometers) inside, they can determine where the actual edge of totality passes to within a hundred meters. This translates into the ability to measure the angular radius of the Sun to an accuracy of 0.04 arc second or about 20 miles (30 kilometers). The Sun in 1983 was, according to their measurements, about 0.2 arc second larger than it was in 1979, which means that the Sun had increased in radius by about 90 miles (140 kilometers). However, the Sun seemed to be 0.5 arc second smaller in 1979 than it was in 1715 or even 1925, a decrease in the solar radius by about 230 miles (375 kilometers).[1]

It is a little surprising to think that our Sun might be expanding, shrinking, or pulsating. But then, it was a shock to people almost four centuries ago when Galileo used his telescope on the Sun, a celestial body thought to be "pure" and immune to change, and found spots on the Sun that appeared, changed, disappeared, and showed that the Sun was rotating. Over the centuries, scientists recognized more and more changes on the Sun: the shape of the corona, prominences, flares, the sunspot cycle. What is most surprising is not that the Sun changes but that changes on the Sun are not reflected more dramatically (and catastrophically) in short-term climate changes on Earth.[2]

Do the Dunhams, Fiala, and their colleagues regret missing the "main attraction" of a total eclipse? No, says Dunham, he actually prefers the view from the edge and chides "central line prejudice." He sacrifices a long look at the corona at the center of the eclipse path, but he gains a maximum-duration view of Baily's beads and the chromosphere—far longer than the few seconds that those near the central line of the eclipse see them. He also enjoys watching the shadow bands, which he sees without special effort at virtually every eclipse.

As of 2007, IOTA teams have successfully observed 16 total and annular eclipses in this fashion and, funds permitting, they plan to collect data at many more.

If this variation in the diameter of the Sun is real, it may be a cyclic pulsation tied to the sunspot cycle of approximately 11 years. Why, the Sun's diameter would vary, Dunham does not know. He refers theoretical questions to others, such as Sabatino Sofia. Dunham concentrates on refining his equipment and techniques to hold experimental error to low enough levels so that his data will have clear meaning.

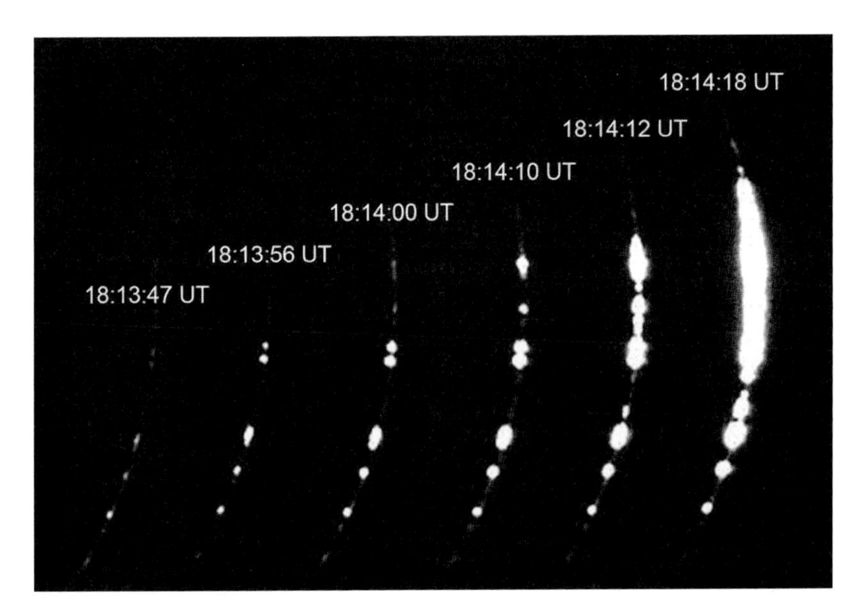

Baily's beads evolving over a 31-second period as the Sun emerges from total eclipse on February 26, 1998. This video sequence was made from the southern edge of totality in Curaçao as part of the International Occultation Timing Association's effort to measure the diameter of the Sun to see if it is changing. [©1998 Richard Nugent]

The task is simple in theory but very complicated in practice. The raw data—what was seen from each precisely marked viewing position at precisely what time—are just the start of the project. To compute the angular size of the Sun to see if it has changed, one must take into consideration a host of factors, the most complex of which is the landscape of the Moon. Because of the mountains and valleys on the Moon, its limb is jagged, producing Baily's beads as totality nears and the only light from the Sun's photosphere that reaches the Earth is passing through the deepest valleys. But because the Moon has an elliptical and inclined orbit, the features visible along the Moon's edge change from one eclipse to the next.[3]

For the radius of the Sun to be measured accurately using the Moon's shadow during an eclipse, the exact height of the north and south limb features of the Moon as they appear during each specific eclipse must

The Corona, Eclipses, and Modern Solar Research

by Jay M. Pasachoff

The disk of the Sun is visible from the ground every clear day, but the solar corona is most clearly visible from the ground only during total eclipses. Scientists are interested in the corona for four major reasons.

First, we learn about how the Sun works. All the energy that leaves the Sun passes through the corona. Why the temperature of the corona is as high as 2 million °C is still a major question, with theories of coronal heating via magnetic fields currently dominating.

Second, we study the solar corona to learn more about the Earth. The corona continually expands into space in the form of the solar wind, which envelops our planet. Thus the Sun causes many effects on Earth: the aurorae, short-wave and CB communication blackouts, and surges on power lines that lead to outages. Changes in solar radiation may affect the Earth's climate on a timescale short enough to detect.

Third, the Sun is a rather typical star, so by studying the Sun we gain detailed close-up information we can apply to the distant stars that are mere pinpoints of light in our largest telescopes. The Sun is thus a key to understanding the universe.

Fourth, the Sun is a celestial laboratory where we learn basic laws of physics. The conditions on the Sun are often not duplicable on Earth. For example, the density of the solar corona is so low that it would be considered a fantastic vacuum in a laboratory on Earth. The corona is a hot plasma which the solar magnetic field directs into the beautiful streamers seen during eclipses. Thus the corona reveals the behavior of hot gas held in a magnetic field. This knowledge is useful to magnetic fusion research which may one day provide energy on Earth. In the laboratory, we have trouble holding plasma together long enough for protons and deuterons to come sufficiently close together to fuse, releasing energy according to Einstein's formula $E = mc^2$. Positively charged particles

A stunning composite of the total solar eclipse of February 28, 1998 was produced using 15 exposures. Features on the dark side of the Moon are also visible. [Olympus OM1 SLR, 55 mm f/8 fluorite refractor, fl = 440 mm, f/8, exposures: 1/250 s to 1 s, Kodak Ektar 100, ©1998 Christian Viladrich]

The Altiplano of Chile offered a unique landscape complete with volcanoes for the total solar eclipse of November 3, 1994. [35 mm SLR, 28 mm lens, f/2.8, 2 s, Ektachrome 100 slide film, ©1994 Alan Dyer]

Four people watch totality from the Lyndhurst airstrip, Australia during the total solar eclipse of December 4, 2002. [Nikon F3 SLR, 28 mm Nikkor, f/2.8, 1/125 s, Kodak CG400 negative, ©2002 David Makepeace]

Despite its short duration (30 seconds) and low altitude, the December 3, 2002 total solar eclipse was captured in a series of exposures from Nullarbor, Australia. It required twelve hours of work to combine seven exposures into a corona composite with Adobe Photoshop. [Canon EOS D60 DSLR, 100–400mm Canon F4.5–5.6 L IS, f/8, exposures: 1/500 s to 8 s, ISO 100, ©2002 Fred Bruenjes]

A huge baobab tree frames the corona during the total solar eclipse of December 4, 2002 from Messina, South Africa. [Nikon F100 SLR, 28 mm Nikkor lens, f/5.6, 1/4 s, Fujichrome 100, ©2002 Kris Delcourte]

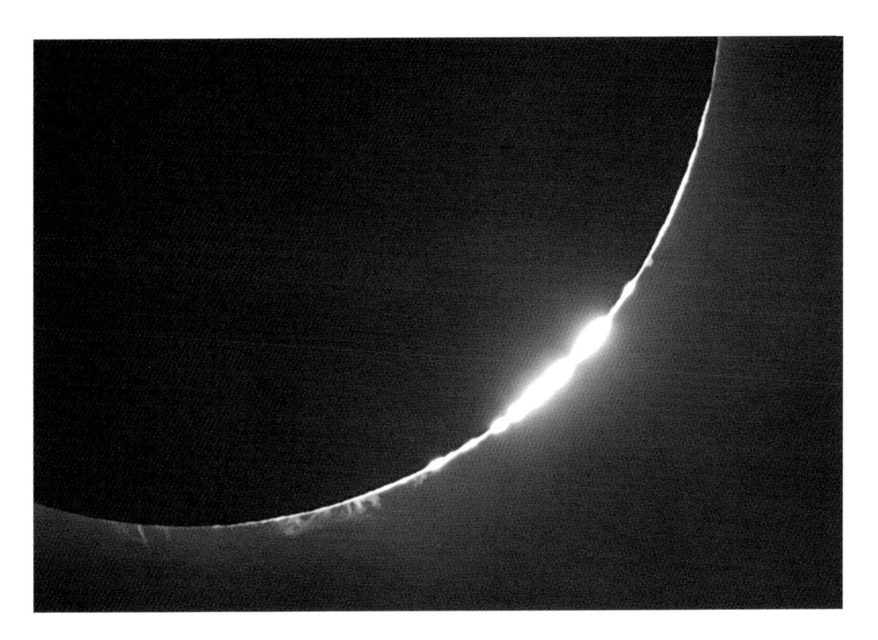

Thin clouds do not prevent views of second contact during the total eclipse of December 4, 2002 from Navashila, Mozambique. [Nikon D1H DSLR, Astrophysics 105 EDT refractor, fl = 620 mm, f/5.9, Nikon TC-300 teleconverter, efl = 1315 mm, f/13, 1/125 s, ISO 200, ©2002 Friedhelm Dorst]

The 2003 total solar eclipse was photographed over Antarctica aboard a Qantas charter flight out of Australia. [Sony DSC-V1 digital camera, 7 mm lens, f/2.8, 1/4 s, ISO 100, ©2003 Alan Dyer]

From the frigid landscape of Antarctica, an eclipse watcher is dramatically captured against the total eclipse of November 23, 2003. [Canon EOS D60 DSLR, 100–400mm Canon F4.5–5.6 L IS, f/8, 1/60 s, ISO 100, ©2003 Fred Bruenjes]

be calculated to high accuracy; otherwise the fraction-of-a-mile uncertainty in the bead-creating diameter of the Moon washes out the accuracy in the measurement of the Sun's radius.

Many astronomers believe the Sun's diameter can be determined more precisely by instruments in space Measurements by the orbiting Solar and Heliospheric Observatory (SOHO) contradict the large diameter changes seen in grazing eclipse observations.[4] However, many of the IOTA eclipse measurements cover time periods different from the SOHO data.

Undiscouraged, Dunham and his colleagues continue their grazing eclipse measurement. If the Sun actually does pulsate slightly in the course of a sunspot, and hence magnetic, cycle, it will probably take another decade or two to prove the point—a long-term project.[5]

repel each other strongly. In a hot gas, particle velocities are so high that protons and deuterons approach each other closely despite their mutual repulsion. But no material container can hold such hot gas, so magnetic fields are used to constrain it. Astronomical studies of the corona help physicists and engineers with this important problem.

The corona is so important to science that no method of observing it should be ignored. Solar telescopes in space, such as the Hinode ("Sunrise") and two STEREO (Solar TErrestrial RElations Observatory) satellites launched in 2006, TRACE (Transition Region and Coronal Explorer) from 1998, and SOHO (Solar and Heliospheric Observatory) from 1995 observe the Sun in parts of the spectrum that do not reach the ground and have provided magnificent new insights. But to limit the scattering of sunlight, space-borne coronagraphs to date have had to block out the innermost corona. And ground-based observations of the corona outside of eclipses cannot see the corona as far from the Sun as can be seen at a total eclipse. Thus, eclipse observations are essential supplements to ground- and space-based coronal observations if the full picture is to be grasped.

The traditional advantages of eclipses over space observations for solar research are flexibility and cost. New observations and instruments can be incorporated into an eclipse expedition on short notice. State-of-the-art and bulky equipment can be transported to eclipse sites far more cheaply than to space. Eclipse equipment does not have to meet launch standards of sturdiness. Further, eclipse instruments can be mounted on steady bases, with the Earth as a platform, and can be adjusted at the last minute by qualified scientists.

The SOHO solar satellite cost about a billion dollars. For less than one-tenth of one percent of that cost, a ground-based eclipse expedition can be mounted to explore current astrophysical problems using the latest equipment. Even allowing for some of the eclipses to be clouded out, eclipses can be a very cost-efficient way of doing forefront astronomy research.

Jay M. Pasachoff, Ph.D. is Field Memorial Professor of Astronomy and Director of the Hopkins Observatory at Williams College in Massachusetts.

Across the Spectrum

Dunham and IOTA members are by no means the only astronomers who still see important scientific value in solar eclipses. In fact, solar eclipses provide so many unique opportunities to study the Sun's atmosphere that this chapter can present only a small sampling of the research in progress.

Spectroheliographs and coronagraphs allow photographic inspection of many aspects of the Sun's chromosphere, prominences, and lower corona outside of an eclipse. But observing in special wavelengths or by placing an occulting disk at the focal plane to block the face of the Sun cannot compensate for the turbulence in the Earth's atmosphere that causes the sunlight to be refracted so that it comes in from very slightly different angles. This deflection of the light blurs the image enough to suppress viewing of fine detail in the corona and prominences altogether, leaving only the most conspicuous solar atmospheric features visible.

For more than a century, astronomers have used occultation by the sharp lunar limb to detect detail in sources that is much too fine to resolve with the most powerful telescopes. What appears to the telescope as a single star can be identified as a close binary star system as the Moon passes in front of it.

The basis of this technique is to measure the "occultation curve" of the source—how rapidly the intensity of the source falls off as the solid surface of the Moon occults it. If the source is relatively broad and diffuse, the intensity decreases relatively gradually, over many seconds, as the Moon slides in front of it and covers it up. If the source is extremely sharp and fine, the intensity drops to zero in a few seconds or less. Eclipses allow scientists to apply the occultation technique directly to the solar limb, resolving the density and temperature profile of the solar chromosphere in fine detail.

Alan Clark and John Beckman pioneered this technique in the infrared wavelengths, observing solar eclipses with telescopes aboard high-flying aircraft. Charles Lindsey and Eric Becklin refined their techniques using NASA's Kuiper Airborne Observatory, a military cargo jet converted to carry a 36-inch (0.9-meter) infrared telescope.

Lindsey, Becklin, and their colleagues from the United States, the United Kingdom, and Canada applied the same technique to the spectacular eclipse of July 11, 1991, which passed directly over Hawaii's Mauna Kea Observatory, perched at an elevation of 13,860 feet (4,225 meters). Because of exceptionally dry conditions atop high mountains, infrared wavelengths can be recorded there without the use of aircraft. These observations were made with the 50-foot-diameter (15-meter-diameter) James Clerk Maxwell Telescope, a radio

The eclipse of July 11, 1991 gave astronomers a rare opportunity because its path of totality crossed Hawaii and the great observatory complex atop Mauna Kea. [Nikon F4S, fl = 58 mm, f/1.2, 1/8 s, ISO 100. © 1991 Serge Brunier]

observatory operated on Mauna Kea by the United Kingdom, the Netherlands, and Canada.

Occultation curve observations made during solar eclipses in the 1980s and early 1990s have now provided scientists with remarkably detailed profiles of the structure of the solar chromosphere in radiation ranging from 30 out to 1,200 microns (1.2 millimeters), wavelengths that are about 40 to 1,600 times longer than the reddest light the human eye can see.[6]

The Baked Alaska Problem

Spectroscopy of the lower atmosphere of the Sun reveals the presence of carbon monoxide (CO)—a molecule. That is weird. Temperatures near the surface of the Sun should instantly split carbon monoxide into carbon and oxygen unless this molecular gas has a temperature as low as sunspots, some 2,700 °F (1,500 °C) cooler than the 10,000 °F (5,500 °C) photosphere.

Exactly where is this "cold" material in the solar atmosphere and how high does it extend? Attempts to pin down its precise vertical extent had been thwarted by fluctuations in the solar image caused by the Earth's turbulent atmosphere. This turbulence causes stars to appear to twinkle and the Sun's disk to appear to quiver.

Alan Clark thought it might be possible to avoid this "seeing" problem and pinpoint the position of the carbon monoxide by using the geometry of a solar eclipse. He and Rita Boreiko made initial observations at eclipses in the 1980s by flying in a small Learjet above most of the infrared-radiation-absorbing water vapor in the Earth's atmosphere. They got mixed results.

But when 1991 brought a total eclipse across the summit of Mauna Kea in Hawaii, Clark and David Naylor exploited the opportunity by using infrared photometry to show that the carbon monoxide lay in a narrow band above the photosphere.

The annular eclipse of May 10, 1994, however, provided Clark with an almost ideal opportunity to pinpoint the height of this surprisingly cold atmospheric component. The Moon's shadow swept close to the world's largest solar telescope, the McMath–Pierce instrument on Kitt Peak in southern Arizona. Furthermore, this telescope was equipped with a powerful infrared spectrograph and imaging camera. As the Moon progressively eclipsed the solar atmosphere, Clark, Charles Lindsey, Douglas Rabin, and William Livingston were able to watch specific spectral lines of carbon monoxide molecules as they changed from absorption lines when seen against the hotter background of the solar disk to emission lines when seen against the background of space at the solar limb. They could determine the position of this process to within about 50 miles (80 kilometers). Such observations are relatively unhindered by "seeing" fluctuations because the obscuring Moon is above the Earth's atmosphere. Most of the carbon monoxide was concentrated within 280 miles (450 kilometers) of the Sun's surface, although very small amounts were detected up to about 625 miles (1,000 kilometers).

The remaining problem is to explain how this cool layer can exist, in which carbon monoxide survives dissociation into atoms while surrounded by fierce temperatures. Clark calls this the "baked Alaska problem." How can the ice cream (the carbon monoxide) remain frozen within the cake while in an oven (the chromosphere) hot enough to bake the meringue?

The present idea is that a network structure of hot, concentrated plasma containing magnetic fields is spread like a fishnet across the solar surface. The carbon monoxide appears to survive in cold pools within the cells of this network, cooling itself by emitting intense infrared radiation by which it is detected along the solar limb. It survives only up to the altitude at which the network magnetic fields spread out to cover the solar surface uniformly. These measurements using eclipses thereby establish the "canopy height" for the chromospheric network magnetic fields.

Thus, eclipses provide the resolution necessary to help clarify the contribution of magnetic fields to the structure of the Sun's lower atmosphere.[7]

More Information from the Infrared

Jeffrey Kuhn and his colleagues also use eclipses to make observations of the solar atmosphere in infrared wavelengths. For the 1998 eclipse, they instrumented a C-130 cargo plane with a hole in its roof through which they could track the Sun. This National Center for Atmospheric Research aircraft allowed them to climb to 18,000 feet (5,500 meters), above much of the water in the atmosphere that absorbs the Sun's infrared radiation. Kuhn's team flew from Panama out over the Pacific Ocean to intercept the eclipse. Then they turned around and by flying eastward along the eclipse path, they were able to extend the duration of totality from 4 minutes to 4 minutes 40 seconds. Through these observations, Kuhn and his coworkers discovered a new line in the spectrum of the Sun created by silicon with 8 of its 14 electrons missing. This spectral line is strong enough to use in future eclipses as a probe to trace and measure the magnetic fields of the corona.[8]

Infrared studies of the Sun's corona during eclipses have also allowed Kuhn and his colleagues to detect fine dust particles deposited in the inner solar system by outgassing comets and colliding asteroids. Sunlight reflecting from this dust provides the faint, hazy zodiacal light that can be seen on Earth under dark-sky conditions rising before the Sun or setting after the Sun. The dust that reflects the zodiacal light lies primarily in and near the plane of the solar system and gradually spirals in closer to the Sun. With their infrared measurements during eclipses, Kuhn and his co-workers have detected zodiacal dust within the corona, but, as it spirals into the Sun, this dust does not pile up in rings in the outer corona, as some astronomers had suspected.[9]

Desert Observations

Some infrared research can be carried out from the ground if the amount of water vapor in the air can be minimized. The problem with water is that it absorbs infrared radiation ferociously well, which is why a cup of water for tea or coffee heats up so rapidly in a microwave oven. It's absorbing infrared wavelengths.

To avoid water vapor in the atmosphere, infrared astronomers either take to the air and fly above most of it (as Kuhn did for the 1998 eclipse) or they fire their experiments into space, above all of the water vapor.

If infrared astronomers must stay on the ground, they go to high mountaintops, such as 13,796-foot (4,205-meter) Mauna Kea in Hawaii, or to the driest of deserts. Total eclipses rarely have the courtesy to pass over mountaintop observatories, although one passed over Mauna Kea in 1991. On the other hand, deserts are plentiful.

Kuhn, Shadia Habbal, and their colleagues at the University of Hawaii were on the ground in Libya for the 2006 eclipse, continuing

Observing an Eclipse from Mauna Kea

by Drake Deming

In 1991, a total solar eclipse occurred over Mauna Kea in Hawaii. The summit of Mauna Kea is home to some of the largest astronomical telescopes in the world. My colleagues and I were granted time on the 3-meter NASA Infrared Telescope Facility (IRTF) to investigate a mysterious infrared emission line that occurs in the solar spectrum near 12 microns. The long wavelength of this line gives it a very large Zeeman splitting in the presence of solar magnetic fields. But to be useful as a magnetic sensor, we had to know the height in the solar atmosphere where the line originates. The eclipse provided a unique opportunity to measure the height of the line-forming layer by noting the exact time when the Moon blocked the line emission. To make rapid measurements in the infrared where the Sun is not as bright as in visible light we needed a large infrared telescope like the IRTF.

We planned to begin the observations just prior to totality, when there would still be significant amounts of sunlight visible. We had to block the Sun's strong visible light emission and just allow the infrared light to reach the focal plane of the telescope. The best way to do that would be to block the visible light using a filter. But that filter would have to be placed in front of the telescope, so it would have to be 3 meters in diameter! We planned to use a sheet of plastic—if we could find a plastic that was opaque in the visible wavelengths but transmitted infrared radiation at 12 microns. It proved difficult to find the right material. Every sample that we measured in our laboratory spectrometer was unsatisfactory. I was inspecting the transmission curves of various samples one day while eating lunch at my desk. Some potato chips I was munching on were packaged in

their infrared- and optical-wavelength study of fine details in the corona that can only be seen from Earth during total eclipses.

"We're in the Sahara Desert," Kuhn says. "It's blisteringly hot. And then we find a source of cool gas—in the outer regions of the Sun."

Well, relatively cool. The high speeds of ionized gases in the corona give it a temperature of several million degrees. But here and there, Kuhn and his collaborators found, the temperatures are hundreds of times lower—only tens of thousands of degrees. At those relatively low temperatures, helium is not ionized. It remains in its neutral state, just mildly excited, and gives off infrared radiation.

The University of Hawaii team can detect this radiation weakly without waiting for an eclipse, but when the Moon covers the face of the Sun, it blocks interfering radiation from the Sun's visible surface and allows Kuhn and his colleagues to pinpoint where in the corona this neutral helium is located so they can try to understand what is happening—how

a slick plastic bag of a kind I hadn't seen before. So I rushed the potato chip bag down to the lab and measured its transmission. It was perfect! After some phone calls to the supermarket and the potato chip company, I located the manufacturer of the plastic bags. It was polypropylene, and they were happy to send me a "small" sample—50,000 square feet—enough to cover every telescope on Mauna Kea! So we made an aluminum frame to hold the polypropylene and went off to the eclipse.

Our eclipse experiment was filmed by WGBH (Boston) for the PBS television series *Nova*. The producer told us they wanted to film "the thrill of victory or the agony of defeat." Our experiment was difficult, but naturally we didn't want to be the "agony of defeat." We had problems with our infrared spectrometer, but after 48 hours of continuous work, we were ready to observe the eclipse. I had programmed our instrument computer to acquire the data automatically, so at the onset of totality, I hit "return" on the computer and went outside to relax and enjoy the eclipse. To my surprise, the *Nova* film crew followed me. I had naively thought they would be more interested in filming the eclipse itself and the instrumentation. Instead, I found that I was watching the eclipse while an enormous lens was focused on me from close range. If you watch that *Nova* episode closely, you will notice that I glance back at the camera briefly, annoyed to be tracked so closely. I thought briefly about trying to escape from the film crew, but then I remembered that the head photographer had successfully filmed giant Australian saltwater crocodiles. I figured that if the crocodiles couldn't escape him, I had no chance! So I tolerated the film crew, and the entire eclipse was a big success. We even published the results,[14] which isn't always accomplished for scientific eclipse experiments. After the eclipse, I slept for two whole days.

Drake Deming, Ph.D. is Chief of the Planetary Systems Laboratory at NASA's Goddard Space Flight Center in Greenbelt, Maryland.

Shadia Habbal, Jeffrey Kuhn, and their University of Hawaii colleagues used an imaging spectropolarimeter to search in infrared wavelengths for relatively cool pockets in the Sun's corona during the 2006 eclipse in Libya. The excited helium they detect may be part of the interstellar gas and dust through which the Sun and planets are passing. [Courtesy of Jeffrey Kuhn]

such a hot charged gas like the corona can have such cool pockets of neutral gas within it.

The instrument they use is an imaging spectropolarimeter. As the name implies, this device provides images, spectrograms, and polarization measurements.

The helium they are detecting could be coming from the Sun. If so, the helium would be hot, ionized, and moving fast like most everything caught in the magnetic fields of the corona, until the helium struck tiny grains of dust in the corona, causing the helium to lose energy and return to its neutral state, at least briefly.

But Kuhn thinks the neutral helium he and his colleagues are seeing is not flowing from the Sun but, instead, is part of the cool interstellar gas through which the Sun and planets are passing in their revolution around the core of the Milky Way Galaxy. As the Sun plows through this tenuous interstellar gas, which contains a small percentage of helium, the helium is heated enough to give off the specific radiation the Hawaiian team has detected. The stimulated helium appears to be in pockets rather than evenly distributed because the corona is of highly uneven density and thus excites the helium only where the density is greater.

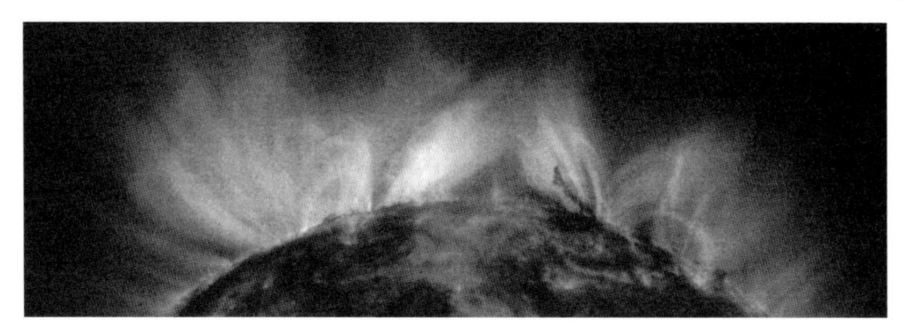

Coronal loops on the Sun, showing the magnetic lines of force, December 30, 1999. [ESA/NASA SOHO EIT]

So what Kuhn may be doing is using interstellar gas as a probe to reveal the structure of the outer corona of the Sun.

Kuhn, Habbal, and their associates hope that they and their imaging spectropolarimeter will be going to a desert in western China for the total eclipse in 2008 to continue their studies.[10]

Threads and Plasmoids

Serge Koutchmy and Laurence November wait for the great natural coronagraph-in-space, the Moon, to eclipse the Sun to permit them a view of the lowest regions of the corona. They hunt for narrow, bright coronal loops indicative of powerful magnetic fields. Studying these loops carefully in ordinary white light, they have observed bright and dark "threads" within the loops—and have found that the dark threads must be essentially vacuums, void of material. Full evacuation requires a magnetic field of a specific strength. Thus the evacuated dark threads provide November with a new perspective on how the magnetic field may determine the temperature of the corona.[11]

When the total solar eclipse of July 11, 1991 passed over Hawaii and the huge observatory complex atop the extinct volcano Mauna Kea, Koutchmy, November, and their co-workers seized the opportunity to observe the corona with the largest optical telescope ever to record a total eclipse, the 142-inch (3.6-meter) Canada–France–Hawaii Telescope. With so large a telescope, they were able to see details in the corona too small to be seen before. They think they observed a coronal plasmoid, a high-density bubble of ionized gas in the corona, about 900 miles (1,400 kilometers) in diameter, moving at 60 miles per second (100 kilometers per second) outward from a magnetic loop prominence that probably spawned it. During the four minutes of the eclipse, the plasmoid distorted, stretched out, and broke into smaller plasmoids.

Koutchmy, November, and their colleagues suspect that it is these plasmoids, small-scale ejection events, not the large coronal mass ejections, that supply most of the material to the upper corona that the Sun loses to space as the background solar wind.[12]

The Chaos of the Corona

The atmosphere of Earth is controlled by three primary forces: gravity, pressure variations, and the Earth's rotation. The Sun's atmosphere is shaped not only by gravity, pressure, and rotation, but also by a fourth force: magnetism. A total eclipse offers the best chance to see how these magnetic fields sculpt the atmosphere of the Sun. By revealing the corona, totality also provides a chance to study a high-temperature plasma under conditions that cannot be duplicated on Earth: extremely hot ionized gases at extremely low density.

Jay Pasachoff is an astronomer who has long used eclipses to probe the Sun. He is eager to understand how the corona is heated to temperatures more than 300 times hotter than the photosphere. The energy of

A merger of an ESA/NASA SOHO spacecraft image and a total eclipse image taken by the Williams College Eclipse Expedition from Kastellorizo, Greece in 2006. SOHO can observe the outer corona and the corona on the face of the Sun, but the "doughnut" between those images is not visible except during total solar eclipses. Combining the eclipse and space images allows astronomers to trace features in the corona from the Sun's surface until the gas escapes into interplanetary space. [Courtesy of Jay Pasachoff]

the Sun is created at its core, where the temperature is about 27 million °F (15 million °C). By radiation and then convection the photons of light energy make their way to the surface, gradually cooling en route, so that the Sun's photosphere glows with a temperature of only about 10,000 °F (5,500 °C). The base of the chromosphere, the lower atmosphere of the Sun, is cooler still, about 7,200 °F (4,000 °C). But the temperature in the corona exceeds one million degrees. Temperature is a measure of the random motion of atoms and subatomic particles, and the particles in the corona are indeed traveling at great speed. But the corona has a very low density, so the total amount of heat energy in the corona is quite small.

What causes the temperature in the corona to be so much higher than that of the visible surface of the Sun? These surprising temperatures seem to result from processes that involve magnetic fields. The magnetic fields that cause sunspots in the photosphere pervade the corona as well. They are responsible for streamers and other coronal features; they also shape the prominences and filaments in the chromosphere.

A favored explanation for the high temperature of the corona is that magnetic waves (for example, Alfvén waves), excited by motion in the convective region beneath the photosphere, propagate energy into the corona, where it is then dissipated as heat. Such waves might manifest themselves as local oscillations in coronal brightness with periods in the range of about 1/10 to 10 seconds.

Jay Pasachoff and his Williams College colleagues and students search the corona for evidence of magnetic waves during total eclipses, such as the one on March 29, 2006 on the Greek island of Kastellorizo. Some coronal-heating theories predict the existence of waves along coronal loops. If magnetic waves are present, observers of the loops might see rapid oscillations in the density of the plasma (ionized gases)

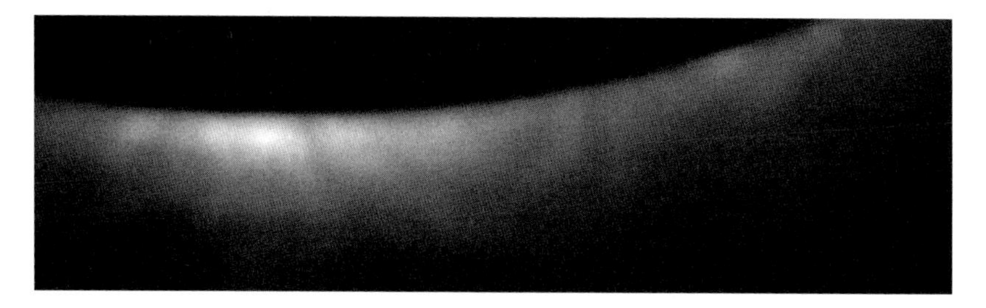

During the total eclipse of 2006, the Williams College Eclipse Expedition photographed the solar limb at a wavelength of 530.3 nanometers, the coronal green line created by iron that has lost 13 of its 26 electrons because of the high temperature of the corona. Pictures were taken 5 times per second to determine the oscillation periods of coronal loops at the edge of the Sun. Their oscillation may help solve the mystery of how the corona is heated. [Courtesy of Jay Pasachoff]

as the waves pass through the gases. Because the coronal gases are so hot, they give off light at distinctive wavelengths (such as a green spectral line that iron emits when half its 26 electrons are missing).

The coronal plasma is quite rarefied, however, so the glow is weak. But as waves pass through a point within these gases, the waves should briefly increase the gas density at that point, thereby briefly increasing the brightness of the gas, and this fluctuation in brightness would give evidence that, indeed, waves are present.

Pasachoff, his colleagues, and his students use the total phase of a solar eclipse to make a sequence of images of coronal loops using the light of the coronal green line. For the 2006 eclipse, they used a new detector that allowed them to simultaneously observe a corresponding coronal red line caused by iron atoms that have lost 9 of their 26 electrons. They watch for periodic brightness variations in small regions of the loops that would suggest the passage of waves. They think that at prior eclipses they detected oscillations with periods in the 0.2- to 5-second range. The analysis of the new data continues.

Pasachoff uses every accessible total eclipse to search for these oscillations in the corona of his favorite star. He hopes this research will help to identify what kind of wave is responsible for the surprisingly high temperatures of the corona and how that heating occurs.[13]

Notes and References

Epigraph: Robert Louis Stevenson: "My Shadow," lines 1–2, in *A Child's Garden of Verses* (New York: C. Scribner's Sons, 1905), page 23.

[1] Sabatino Sofia, David W. Dunham, Joan B. Dunham, and Alan D. Fiala: "Solar Radius Change Between 1925 and 1979," *Nature*, volume 304 (August 11, 1983), pages 522–526.

[2] Some changes observed on the Sun may have been reflected in modest climate changes on Earth within historical times. From 1645 to 1715, there were very few sunspots. During this period, called the Maunder minimum, temperatures in Europe were lower than usual, leading some modern scientists to refer to these seven decades as the Little Ice Age.

[3] The Moon appears to keep the same side always facing the Earth because it rotates once on its axis as it makes one revolution around our planet. However, the Moon's orbital speed is not constant, because its orbit is elliptical. The Moon moves faster when closer to the Earth and slower when farther away. As a result, the Moon does not quite keep one side precisely aligned with Earth. This slight misalignment provides different horizons along the eastern and western edges of the Moon.

The 5° inclination of the Moon's orbit to the ecliptic means that as the Moon approaches the Sun for an eclipse, it is ascending or descending through the Sun–Earth plane. Because observers on Earth see the Moon crossing the Sun slightly top or bottom first, this perspective foreshortens or extends the view across the Moon's polar regions and changes which features cause Baily's beads.

A lesser libration effect is the view from different locations on Earth. Depending on our latitude, we see a little over or under the Moon's poles. From the east or west, we see the polar features at slightly different angles. This change of perspective with each eclipse and each position along the eclipse path causes the apparent positions and angular heights of features to change minutely.

In addition, the Moon actually wobbles slightly due to the tidal effects of the Earth and Sun, adding in a complex way to this libration effect.

[4] M. Emilio, J. R. Kuhn, R. I. Bush, and P. Scherrer: "On the constancy of the Solar Diameter," *Astrophysical Journal*, Volume 543, Issue 2 (10 November 2000), pages 1007–1010.

[5] Dunham hopes to compare IOTA measurements of the Sun's diameter made during eclipses with those made by the French space agency's Picard satellite, scheduled for launch in 2008. Picard is specifically designed to measure the diameter of the Sun.

[6] C. Lindsey, E. E. Becklin, F. Q. Orrall, M. W. Werner, J. T. Jefferies, and I. Gatley: "Extreme Limb Profiles of the Sun at Far-Infrared and Submillimeter Wavelengths," *Astrophysical Journal*, volume 308 (September 1, 1986), pages 448–458; and T. L. Roellig, E. E. Becklin, J. T. Jefferies, G. A. Kopp, C. A. Lindsey, F. Q. Orrall, and M. W. Werner: "Submillimeter Solar Limb Profiles Determined from Observations of the Total Solar Eclipse of 1988 March 18," *Astrophysical Journal*, volume 381, November 1, 1991, pages 288–294; and C. Lindsey, J. T. Jefferies, T. A. Clark, R. A. Harrison, M. Carter, G. Watt, E. E. Becklin, T. L. Roellig, D. C. Braun, D. A. Naylor, and G. J. Tompkins: "Extreme-Infrared Brightness Profile of the Solar Chromosphere Obtained During the Total Eclipse of 1991," *Nature*, volume 358, July 23, 1992, pages 308–310. See also Peter Foukal: "Darkness Can Illuminate," *Nature*, volume 358, July 23, 1992, pages 285–286.

[7] T. A. Clark, C. A. Lindsey, D. M. Rubin, and W. C. Livingston: "Eclipse Measurements of the Distribution of CO Emission Above the Solar Limb," in J. R. Kuhn and M. J. Penn: *Infrared Tools for Solar Astrophysics, What's Next?*, Proceedings of the National Solar Observatory/Sacramento Peak Observatory Workshop, September 1994, pages 133–138; also Alan Clark, personal communication, October 20 and 24, 1998.

[8] This collaboration included scientists from the Max Planck Institute in Germany, Rhodes College, the National Solar Observatory, Michigan State University, and the High Altitude Observatory.

[9] J. R. Kuhn, M. J. Penn, and I. Mann: "The Near-Infrared Coronal Spectrum," *Astrophysical Journal*, volume 456, January 1, 1996, pages L67-L70; J. R. Kuhn, H. Lin, P. Lamy, S. Koutchmy, and R. N. Smartt: "IR Observations of the K and F Corona During the 1991 Eclipse," in D. M. Rabin et al., editors: *Infrared Solar Physics* (Netherlands: International Astronomical Union, 1994), pages 185–197; P. Lamy, J. R. Kuhn, H. Lin, S. Koutchmy, and R. N. Smartt: "No Evidence of a Circumsolar Dust Ring

from Infrared Observations of the 1991 Solar Eclipse," *Science*, volume 257, September 4, 1992, pages 1377–1380.

[10] This section is based on correspondence and conversations with Jeffrey Kuhn, spring 2007.

[11] Laurence J. November and Serge Koutchmy: "White-Light Coronal Dark Threads and Density Fine Structure," *Astrophysical Journal*, volume 466, July 20, 1996, pages 512–528; Serge Koutchmy and Michaël M. Molodensky: "Three-dimensional image of the solar corona from white-light observations of the 1991 eclipse," *Nature*, volume 360, December 24/31, 1992, pages 717–719.

[12] S. Koutchmy, O. Bouchard, S. Grib, L. November, J-C. Vial, P. Gouttebrose, V. Koutvitsky, M. Molodensky, L. Solov'iev, and I. Veselovsky: "About Small Plasmoids Propagating in the Solar Corona," *Proceedings of the Third SOHO Workshop—Solar Dynamic Phenomena and Solar Wind Consequences* (September 26–29, 1994) (European Space Agency publication SP-373, December 1994), pages 139–142.

[13] Report to the 1998 Keck Northeast Astronomy Consortium Student Research Symposium by Kevin D. Russell, with special reference to work by Tim McConnochie, supplied by Jay M. Pasachoff, October 16, 1998.

J. M. Pasachoff, B. A. Babcock, K. D. Russell, T. H. McConnochie, and S. Diaz: "A Search at Two Eclipses for Short-Period Waves that Heat the Corona," *Solar Physics*, volume 195, number 2 (August 2000), pages 281–298.

J. M. Pasachoff, B. A. Babcock, K. D. Russell, and D. B. Seaton: "Short-Period Waves that Heat the Corona Detected at the 1999 Eclipse," *Solar Physics*, volume 207, number 2 (June 2002), pages 241–257. Also www.arxiv.org:astro-ph 0202237.

[14] D. Deming, D. E. Jennings, G. McCabe, R. Noyes, G. Wiedermann, and F. Espenak, "Limb Observations of the 12.32 Micron Solar Emission Line During the July 1991 Total Eclipse," *Astrophysical Journal Letters*, volume 396 (1 September 1993), L53.

10
Observing a Total Eclipse

> Now eclipses are elusive and provoking things... visiting the same locality only once in centuries. Consequently, it will not do to sit down quietly at home and wait for one to come, but a person must be up and doing and on the chase.
>
> Rebecca R. Joslin (1929)

A total eclipse of the Sun. What is this sight that lures people to travel great distances for a brief view at best and a substantial possibility of no view at all, with no rain check? And how do you get the most out of the experience of a total eclipse?

In these pages, 20 astronomers, professionals and amateurs, plus some eclipse seekers from earlier times share their experiences with you. Among them, they have witnessed over 200 total eclipses.

"It's like a religious experience," says Jay Anderson, "the anticipation as the time until totality is counted in days, then hours, then minutes. It's the perfect build-up. Spielberg couldn't do it better. It's an intensely moving event."

Steve Edberg agrees: "It is the intensity of the event. You grab as much as you can. I like action in the heavens and you can't get much better than this."

Roger Tuthill had a ritual he followed the day before an eclipse. He walked around inside the path of totality observing the people who live there. They have no idea what they are about to experience and there are no words to adequately prepare them.

First Contact

There is a special feeling from the instant when the Moon begins to slide in front of the Sun. In less than a minute, observers with small telescopes see the first tiny "bite" out of the western side of the Sun. For Virginia Roth, "That's when the magic starts."

Panel of Eclipse Veterans

Jay Anderson, meteorologist, University of Manitoba, Winnipeg
John R. Beattie, typesetter, New York City
Richard Berry, former editor-in-chief, *Astronomy*; astronomy author
Dennis di Cicco, senior editor, *Sky & Telescope*, Cambridge, Massachusetts
Stephen J. Edberg, astronomer, Jet Propulsion Laboratory, Pasadena, California; astronomy author; executive director of the annual RTMC Astronomy Expo conference of amateur astronomers
Alan D. Fiala, chief, Nautical Almanac Division, U.S. Naval Observatory, Washington, D.C., now retired
Ruth S. Freitag, senior science specialist, Library of Congress, Washington, D.C.; astronomy author
Joseph V. Hollweg, professor of astronomy, University of New Hampshire, Durham
George Lovi, astronomy author/columnist, Lakewood, New Jersey (George died in 1993)
Frank Orrall, professor emeritus of physics and astronomy, Institute for Astronomy, University of Hawaii, Honolulu (Frank died in 2000)
Jay M. Pasachoff, Field Memorial professor of astronomy and director of the Hopkins Observatory, Williams College, Williamstown, Massachusetts; astronomy author
Leif J. Robinson, editor in chief, *Sky & Telescope*, Cambridge, Massachusetts; now editor emeritus; astronomy author
Virginia and Walter Roth, proprietors, Scientific Expeditions, Inc., Venice, Florida (now deceased)
Roger W. Tuthill, president, Roger W. Tuthill, Inc., telescope accessories, Mountainside, New Jersey (Roger died in 2000)
Jack B. Zirker, astronomer, National Solar Observatory, Sunspot, New Mexico, now retired

First contact remains a very special moment for Jay Pasachoff. As an astronomer, he can more readily than most appreciate all the factors that go into predicting precisely when an eclipse will occur and exactly where on Earth it will be seen. And when the call "First Contact" comes right on schedule, he always finds this ingenuity of mankind astounding.

Even a century and a half ago, this commencement of a rare event was already exerting a powerful effect on its beholders. French astronomer François Arago observed the 1842 eclipse from Perpignan in southern France amid townspeople and farmers who had been educated about the eclipse and who were watching the sky intently. "We had scarcely, though provided with powerful telescopes, begun to perceive a slight indentation in the sun's western limb, when an immense shout, the commingling of twenty thousand different voices, proved that we had only anticipated by a few seconds the naked eye observation of twenty

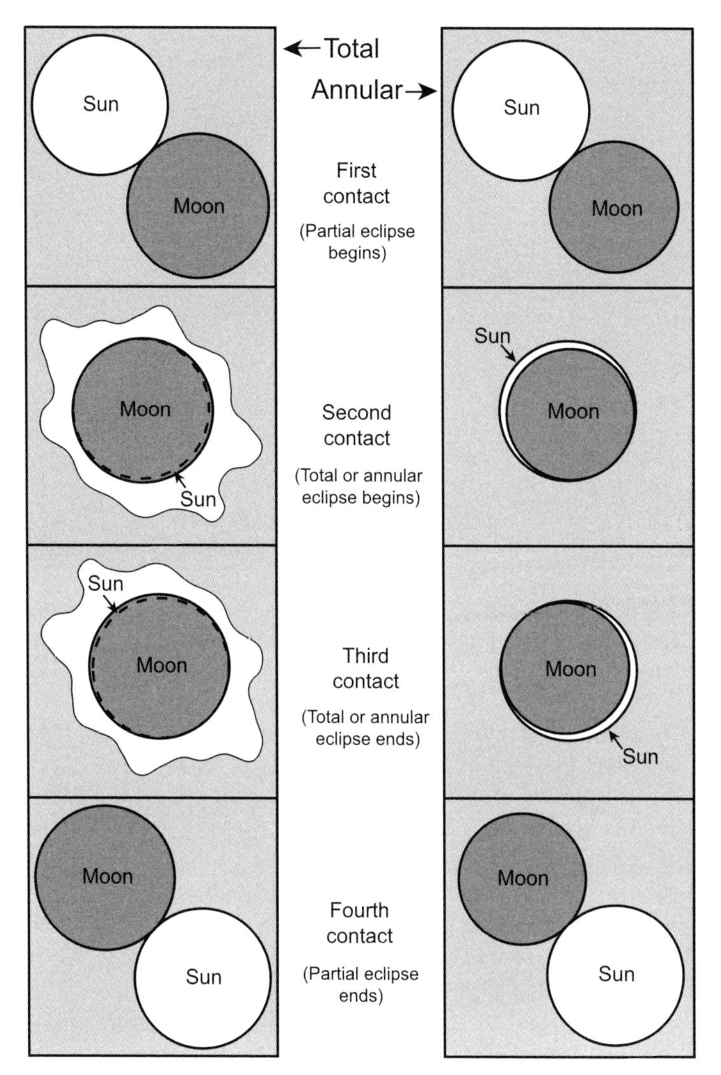

Points of contact in a total solar eclipse *(left)* and an annular eclipse *(right)*. In a partial eclipse, there are only first and fourth (last) contacts.

thousand astronomers equipped for the occasion, and exulting in this their first trial."[1]

The Crescent Sun

Solar physicist Joe Hollweg likes to view the partial phases leading up to a total eclipse with a patch over one eye so that it becomes dark-adapted.[2]

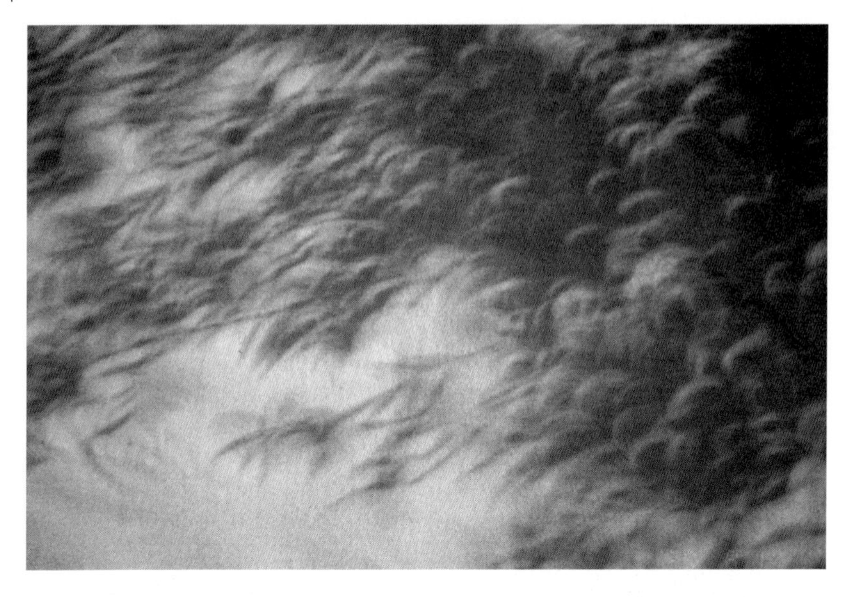

A tree's shadow reveals crescent images of the partially eclipsed Sun created by holes in the foliage. [©1999 Greg Babcock]

Totality is about as bright as a full-moonlit sky, and a dark-adapted eye makes for better viewing of totality.

The partial phase of a total eclipse has a power all its own, as the Moon steadily encroaches upon the Sun, covering more and more of its face. A partial eclipse close to total visited a Russian monastery in medieval times, near sunset on May 1, 1185. A chronicler recorded: "The sun became like a crescent of the moon, from the horns of which a glow similar to that of red-hot charcoals was emanating. It was terrifying to men to see this sign of the Lord."[3]

As the partial phase proceeds, a thousand tiny images of the crescent Sun may be visible on the ground beneath trees as the spaces between the leaves create pinhole cameras to focus the crescent image of the waning Sun.

Yet, as George B. Airy, Astronomer Royal of England, learned when he saw his first total eclipse in 1842: "[N]o degree of partial eclipse up to the last moment of the sun's appearance gave the least idea of a total eclipse..."[4]

The Changing Environment

Every eclipse veteran urges newcomers to pause as totality approaches to look around at the landscape and notice the changes in light levels and color. Many people are surprised how little the landscape darkens until the last ten minutes or so before the eclipse becomes total. All the better for the drama, because once the light begins to fade, it sinks quite noticeably. Usually, everyone becomes quite silent. You can feel the tension and rising emotion. A primitive portion of your brain tugs at you to say that something peculiar is going on, that it ought not to be growing dark in the midst of the day.

Dennis di Cicco remembers that during the 1976 eclipse in Australia, the birds responded to the fading light by raising a racket and going to roost. In 1973, he witnessed an annular eclipse in Costa Rica that took place shortly after sunrise. The cows grazing around him ignored the partial phases of the eclipse, but as the eclipse reached maximum, the cows ceased grazing, formed a line, and marched back to their barn.

Walter Roth recalls watching the 1973 eclipse from a game preserve in Africa. As the light faded in the minutes just before totality, the birds flocked into the trees, complaining madly. Elephants, which had been grazing peacefully, milled around nervous and confused.

"As the light fades, the Sun is a thin crescent and shrinking quickly," says Richard Berry. "The quantity and quality of the light have begun to change noticeably. The temperature is dropping; the air feels still and strange."

Shadow Bands

As the eclipse nears totality and shortly after it emerges from totality, shadow bands—faint undulations of light rippling across the ground at jogging speed—are sometimes visible. They are one of the most peculiar and least expected phenomena in a total eclipse. Many eclipse veterans have never seen them; some do not want to take time to try because they occur in the last moments before totality as Baily's beads and other beautiful sights are visible overhead. Other veterans consider shadow bands one of the true highlights of a total eclipse. They resemble the graceful patterns of light that flicker or glide across the bottom of a swimming pool.

Shadow bands occur when the crescent of the remaining Sun becomes very narrow so that only a thin shaft of sunlight enters the atmosphere of Earth overhead. There it encounters currents of warmer and cooler air which have slightly different densities. The different densities act as very weak lenses to bend the light passing from one parcel to the next.[5] It is the focusing and defocusing of light by these ever-present air currents in motion that causes stars to twinkle. The Sun would twinkle too if it were a starlike dot in the sky. And so it does, near the total phase of a solar eclipse. When the Sun has narrowed from a disk to a sliver, the light from the sliver twinkles in the form of shadow bands rippling across the ground.

Catching Shadow Bands

by Laurence A. Marschall

Even though shadow bands are only visible for a few fleeting minutes, it is possible to catch them in flight if you prepare in advance. Get a large piece of white cardboard or white-painted plywood to act as a screen—the bands are subtle and can be more easily seen against a clean, white surface. A large white sheet staked to the ground may be more portable and will serve in a pinch, but ripples in the sheet can mask the faint gradations of the shadow bands.

Lay out on the screen one or two sticks marked with half-foot intervals (yardsticks will do nicely). Orient the sticks at right angles to one another so that at the first sign of activity you can move one stick to point in the direction that the shadow bands are moving. Then, using the marks on that stick, make a quick estimate of the spacing between the bands (typically 4 to 8 inches; 10 to 20 centimeters). Finally, using a watch, make a quick timing of how long a bright band takes to go a foot or a yard. Jot down the figures or, better yet, dictate your measurements into a small tape recorder. If you practice this procedure before the eclipse, you will be able to see the shadow bands and then swing your attention back to the sky to catch Baily's beads and the onset of totality.

The second stick, by the way, is reserved for measuring shadow bands *after* totality, if they are visible.

Later, when the excitement of the eclipse is over, you can take stock of the data. Do the shadow bands seem to move at all? At some eclipses, especially when the air is very still, they just shimmer without going anywhere. If they move, how fast? Typical speeds are about 5 to 10 miles per hour (8 to 16 kilometers per hour). Do they change directions after eclipse? Usually they do, unless you happen to be standing directly along the central line of totality.

Dr. Laurence A. Marschall is W.K.T. Sahm Professor of Physics at Gettysburg College in Pennsylania

In 1842, George B. Airy, the English Astronomer Royal, saw his first total eclipse of the Sun and recalled shadow bands as one of the highlights: "As the totality approached, a strange fluctuation of light was seen... upon the walls and the ground, so striking that in some places children ran after it and tried to catch it with their hands."[6]

The Approach of Totality

During the final one to two minutes before totality, the Sun is about 99% covered and the light is fading rapidly. In these brief moments, the diamond ring, Baily's beads, and the corona all appear.

Shadow bands on an Italian house in 1870. [Mabel Loomis Todd: *Total Eclipses of the Sun*]

Diamond ring effect at the total solar eclipse of June 21, 2001 from Chisamba, Zambia. [Nikon F2 SLR, 60 mm f/15 Unitron refractor, f/15, 1/125 s, ISO 200. © 2001 Michael F. Barrett]

Looming on the western horizon, growing ever larger, is the Sun-cast dark shadow of the Moon. And it is coming toward you. Alan Fiala describes its appearance as the granddaddy of thunderstorms, but utterly calm. If you are observing from a hill with a view to the west, the approach of the Moon's shadow can be quite dramatic, even chilling. In the words of astronomer Mabel Loomis Todd a century ago, it is "a tangible darkness advancing almost like a wall, swift as imagination, silent as doom."[7] And this is how astronomer Isabel Martin Lewis in 1924 described the onrush of the Moon's shadow at the onset of totality: "[W]hen the shadow of the moon sweeps over us we are brought into direct contact with a tangible presence from space beyond and we feel the immensity of forces over which we have no control. The effect is awe-inspiring in the extreme."[8]

As the shadow races east toward you, it accelerates. Ten seconds before totality, you feel as if you are being swallowed by some gigantic whale. The soft white glow of the corona begins to silhouette the Moon's dark disk while a dazzling beacon of sunlight clings to one edge of the Moon—the diamond ring effect. A few seconds before totality begins, the ends of the remaining sliver of the Sun fracture into Baily's beads. Each bead lasts only an instant and flickers out as new ones form. As the length of the Sun's sliver shortens, the two separated groups of beads converge and combine, and suddenly only one remains. "For one fleeting moment this last bead lingers, like a single jewel set into the arc that is the lunar limb," says John Beattie. And then it is gone.

The eclipse is total.

Remotely Close[a]

by Patricia Totten Espenak

In another life
the Moon
touched us
with her dark shadow,

Made invisible
impressions,
creased
our minds.

And now that image,
impossible to imagine,
draws us to
the next chosen place.

Where the Moon's
shadow sword
will nick
the Earth again.

[a] After three failed attempts spanning 25 years, Pat Espenak witnessed her first total eclipse of the Sun in India on October 24, 1995

Corona Emerging/Second Contact

It is difficult to find words that do justice to the corona, the central and most surprising of all features of a total eclipse. The Sun has vanished, blocked by the dark body of the Moon. A black disk surrounded by a white glow. "It is the eye of God," says Jack Zirker.

In the bitter cold of the high plateau in Bolivia, astronomer Frank Orrall was part of a research team studying the total eclipse of 1966. In reviewing his observing notes, he realized that he had written "The heavens declare the glory of God." "My notebooks normally do not say such things," he mused.

Even the color of the corona is difficult to capture. "I prefer 'pearly,' " says Ruth Freitag, "because it conveys a luminous quality that 'white' lacks." "A silver gossamer glow," suggests Steve Edberg. "Remember," said George Lovi, "the human eye is the only instrument that can see the corona in all its splendor."

During Totality

In the midst of totality, it is particularly impressive to look around at the nearby landscape and the distant horizon. Depending on your position within the eclipse shadow, the size of the shadow, and cloud conditions away from the Sun's position, the light level during totality can vary greatly from eclipse to eclipse and from place to place within an eclipse. On some occasions, totality brings the equivalent of twilight soon after

The glorious corona is visible for just a few short minutes during the total solar eclipse of June 21, 2001 from Zambia. [Nikon D1H DSLR, Astrophysics 105 EDT refractor, fl = 620 mm, Nikon 1.4 × -converter TC-14, efl = 917 mm, f/8.7, composite of 21 exposures: 1/4000 s to 2 s. © 2002 Friedhelm Dorst and Miloslav Druckmüller]

sunset; on others it is dark enough to make reading difficult. Yet it is never as black as night.

The color that descends upon everything is hard to specify. Most veteran observers describe it as an eerie bluish gray or slate gray, and then apologize for the inadequacy of the description. The Sun's corona contributes some brightness, but only about as much as a full moon. Primarily, the light that brightens your location within the shadow of the Moon is light reflected from a short distance away, where the Sun is not totally eclipsed.

Looking around the horizon in the midst of totality, you have the best impression of standing in the shadow of the Moon. You also have a renewed appreciation for the power of the Sun. There it is, its face completely blocked by the Moon. How big in the sky is that body whose overwhelming brightness creates the day and banishes the stars from view? Reach out your arm toward the eclipsed Sun, just as you did to measure the Moon. Squeeze the darkened Sun between your index finger and your thumb until it just fits between them. It is the size of a pea. Yet that little spot in the sky, darkened now, is usually enough to blanket the Earth in light and warmth and completely dazzle the eye. For just

The View from the Edge

by Alan D. Fiala

The advice in this chapter assumes that the observer is near the central line of an eclipse. But observing a total eclipse from the center of its path is not the only possibility.

At any total eclipse, if you move toward either edge of the path of totality, you gain in the duration of Baily's beads activity. The number of beads you see depends on whether you go toward the southern or northern limit of the eclipse path. The southern edge provides more beads because the terrain of the Moon's southern limb is much rougher. The bead activity at the edges of the path is always caused by the same lunar features, so you can control the amount of bead activity you see by how close you come to the correctly predicted limit.

If you observe from the central line of the eclipse, the beads are confined to the eastern limb of the Moon as totality approaches and the western edge of the Moon after the end of totality. The mountains and valleys along the eastern and western limbs are strongly affected by lunar librations,[a] which change the view significantly from one eclipse to another and determine whether or not any Baily's bead activity occurs at second or third contact.

If you want to study the inner corona or the corona near one pole (where the corona tends to appear more brush-like), once again you gain by going toward the edge of the path of totality. In this way, you control what portion of the inner corona you observe.

a few minutes its face is hidden. Daytime has become night and the temperature is falling. Do you feel something of what ancient people must have felt as they watched the Sun, upon which they depended for warmth and light and life itself, disappear?

Leif Robinson emphasizes that a total eclipse cannot be experienced vicariously. Even the best photography and video recordings are pale reflections of the event. Taking pictures during an eclipse is fine, but be sure you *look* at what is truly a visual spectacle. Do not miss the ambience of the moment either, he advises. Look at other people to see how they are reacting. Notice the changing colors in the sky. He does not, however, recommend trying to see planets or stars in the sky during a total eclipse. "Time during totality is too precious to spend straining to see stars when you can see those same objects much better in

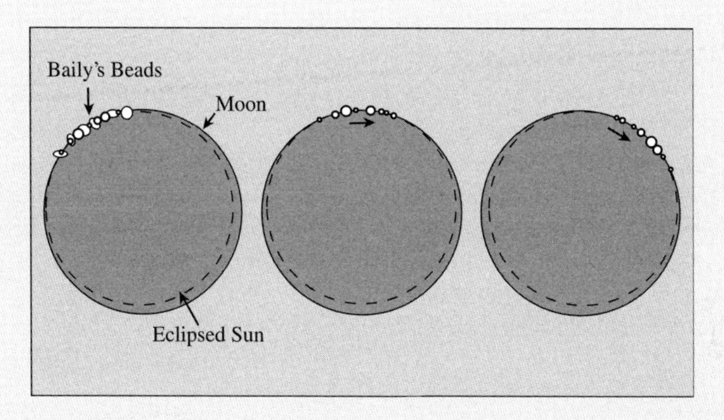

Migration of Baily's beads as seen from the edge of the path of totality. As the Moon moves across the face of the Sun, different valleys on the Moon allow light from the Sun's surface to pass, creating different beads.

The central line is not the only place from which to enjoy a total eclipse of the Sun.

---

[a] The Moon rotates once on its axis while it revolves once around the Earth. The result is that we always see the same face of the Moon; the other half is always turned away from us. Almost, but not quite. Instead of seeing only 50% of the Moon, over time we actually see 59% percent due to librations. In this case, longitudinal and diurnal librations are at work: (1) The speed of the Moon's spin is constant but the speed of the Moon in its elliptical orbit varies, which allows us to see a little beyond the eastern and western limbs of the Moon in the course of a month; (2) When the Moon is near the horizon, we are looking at it a little from the side, so we are peeking a bit beyond the eastern or western limb seen when the Moon is high in the sky.

Dr. Alan D. Fiala was chief of the Nautical Almanac Division, U.S. Naval Observatory before he retired.

the nighttime sky." Steve Edberg feels differently: "One of my strongest memories from eclipses is seeing stars during the day."

The transformation in the appearance of the Sun from a bright crescent to a dark disk surrounded by ghostly light was recorded by François Arago, the dean of French astronomy a century and a half ago, as he watched the eclipse of July 8, 1842 with a group of astronomers and nearly 20,000 townspeople in southern France.

> *[W]hen the sun, being reduced to a narrow filament, began to throw only a faint light on our horizon, a sort of uneasiness took possession of every breast, each person felt an urgent desire to communicate his emotions to those around him. Then followed a hollow moan resembling that of the distant sea after a storm, which increased as the slender crescent diminished. At last, the crescent disappeared, darkness instantly followed, and this phase of the eclipse was marked by absolute silence... The magnificence of the phenomenon had triumphed over the petulance of youth, over the levity affected by some of the spectators as indicative of mental superiority, over the noisy indifference usually professed by soldiers. A profound calm also reigned throughout the air: the birds had ceased to sing.*
>
> *After a solemn expectation of two minutes, transports of joy, frenzied applauses, spontaneously and unanimously saluted the return of the solar rays. The sadness produced by feelings of an undefinable nature was now succeeded by a lively satisfaction, which no one attempted to moderate or conceal. For the majority of the public the phenomenon had come to a close. The remaining phases of the eclipse had no longer any attentive spectators beyond those devoted to the study of astronomy.*[9]

How do you take it all in? It is not possible. There is too much to see—and feel. "Everybody, myself included, tries to do too much," says Steve Edberg. "Save time near the beginning, middle, and end of totality just to stare," urges Jay Anderson. "Make a deliberate effort to store the sights in your mind." Mabel Loomis Todd wrote, a century ago, that "When Dr. Peters of Hamilton College was asked what single instrument he would select for observing an eclipse, he replied, 'A pillow.' "[10]

The End of Totality

All too soon, no matter what the duration of totality (and it can't exceed 7 minutes 32 seconds[11]), the Moon's shadow moves on and a bright flash of sunlight appears from the western edge of the Moon.

Rebecca Joslin and her college astronomy teacher traveled from the United States to Spain for an eclipse in 1905, only to be clouded out. So

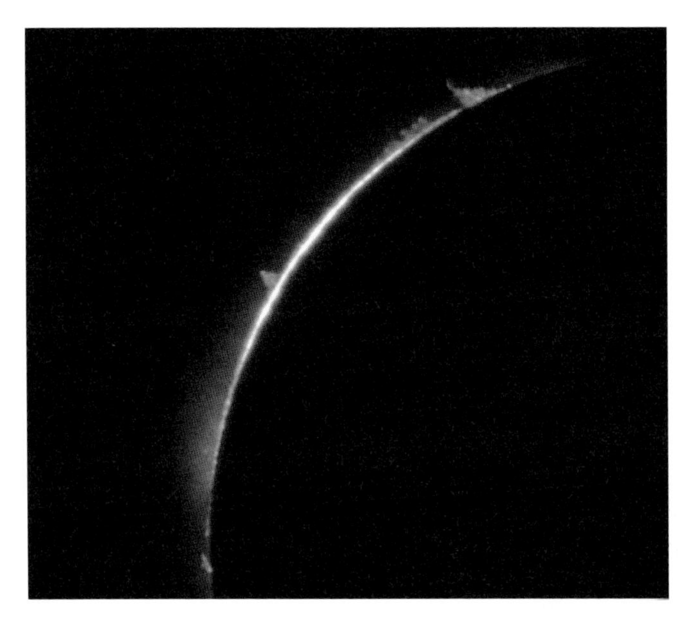

Brilliant red prominences and the chromosphere are seen at the total solar eclipse of March 29, 2006 from Jalu, Libya. [35 mm SLR, Borg 100ED f/6.4 refractor, fl = 640 mm, Vixen GP mount, 2 × teleconverter, f/12.8, 1/250 s to 1 s, Fuji Velvia 50 slide film. © 2006 Dave Kodama]

they shifted their attention to nature around them until light pierced the cloud to tell them that the unseen total phase of the eclipse was over.

> *But we hardly had time to draw a breath, when suddenly we were enveloped by a palpable presence, inky black, and clammy cold, that held us paralyzed and breathless in its grasp, then shook us loose, and leaped off over the city and above the bay, and with ever and ever increasing swiftness and incredible speed swept over the Mediterranean and disappeared in the eastern horizon.*
>
> *Shivering from its icy embrace, and seized with a superstitious terror, we gasped, "WHAT was THAT?"... The look of consternation on M's face lingered for an instant, and then suddenly changed to one of radiant joy as the triumphant reply rang out, "THAT was the SHADOW of the MOON!"*[12]

"I doubt if the effect of witnessing a total eclipse ever quite passes away," wrote Mabel Loomis Todd. "The impression is singularly vivid and quieting for days, and can never be wholly lost. A startling nearness to the gigantic forces of nature and their inconceivable operation seems to have been established. Personalities and towns and cities, and hates and jealousies, and even mundane hopes, grow very small and very far away."[13]

"Beware of post-eclipse depression," warns Steve Edberg. One minute after third contact, with the passage of the Moon's shadow, the disappearance of the shadow bands, and the reappearance of the

The Personalities of Eclipses

by Stephen J. Edberg

Weather, geography, and companionship profoundly affect the experience of an eclipse. But each eclipse has its own intrinsic differences from all others and these differences continue to lure eclipse veterans. They use them in planning their observations.

One factor is the magnitude of the total eclipse, the degree to which the disk of the Moon more than covers the disk of the Sun. When the angular size of the Moon is great, the Moon at mid-totality will mask not only the Sun's photosphere but also its chromosphere and innermost corona. Except at the beginning and end of totality (or near the eclipse path limits), the stunning fluorescent pink of the prominences will be hidden from view, unless an absolutely gigantic prominence happens to be present on the Sun's limb (as there was for the 1991 eclipse).

However, because the relatively bright inner corona is blocked from view, a large magnitude eclipse is the best time to observe the full extent and detail in

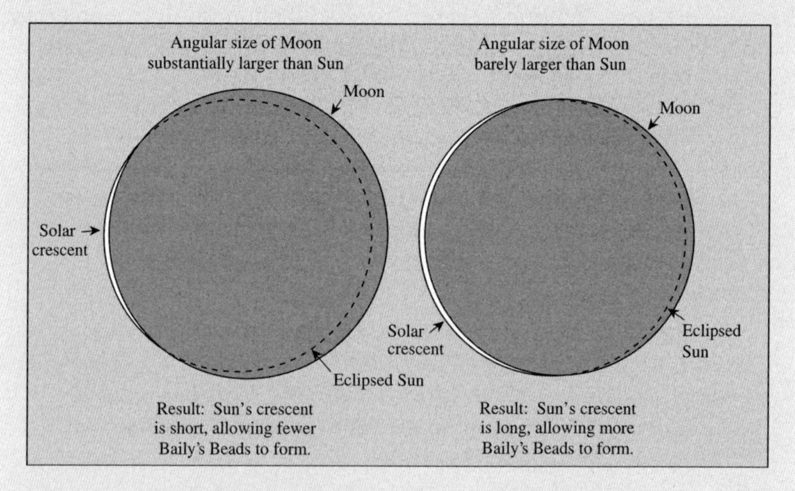

Left: When the Moon's angular size is large, the Sun's crescent is short, reducing the span of Baily's beads. ***Right***: When the Moon just barely covers the Sun, so that the disks appear nearly the same size, the Sun's crescent is much longer and the span of Baily's beads is greater.

crescent Sun, you feel exhausted: worn out by totality and the excitement of getting ready for it. Most observers are too tired to watch after third contact, and what follows is anticlimactic anyway. The cure for blue sky blues? "Socialize," says Edberg. "Ask people what they saw. Share war stories."

"The end of totality is a time for celebration," says Jay Anderson. "In the wake of such a powerful shared experience, conversations become animated and casual acquaintances tend to become lifelong friends."

the corona. Because the Moon appears larger, its shadow is wider. Thus the total eclipse lasts longer and the darkness is deeper, allowing the full majesty of the corona to shine through, as well as any planets and bright stars that happen to be above the horizon.

The Moon's apparent motion over the Sun's disk is easier to photograph during long eclipses. One photo taken at the onset of totality and a second taken just before the end maximizes the change in the Moon's position with respect to the Sun. When viewing these pictures as a stereo pair, the Moon appears to actually hang in front of the corona and prominences.

By contrast, in a smaller magnitude total eclipse, the Moon's disk is not big enough to mask the lower corona and prominences may be seen all the way around the disk of the Sun during most of totality. But this eclipse will be comparatively brief and the full extent of the corona may not be evident. A total eclipse with less obscuration often provides a better display of Baily's beads and the diamond ring effect. Because the Moon's disk is nearly the same apparent size as the Sun's disk, as totality nears, the crescent of the Sun is long and narrow, allowing Baily's beads to glimmer like jewels on a necklace. When the Moon's apparent disk is large, the length of the Sun's crescent is greatly shortened as it narrows. There may be only one or two Baily's beads.

As veteran observers plan for upcoming eclipses, they also take into consideration the sunspot cycle. At sunspot maximum, the corona is brighter, rounder, and larger. Prominences also tend to be more numerous. At sunspot minimum, the corona is fainter and broader at the equator than at the poles. Brush-like coronal features projecting from the poles are more noticeable.

A third factor used by eclipse followers is the proximity of the shadow path to the Earth's equator, where the rotation of the Earth causes the ground speed of the Moon's shadow to be slowest, making the duration of totality longest. Eclipses with six to seven minutes of totality are almost always found between the Tropic of Cancer and the Tropic of Capricorn. The rotation of the Earth extends not just the period of totality but all aspects of the eclipse, so that, near the equator, the partial phases of the eclipse last longer. The last crescent of the Sun is covered more slowly, so Baily's beads and the diamond ring effect, while still brief, last a little longer, and the view of the prominences at the limb of the Sun is prolonged.

Stephen J. Edberg is a Jet Propulsion Laboratory scientist, author, and executive director of the RTMC Astronomy Expo, a major annual conference of amateur astronomers.

My Favorite Eclipse

by Ken Willcox

Every total eclipse is different—and wonderful—but if you see more than one, you probably have a favorite, for some special reason. My favorite was the eclipse of November 3, 1994 on the Altiplano in Bolivia.

The Altiplano, or Puno, is a wide flat desert high in the Andes Mountains. It was there we had come, 110 of us from all over the United States and Europe, to a small plot of land 12,516 feet (3,815 meters) above sea level, hoping for a spectacular view of a total eclipse of the Sun.

The Altiplano site was selected because it provided the best chance of clear skies along the eclipse path. But the price of clear skies was the major logistical problem of transporting 110 people safely and comfortably to a remote high-altitude viewing site in a foreign land.

On Wednesday, November 2, we boarded our own private train in La Paz, with a dining car and also a baggage car to accommodate the array of equipment brought to record three precious minutes of time. At 4:50 P.M., our chartered narrow-gauge train began its spiraling climb out of the trench in which La Paz lies, then turned south toward our destination 200 miles (320 kilometers) away, a spot on the central line south of Sevaruyo. In the middle of the night, using the global positioning system (GPS), Jim Zimbleman of the Smithsonian Institution and I navigated the train to a precise spot along the tracks selected two years earlier.

As we approached our site about 4 A.M., we were met by 70 gun-toting soldiers provided by the Bolivian army. Under floodlights, before dawn, in the middle of nowhere high in the Andes, amateur and professional astronomers and Bolivian soldiers dragged equipment from the train and prepared to record an extraordinary celestial event. The soldiers set up a perimeter to keep out local people who might wander into the forest of telescopes that had suddenly sprouted in the desert.

One hour after sunrise, the clouds began breaking up. Only high, scattered cirrus remained at first contact at 7:19 A.M., and those too were vanishing.

Charles Piazzi Smyth, Astronomer Royal of Scotland, saw his first total eclipse in 1851, and recognized its ability to overwhelm an observer, distracting him from his carefully planned research. "Although it is not impossible but that some frigid man of metal nerve may be found capable of resisting the temptation," he wrote, "yet certain it is that no man of ordinary feelings and human heart and soul can withstand it."[14]

Confessions of Eclipse Junkies

What happens if you are clouded out, as happens even with good planning about one out of every six times? Unlike a rocket countdown, which can be "T minus 30 minutes and holding" while the clouds clear,

Approximately a half hour before totality, the wind began to increase from the east, rather than decrease as it usually does preceding a total eclipse of the Sun. The heat from the Sun falling on the Altiplano and its surrounding mountains was being extinguished by the increasing eclipse. The cooling was most rapid in the mountains and the heavier cold air rushed down the closer eastern slopes and onto the Altiplano.

A family of Aymara Indians, descendants of the Incas, had been invited to join us and were given solar filters to view the partial phases. This Bolivian family huddled around their father as totality approached. We kept glancing at them to watch their response to the disappearing Sun and the onset of totality. A few minutes before totality, the father made his three boys look down at the ground until the eclipse was over. The father feared the souls of the boys would be unalterably affected.

"Here it comes!" someone shouted. "Where?" "Over there," pointing to the northwest. "Oh yes! I see it!" "It's getting dark now... it's getting real dark now... it's getting really dark now... it's really getting... *oh my God*!" Cheers and gasps accompanied the beginning of totality. It took everything I had to keep my mind focused on the task at hand, and even that didn't work when the lady behind me broke down crying. Her husband wrapped his arms around her.

As totality ended, I shot a last few exposures, blinded by my own tears, gave up, and turned around and photographed the couple behind me.

As totality began, astronomer Chris Halas of Lincoln, Massachusetts felt as if her soul had been raptured into the presence of God, only to fall back to Earth at the end of totality. That comes closest to describing what it is like to experience a total eclipse of the Sun—except you don't want it to end, you don't want to come back to Earth.

The Aymara Indians were right. All our souls were unalterably affected by that eclipse. But it was—and is—a sublime experience, one that should be sought, not feared.

an eclipse countdown is inexorable, and it is not always possible to race to another location where there is a break in the clouds.

"My first eclipse was in Maine in the summer of 1963," Joe Hollweg recalls. "I was working in New York City, but I made a trip to Maine just for the eclipse. The sky was partly cloudy, with lots of small cottony fair weather clouds. What a disappointment when one of those little puffs moved over the Sun just at the start of totality! I got to see Baily's beads, and that was all. A few hundred yards away people were in the clear, and they were cheering with excitement. However, I do remember seeing the Moon's shadow racing across those puffy clouds, both at the start of the eclipse and at the end. The motion of the shadow seemed incredibly rapid, perhaps because the clouds weren't very high. That was the first time I had some sense of how fast 1,000 miles per hour really is."

There were high hopes for the total eclipse of August 19, 1887, but clouds spoiled the view along almost the entire path from Germany through Russia to Japan. Someone posted a public notice in Berlin stating that, on account of the weather, the eclipse had been postponed to another day.[15]

Bad luck plagued Rebecca Joslin in her eclipse chasing. An amateur astronomer, she started chasing eclipses as a college student in 1905. Going to Spain in that era meant weeks of travel. The weather was perfect until about 10 minutes before totality, when a single cloud appeared, blotted out the Sun for the entire 3½ minutes of totality, and then moved on. She tried again in August 1914, sailing first for England, with reservations on a ship that would take her to a viewing site in Norway. While she was en route to England, World War I broke out, her England-to-Norway cruise was canceled, and she watched a partial eclipse from England under absolutely clear skies.

It was not until January 24, 1925, 20 years after her first attempt, that Joslin finally saw a total eclipse. This time the eclipse came almost to her. She had only to cross from Massachusetts into Connecticut. She had previously been wiped out by one cloud and one world war. Nature didn't make it easy on her this time either. The temperature at her site was –2 °F (–19 °C). One brief eclipse out of three. She gathered her travels and experiences into a book called *Chasing Eclipses*. Because of her

Clouded out. Glum observers at the August 9, 1896 eclipse in Lapland see only the twilight glow on the horizon in this painting by Lord Hampton. [Annie S. D. Maunder and E. Walter Maunder: *The Heavens and Their Story*]

keen eye for people, places, history, and irony, she salvaged quite a bit from misfortune.

All the veterans agree: Plan your eclipse trip so that you see things, go places, and meet people that will shine in your memory even if the corona does not.

Crucial to the enjoyment of a solar eclipse is eye safety. You wouldn't stare directly at the Sun during a normal day, so you shouldn't stare at the Sun when it is partially eclipsed without proper eye protection (described in the next chapter). However, when the Sun is totally eclipsed—no portion of its disk is showing—it is perfectly safe to look at the Sun without any eye protection. Jay Pasachoff is irritated by governments and news media in foreign lands and even in this country that mislead the public by implying that the Sun gives off "special rays" at the time of an eclipse. Instead of teaching simple safety procedures, they frighten people, thereby depriving them of a rare and magnificent sight.[16]

Alan Fiala recalls the 1980 eclipse in India where pregnant women were instructed to remain indoors during the eclipse. For the 1992 eclipse, Roger Tuthill rented a jumbo jet in Brazil to meet and chase the shadow over the Atlantic, using the speed of the aircraft to expand three minutes of totality into six. To serve the 50 passengers aboard the DC-10, the airline assigned 12 flight attendants. There were plenty of

The eerie twilight of totality silhouettes astronomers as they quickly make their measurements during the total solar eclipse of February 16, 1980 near Hyderabad, India. [Nikon FE SLR, 24 mm lens, f/2.8, 1 s, ISO 200 film. ©1980 Jay M. Pasachoff]

windows for them to watch the eclipse, but almost all the stewardesses refused to look. They feared that if they saw the Sun in eclipse, they could never become pregnant.

The pilot, however, was not afraid to look. He was wildly enthusiastic. Even after the Moon's shadow outran the jet and totality was over, the pilot pressed on eastward toward Africa because he was so

Stages of a Total Eclipse

First Contact The Moon begins to cover the western limb of the Sun.

Crescent Sun Over a period of about an hour, the Moon obscures more and more of the Sun, as if eating away at a cookie. The Sun appears as a narrower and narrower crescent.

Light and Color Changes About 15 minutes before totality, when 80% of the Sun is covered, the light level begins to fall noticeably, and then it falls with increasing rapidity and the landscape takes on a metallic gray-blue hue.

Gathering Darkness on the Western Horizon About five minutes before totality, the shadow cast by the Moon becomes visible on the western horizon as if it were a giant but silent thunderstorm.

Animal, Plant, and Human Behavior As the level of sunlight falls, animals may become anxious or behave as if nightfall has come. Some plants close up. Notice how the people around you are affected.

Temperature As the sunlight fades, the temperature may drop perceptibly.

Shadow Bands A minute or so before totality, ripples of light may flow across the ground and walls as the Earth's turbulent atmosphere refracts the last rays of sunlight.

Diamond Ring Effect About 15 seconds before totality begins, the corona first becomes visible with one dazzlingly bright jewel of sunlight along one edge of the Moon.

Corona As the diamond ring fades, the corona becomes more prominent.

Baily's Beads About five seconds before totality, the Moon has covered the entire face of the Sun except for a few rays of sunlight passing through deep valleys at the Moon's limb, creating the effect of jewels on a necklace.

Shadow Approaching While all this is happening, the Moon's dark shadow in the west has been growing. Now it rushes forward and envelops you.

Second Contact Totality begins. The Sun's photosphere is completely covered by the Moon.

Prominences and the Chromosphere For a few seconds after totality begins, the Moon has not yet covered the lower atmosphere of the Sun and a thin strip

fascinated by the fast-moving pillar of darkness ahead of him. "It's even more exciting than an engine fire," he explained.

George Lovi, Alan Fiala, and others remember the 1983 eclipse in Indonesia. Before the eclipse, the streets were teeming with people. On eclipse morning, however, there was scarcely anyone outdoors. Fire sirens

of the vibrant red chromosphere is visible at the Sun's eastern limb. Stretching above the chromosphere and into the corona are the vivid red prominences.

Corona Extent and Shape The corona and prominences vary with each eclipse. How far (in solar diameters) does the corona extend? Is it round or is it broader at the Sun's equator? Does it have the appearance of short bristles at the poles? Look for loops, arcs, and plumes that trace solar magnetic fields.

Planets and Stars Visible Venus and Mercury are often visible near the eclipsed Sun, and other bright planets and stars may also be visible, depending on their positions and the Sun's altitude above the horizon.

Landscape Darkness and Horizon Color Each eclipse, depending mostly on the Moon's angular size, creates its own level of darkness. At the far horizon all around you, beyond the Moon's shadow, the Sun is shining and the sky has twilight orange and yellow colors.

Temperature Is it cooler still? A temperature drop of about 4 °F (2 °C) is typical.

Animal, Plant, and Human Reactions What animal noises can you hear? How do *you* feel?

End of Totality Approaching The western edge of the corona begins to brighten and the reddish prominences and chromosphere appear.

Third Contact One bright point of the Sun's photosphere appears along the western edge of the Moon. Totality is over. The stages of the eclipse repeat themselves in the reverse order.

Baily's Beads The beads grow and merge into a brilliant solar crescent.

Diamond Ring Effect and Corona As the diamond ring brightens, the corona fades from view. Daylight returns.

Shadow Rushes Eastward

Shadow Bands

Crescent Sun

Recovery of Nature

Partial Phase

Fourth Contact The Moon no longer covers any part of the Sun. The eclipse is over.

wailed like an air raid warning for people to take cover. School children, on government orders, were kept indoors, forbidden to see the eclipse, except perhaps on television. A stunning natural event that comes to you, if you are lucky, once in five lifetimes, was denied to them.

On Java for that 1983 eclipse, Jay Anderson recalls soldiers patrolling the streets to discourage unauthorized observers, although the citizens were so frightened by government warnings that almost all stayed indoors. As the Sun was gradually disappearing, the soldiers near him began to be caught up in the excitement. Anderson offered them a view through his telescope, but the warnings they had received overtook them and they refused. Several observers gave the officers a few minutes of careful explanation before the officers nervously took a peek. "After their first look all apprehension disappeared, and they participated in the event as fully as we did. Ten minutes before totality, the officer in charge was interrupted by a call on his walkie-talkie from his superior at headquarters:

'How are things going out there?'

'O.K. There are no problems.'

'It's time to come in now. Collect your men and bring them back to the barracks. We are supposed to have all troops back before the eclipse begins.'

'There are no problems here; everyone is safe; the tourists are all working on their equipment. Are you sure you want us in?—I see no problems.'

'Orders are to bring everyone in, and leave the tourists to themselves. Bring the men in.'

'I'm sorry, I can't hear you. I'm having trouble with the radio. What did you say?'

At this point the officer turned off the radio and put it back in his pocket. The entire troop stayed to watch the eclipse with us."

Even more memorable to Anderson were two dozen Indonesian students from a local college. They were studying English and attached themselves to Anderson's group to practice their speaking skills. As totality neared, the students became extremely nervous. To reassure them, Anderson and his fellow observers held hands with them during totality—the most magical moment in all the eclipses he has seen.

But less developed countries aren't the only places where superstition and misinformation reign supreme and deprive people of a rare gift of nature. The eclipse of 1970 was the first and most impressive of many George Lovi had seen. It was visible in the eastern United States and he was watching from Virginia Beach. The news media had trumpeted the event, but laid great emphasis on the dangers. Around him were thousands of people, most of them casual observers, watching as the crescent Sun thinned and vanished. Instantly a cry went up, spreading from group to group: "Look away! Look away!" Lovi knew of other curious people who traveled substantial distances to reach Virginia Beach and then, fearful, stayed in their motel rooms and watched the event on television.

Dennis di Cicco was in a cornfield in North Carolina for the 1970 eclipse and recalls that in the midst of totality, a car came up the road with its lights on and drove right by without stopping, its passengers oblivious or impervious to the wonder above them. Di Cicco also remembers Australia in 1976, where government warnings that people should stay in their houses to avoid the dangers of the eclipse were so intense that he and other observers feared that local citizens or the police might stop them from watching the eclipse out of concern that they were hurting themselves.

Steve Edberg was in Winnipeg, Canada for the eclipse of 1979. Students had to have notes from their parents to be allowed outside during the eclipse. All the others were confined to their classrooms where the shades were drawn and the students watched the eclipse on television. Jay Pasachoff recalls that "One school in Winnipeg even asked for permission to ignore fire alarms if any sounded during the eclipse, lest the students rush outside and be blinded." On a more rational note, Anderson remembers that many parents kept their children home from school on eclipse day and took the day off themselves so that they could witness the eclipse as a family.

Sometimes superstitions and misinformation about eclipses can be tragic. In the aftermath of the 1998 eclipse in the Caribbean, an Associated Press wire story reported:

> *They were frightened by an eclipse—and apparently, they were frightened to death.*
>
> *Four members of a Haitian family have been found dead in their homes. Officials say the four may have been killed by an overdose of sleeping pills they took to alleviate their fears of last Thursday eclipse. They also may have suffocated. They'd plugged all the openings to their home with rags, to keep the Sun out.*
>
> *Radio broadcasts in Haiti say another young girl suffocated Thursday in a home that had been sealed. Thousands of Haitians were afraid that the eclipse would blind them or kill them.*
>
> *The government declared a national holiday, in hopes of preventing panic. Police ordered pedestrians off the street during the eclipse, yelling, "Go home! It's dangerous to be out!"*[17]

The great astronomy popularizer Camille Flammarion, quoting François Arago, tells of the eclipse of May 22, 1724, visible from Paris. Supposedly, a marquis and some aristocratic lady friends were invited to observe the eclipse at the Paris Observatory. However, the ladies fussed so long with their gowns and hair styles that the party arrived

Equipment Checklist for Eclipse Day

Checklist of your intended activities during eclipse

Camera equipment

Binoculars and/or telescope

Solar filters for eyes

Solar filters for binoculars and/or telescope

Piece of cardboard to use as a pinhole camera for indirect viewing of partial phases

Portable seat or ground covering

Flashlight with new batteries

Pencil and paper to record impressions or to sketch (also to take down the names and addresses of fellow observers)

Suitable clothing and hat (you will be outside for several hours)

Sunglasses (*not* for direct viewing of partial phases)

Bug repellent, sunscreen lotion, basic first aid kit)

Snacks and a canteen of water

Tape recorder for your comments and impressions or to capture reactions of people or wildlife near you

Tape recorder with earphones and prerecorded tape timed to cue you about what you want to do next (to run from about 5 minutes before totality until 5 minutes after totality)

late, a few minutes after totality had ended. "Never mind, ladies," said the marquis, "we can go in just the same. M. Cassini [the observatory director] is a great friend of mine and he will be delighted to repeat the eclipse for you."[18]

Notes and References

Epigraph: Rebecca R. Joslin: *Chasing Eclipses: The Total Solar Eclipses of 1905, 1914, 1925* (Boston: Walton Advertising and Printing, 1929), pages 1–2.

[1] François Arago: *Popular Astronomy*, volume 2, translated by W. H. Smyth and Robert Grant (London: Longman, Brown, Green, Longmans, and Roberts, 1858), page 360.

[2] Hollweg says he uses a "poor man's version" of a patch. "I stick a clean hanky between my eye and my glasses."

[3] Anton Pannekoek: *A History of Astronomy* (London: G. Allen & Unwin, 1961), page 406.

[4] George B. Airy: "On the Total Solar Eclipse of 1851, July 28," page 1 in Bernard Lovell, editor: *Astronomy*, volume 1, The Royal Institution Library of Science (Barking, Essex: Elsevier Publishing, 1970).

[5] Johana L. Codona; "The Enigma of Shadow Bands," *Sky and Telescope*, volume 81, 1991, page 482.

[6] George B. Airy: "On the Total Solar Eclipse of 1851, July 28," in Bernard Lovell, editor: *Astronomy*, volume 1, The Royal Institution Library of Science (Barking, Essex: Elsevier Publishing, 1970), page 4. This speech was given on May 2, 1851, prior to the total eclipse on July 28.

[7] Mabel Loomis Todd: *Total Eclipses of the Sun*, revised edition (Boston: Little, Brown, 1900), page 21.

[8] Isabel Martin Lewis: *A Handbook of Solar Eclipses* (New York: Duffield, 1924), page 62.

[9] François Arago: *Popular Astronomy*, volume 2, translated by W. H. Smyth and Robert Grant (London: Longman, Brown, Green, Longmans, and Roberts, 1858), pages 360–361.

[10] Mabel Loomis Todd: *Total Eclipses of the Sun*, revised edition (Boston: Little, Brown, 1900), page 19.

[11] Jean Meeus, "The maximum possible duration of a total solar eclipse," *Journal of the British Astronomical Association*, volume 113, number 6 (December 2003), pages 343–348. The maximum possible duration of totality actually changes over the centuries due to variations in the eccentricity of the Moon's orbit. The maximum duration peaked at 7 minutes 36.1 seconds around 120 B.C. It is currently 7 minutes 32.1 seconds and will steadily decrease to 7 minutes 0 seconds around the year 6600 before it once again begins to increase.

[12] Rebecca R. Joslin: *Chasing Eclipses: The Total Solar Eclipses of 1905, 1914, 1925* (Boston: Walton Advertising & Printing, 1929), pages 14–15. "M" was Mary Emma Byrd, first director of the Smith College Observatory.

[13] Mabel Loomis Todd: *Total Eclipses of the Sun*, revised edition (Boston: Little, Brown, 1900), page 25.

[14] Mabel Loomis Todd: *Total Eclipses of the Sun*, revised edition (Boston: Little, Brown, 1900), page 174.

[15] Mabel Loomis Todd: *Total Eclipses of the Sun*, revised edition (Boston: Little, Brown, 1900), page 152.

[16] Jay M. Pasachoff hosts a special IAU (International Astronomical Union) web site for eclipses that focuses on public education and eye safey: <www.eclipses.info>.

[17] Associated Press, March 2, 1998, reporting on the eclipse of February 26.

[18] Camille Flammarion: *The Flammarion Book of Astronomy*, edited by Gabrielle Camille Flammarion and André Danjon, translated by Annabel and Bernard Pagel (New York: Simon and Schuster, 1964), page 147.

11
Observing Safely

Shadow and sun—so too our lives are made—
Here learn how great the sun, how small the shade!
Richard Le Gallienne (1920?)

WARNING
Permanent eye damage can result from looking at the disk of the Sun directly, or through a camera viewfinder, or with binoculars or a telescope even when only a thin crescent of the Sun or Baily's beads remain. The 1% of the Sun's surface still visible is about 10,000 times brighter than the full moon. Staring at the Sun under such circumstances is like using a magnifying glass to focus sunlight onto tinder. The retina is delicate and irreplaceable. There is little or nothing a retinal surgeon will be able to do to help you. Never look at the Sun outside of the total phase of an eclipse unless you have adequate protection.

Once the Sun is entirely eclipsed, however, its bright surface is hidden from view and it is completely safe to look directly at the totally eclipsed Sun without any filters. In fact, it is one of the greatest sights in nature.

There are five basic ways to observe the partial phases of a solar eclipse without damage to your eyes.

The Pinhole Projection Method

One safe way of enjoying the Sun during a partial eclipse—or anytime—is a "pinhole camera," which allows you to view a *projected* image of the Sun. There are fancy pinhole cameras you can make out of cardboard boxes, but a perfectly adequate (and portable) version can be made out of two thin but stiff pieces of white cardboard. Punch a small clean pinhole in one piece of cardboard and let the sunlight fall

Pinhole camera made from two pieces of cardboard. Sunlight falls through the pinhole and forms an inverted image of the Sun on the screen. [Drawing by Sheri Flournoy]

through that hole onto the second piece of cardboard, which serves as a screen, held below it. An inverted image of the Sun is formed. To make the image larger, move the screen farther from the pinhole. To make the image brighter, move the screen closer to the pinhole. Do not make the pinhole wide or you will only have a shaft of sunlight rather than an image of the crescent Sun. Remember, this instrument is used with your back to the Sun. The sunlight passes over your shoulder, through the pinhole, and forms an image on the cardboard screen beneath it. Do **not** look through the pinhole at the Sun.

Welders' Goggles

A second technique for viewing the Sun safely is by looking directly at the Sun through a special filter. Welders' goggles or the filters for welder's goggles with a rating of 14 or higher are safe to use for looking directly at the Sun. They are relatively inexpensive and can be purchased from a local welding supply company.

Solar Eclipse Glasses

Perhaps the most satisfying and convenient way of watching an eclipse is with solar eclipse glasses. These devices consist of two solar filters mounted in cardboard frames that can be worn like a pair of eyeglasses. They look a lot like the old 3-D glasses dispensed in movie theaters for 1950s horror films.

The filters in eclipse glasses are made from one of several types of materials. The shiny silver filters are aluminized polyester or Mylar.[1] A problem with this material is that it gives the Sun an unnatural blue color. But even more important, it may have small pinholes that could allow unfiltered sunlight to reach your eyes and damage them. For that reason, aluminized polyester has been largely replaced with safer materials.

Black polymer is an excellent alternative and is composed of carbon particles suspended in a stiff plastic. It produces a more natural yellow image of the Sun. A number of companies specialize in solar eclipse glasses, including American Paper Optics, Rainbow Symphony, and Thousand Oaks Optical. Check the Internet for web sites offering eclipse glasses for sale.

Eclipse glasses in cardboard frames have become very popular for safe eclipse viewing. They can be ordered over the web from manufacturers such as American Paper Optics, Rainbow Symphony, and Thousand Oaks Optical. [© 2001 Johnny Horn]

Advertisements for eclipse glasses may also be found in popular astronomy and science magazines. Check the optical density of eclipse glasses by verifying that they allow you to look comfortably at the filament of a high-intensity lamp.

When using any filter, do not stare for long periods at the Sun. Look through the filter briefly and then look away. In this way, a tiny hole that you miss is not likely to cause you any harm. You know from your ignorant childhood days that it is possible to glance at the Sun and immediately look away without damaging your eyes. Just remember that your eyes can be damaged without you feeling any pain.

Eye Damage from a Solar Eclipse

by Lucian V. Del Priore, M.D., Ph.D.

The dangers of direct eclipse viewing have not always been appreciated, despite Socrates' early warning that an eclipse should only be viewed indirectly through its reflection on the surface of water. A partial eclipse in 1962 produced 52 cases of eye damage in Hawaii, and a total eclipse along the eastern seaboard of the United States produced 145 cases in 1970. As many as half of those affected never fully recovered their eyesight.

There is nothing mysterious about the optical hazards of eclipse viewing. No evil spirits are released from the Sun during a solar eclipse, and there is no scientific reason for running indoors to avoid "the harmful humors of the Sun." Eye damage from eclipse viewing is simply one form of light-induced ocular damage, and similar damage can be produced by viewing any bright light under the right (or should I say the wrong!) conditions.

Light enters the eye through the cornea and is focused on the retina by the optical system in the front of the eye. Any light which is not absorbed by the retina is absorbed by a black layer of tissue under the retina called the retinal pigment epithelium. The retina is the human body's video camera; it contains nerve cells that detect light and send the electrical signal for vision to the brain. Without it, we cannot see. Most of the retina is devoted to giving us low-resolution side vision. Fine detail reading vision is contained in a small area in the center of the retina called the fovea. People who damage their fovea are unable to read, sew, or drive, even though this small area measures only 1/100th of an inch across, and is less than 1/10,000th of the entire retinal area! Unfortunately, this is the precise area that is damaged if we stare directly at the surface of the Sun. Damage has been reported with less than one minute of viewing. The image of the Sun projected onto the retina is about 1/150th of an inch in size, and this is large enough to seriously damage most of the fovea.

Why is sunlight damaging to a structure which is designed to detect light? Bright sunlight focused on the retina is capable of producing a thermal burn, mainly from the absorption of infrared and visible radiation. The absorption of this light raises the temperature and literally fries the delicate ocular tissue. There is no mystery here, as every schoolchild knows that sunlight focused through a magnifying glass can cause a piece of paper to burst into flames. Yet Sun viewing seldom produces a thermal burn; all but the most intoxicated viewer would surely

Camera and Telescope Solar Filters

Many telescope companies provide special filters that are safe for viewing the Sun. Black polymer filters are economical but some observers prefer the more expensive metal-coated glass filters because they produce sharper images under high magnification. Balder Planetarium AstroSolar Safety Film is another alternative. It's a metal-coated resin with excellent optical quality and high contrast. The company even offers instructions on how to make an inexpensive cardboard cell to mount the filter on your telescope, binoculars, or camera.

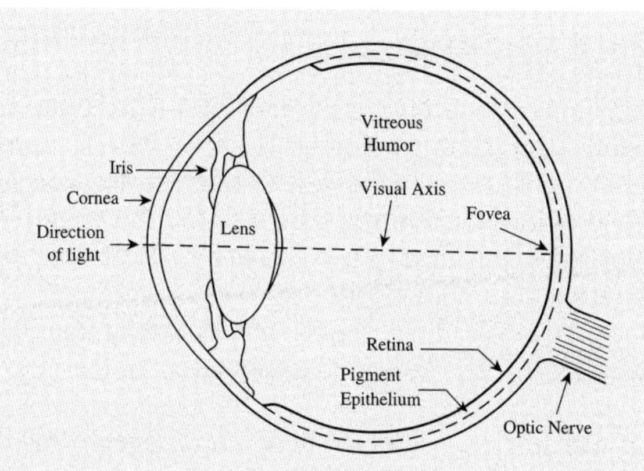

Cross section of the eye. [Drawing by Josie Herr]

turn away before this occurs. Instead, most cases of eclipse blindness are related to photochemically induced retinal damage, which occurs at modest light levels that produce no burn and no pain. Two types of light lesions are recognized clinically and experimentally, and both are probably responsible for the damage observed after eclipse viewing. Blue light (400–500 nanometers) damages the retinal pigment epithelium and leads to secondary changes in the retina, while near ultraviolet light (340–400 nanometers) is absorbed by and directly damages the light-sensitive cells in the outer retina.

Viewing a partial eclipse recklessly is not the only way to produce light-induced retinal damage. Sungazing is a well-known cause of retinal damage even in the absence of an eclipse. Numerous cases of blindness have been reported in sunbathers, in military personnel on anti-aircraft duty, and in religious followers who sungaze during rituals and pilgrimages. The Sun is not even required to produce light damage; other types of bright lights, including lasers and welders' arcs, will have the same effect. The common thread here is clear: direct viewing of bright lights can damage the retina regardless of the source. A solar eclipse merely increases the number of potential victims, and brings the problem to public attention.

Dr. Lucian V. Del Priore, M.D., Ph.D. (physics) is a retinal surgeon, researcher, and professor in the College of Physicians and Surgeons, Columbia University.

Check the Internet or astronomy magazines for dealers offering these filters. Some of the major companies include Meade, Celestron, Orion Telescopes, and Thousand Oaks Optical.

Caution: Do not confuse these filters, which are designed to fit over the front of a camera lens or the aperture of a telescope, with a so-called solar *eyepiece* for a telescope. Solar eyepieces are still sometimes sold with small amateur telescopes. **They are not safe** because of their tendency to absorb heat and crack, allowing the sunlight concentrated by the telescope's full aperture to enter your eye.

Fully Exposed and Developed Black-and-White Film

You can make your own filter out of black-and-white film, but *only* true black-and-white film (such as Kodak Tri-X or Pan-X). Open up a roll of black-and-white film and expose it to the Sun for a minute. Have it developed to provide you with negatives. Use the negatives (black in color) for your filter. It is best to use two layers. With this filter, you can look directly at the Sun with safety.

Remember, however, that if you are planning to use black-and-white film as a solar filter, you need to prepare it at least several days in advance.

Caution: Do not use color film or chromogenic black-and-white film (which is actually a color film). Developed color film, no matter how dark, contains only colored dyes, which do not protect your vision. It is the metallic silver that remains in black-and-white film after development that makes it a safe solar filter.

Eye Suicide

Standard or polaroid sunglasses are not solar filters. They may afford some eye relief if you are outside on a bright day, but you would never think of using them to stare at the Sun. So you cannot use sunglasses, even crossed polaroids, to stare at the Sun during the partial phases of an eclipse. They provide little or no eye protection for this purpose.

Observing with Binoculars

Binoculars were astronomy writer George Lovi's favorite instrument for observing total eclipses. Any size will do. He used 7 × 50 (magnification of 7 times with 50-millimeter [2-inch] objective lenses). "Even the best photographs do not do justice to the detail and color of the Sun in eclipse, and especially the very fine structure of the corona, with its exceedingly delicate contrasts that no film can capture the way the eye can." The people who did the best job of capturing the true appearance of the eclipsed Sun, he felt, were the nineteenth-century artists who photographed totality with their eyes and minds and developed their memories on canvas.

For people who plan to use binoculars on an eclipse, Lovi cautioned common sense. Totality can and should be observed without a filter, whether with the eyes alone or with binoculars or telescopes. But the partial phases of the eclipse, right up through the diamond ring effect, must be observed with filters over the objective lenses of the binoculars. Only when the diamond ring has faded is it safe to remove the filter. And it is crucial to return to filtered viewing as totality is ending and the western edge of the Moon's silhouette begins to brighten. After all, binoculars are really two small telescopes mounted side by side. If observing the Sun outside of eclipse totality without a filter is quickly damaging to the unaided eyes, it is far quicker and even more damaging to look at even a sliver of the uneclipsed Sun with binoculars that lack a filter.

If you don't have solar filters for your binoculars, there is a second way to safely view the partial phases with them. Use the binoculars to

Binoculars can be used to safely project a magnified image of the Sun onto a piece of white cardboard. Never look at the Sun directly through binoculars unless they are equipped with solar filters. [© 2000 Fred Espenak]

project the Sun's image onto a white piece of cardboard. Just hold the binoculars two or three feet from the cardboard using the binoculars' shadow to point them towards the Sun. It takes a little practice but it works great once you get the hang of it. You get two magnified images for the price of one because each half of the binoculars projects a separate image. Since there are no solar filters to remove, you can quickly place the binoculars directly in front of your eyes when totality begins.

Observing with a Telescope

Some observers, including astronomy historian Ruth Freitag, prefer to watch the progress of the eclipse through a small portable telescope, which offers stability and is much less tiring to use for extended periods than binoculars. A telescope also provides more detail at higher powers, if this is desired. The solar filter is removed at totality. When she wants to see a wider view of the corona, she switches to the finder scope. Just make sure that your solar filters are easy enough to remove without bumping the telescope.

A Final Thought

Just remember, Lovi said, "Don't try to do too much. Look at the eclipse visually. Don't be so busy operating a camera that you don't see the eclipse. And don't set off for the eclipse so burdened down by baggage and equipment that you are tired and stressed and too nervous to enjoy the event."

Astronomer Isabel Martin Lewis also warned of the dangers of too many things to do: "A noted astronomer who had been on a number of eclipse expeditions once remarked that he had never SEEN a total solar eclipse."[2]

Notes and References

Epigraph: Richard Le Gallienne: "For Sundials," cited in John Bartlett: editor: *Familiar Quotations*, 13th edition (Boston: Little, Brown, 1955), page 830.

[1] Mylar is a registered trademark of DuPont.

[2] Isabel Martin Lewis: *A Handbook of Solar Eclipses* (New York: Duffield, 1924), page 98.

12
Eclipse Photography

One picture is worth a thousand words.
Anonymous

How do you capture the amazing spectacle of a total eclipse with a camera? Photographing an eclipse really isn't very difficult. It doesn't take a lot of fancy or expensive equipment. You can take a snapshot of an eclipse with a simple camera if you can hold the camera steady or place it on a tripod.

The first step in eclipse photography is to decide what kind of pictures you want. Are you partial to scenes with people and trees in the foreground and a small but distinct eclipsed Sun overhead? Or do you prefer a close-up in which the radiant corona or vivid red prominences of the eclipsed Sun fill the frame? Your decision will determine what kind of equipment you need. Look at the photographs and captions throughout this book. They illustrate some of what can be done with a range of cameras, lenses, telescopes, film, and exposures.

New technologies in cameras and electronics are making eclipse photography easier than ever before. Even beginners can take great eclipse photos with some careful planning. *Planning* is the key. The day of the eclipse is *not* the time to try out a new tripod or lens. You need to be completely familiar with your camera and equipment, and you need to *rehearse* with them weeks before the eclipse. A total eclipse grants you only a few precious minutes and everything must work perfectly. Nature does not provide instant replays.

Film or Digital?

In less than a decade, digital cameras have revolutionized the way we shoot pictures. With the drastic fall in the price of electronic image sensors and memory cards, digital cameras have become much more affordable. Nevertheless, many film cameras will remain in use for years to come. So which is right for you?

Digital photography has more "up front" costs since the cameras are more expensive and you need to buy memory cards and rechargeable batteries. You might even need a new computer and software to edit your photos and create digital slide shows to distribute to friends and family on CD. However, you'll never have to buy film again. Since you can choose which shots to print, you won't have to pay expensive processing fees for every picture "on the roll." With a large memory card, you can shoot hundreds of eclipse images in a few minutes and see your results immediately after totality ends.

On the other hand, you may already own a suitable film camera and accessories. Film can record a tremendous amount of detail at much lower cost than digital. You need a digital camera with at least 10 megapixels to achieve the resolution comparable to a good 35mm negative. Some people prefer the appearance of photos shot on film compared to digital. If you don't want to deal with computers and expensive or intimidating technology, then film may be best for you. No matter which camp you feel more comfortable in, both film and digital cameras are capable of taking excellent eclipse photos.

The following sections point out the differences between film and digital photography and how you can use either to bag that prized eclipse photo.

Choosing Film

If you're using a film camera, what film is best for eclipse photography? That all depends on whether you prefer print film (negatives) or slide film (transparencies).[1] A major advantage of print film is that it is has a wide exposure latitude. You can be off as many as three f-stops (or shutter speeds) from the best exposure and still produce an acceptable photograph.

Slide film is less forgiving. If your exposure is off by more than one f-stop, your photograph is usually ruined. Slides are also less convenient to view than prints because you need a slide projector. Nevertheless, slide film does have some advantages. Only one processing step is required, which helps to retain accurate color. When slides are projected onto a screen, they can be shared with a large audience.

If you were allowed only one roll of film to photograph an eclipse, it is hard to go wrong with the latest ISO 400 film from Agfa, Fuji, Kodak, or Konica. The ISO rating is a measure of the film's sensitivity: how fast it responds to light. Films with an ISO of 100 or less are "slow" and work best in bright light. Films with an ISO of 400 or more are "fast" and can work in dimmer light.

There is a trade-off between film speed (ISO rating) and resolution. As film speed increases, so does the graininess of the film. Fortunately,

This hand-held shot of the diamond ring effect was made from the deck of rocking ship in the Pacific Ocean. An image-stabilizing zoom lens and a DSLR made it possible during the total solar eclipse of April 9, 2005. [Nikon D100 DSLR, Nikkor 80–400 VR zoom lens, fl = 400 mm, f/5.6, 1/500 s, ISO 400. © 2005 Robert Shambora]

film technology has advanced by leaps and bounds. Today's ISO 400 films are finer grained than the best ISO 100 films were only a decade ago.[2]

A greater threat to image sharpness, especially among beginners, is camera vibration caused by wind, flimsy tripods, and nervous photographers. ISO 400 films allow you to use faster shutter speeds, which can help minimize blurring from vibrations. Always be sure to use a 36-exposure roll of film in your camera. Totality is brief and there's no time to change rolls at mid-eclipse.

In general, films that work well for everyday photography also work well for shooting eclipses.

Digital "Film"

Digital cameras use memory cards instead of film to store their images. Most memory cards fall into one of four major types: compact flash (CF), secure digital (SD), Memory Stick (Sony), and xD. Depending on the make and model of a digital camera, it is designed to work with only one of these cards. The cards themselves come in a range of capacities: 128 MB, 256 MB, 512 MB, 1 GB, 2 GB, 4 GB, and 8 GB. MB stands

A Digital Camera Primer

Digital cameras work by converting the images focused on their light-sensitive detectors into electrons. The number of electrons is proportional to the light intensity and this information gets stored on the camera's memory card. The image sensor, either a CCD (charge-coupled device) or CMOS (complementary metal-oxide-semiconductor), is an integrated circuit containing a two-dimensional array or matrix of tiny, light-sensitive picture elements or pixels. Each pixel is equipped with a filter allowing it to measure the light intensity in one of the three primary colors (red, green, or blue) from its a unique position (x-y coordinate) on the sensor.

The camera creates the final image by combining the complementary color information from neighboring pixels. The resulting image contains only one-third as many pixels because each pixel has information in all three colors. When cameras are advertised to have a certain number of pixels, that number refers to the pixels in the final image and not the camera's sensor chip.

Ever since digital cameras hit the marketplace, manufacturers have been engaged in a war of escalating megapixels (millions of pixels). The more pixels a camera offers, the more detail you can capture in the final photo. Early cameras could only deliver images of about half a megapixel (500 KB). For a square image, half a megapixel is equivalent to about 700 × 700 pixels. That's good enough for the web but not for prints.

At a typical viewing distance of 18 inches, a print made from a digital image must contain at least 250 pixels per inch to be of "photographic quality." Less than that and your eye will start to see the individual pixels making up the image. By these criteria, a half-megapixel image can be printed no larger than 2.8 inches square!

Fortunately, the typical digital camera now delivers images containing 6 megapixels. This is equivalent to an image measuring 2000 × 3000 pixels which can be enlarged to an maximum size of 8" × 12" at 250 pixels per inch.

The megapixel war is still in progress, so expect to see cameras with higher pixel counts in the future.

for megabyte (one million bytes) and GB stands for gigabyte (one billion bytes). Larger capacity cards are more expensive but they hold more images.

When a digital camera records an image onto a memory card, it usually writes a JPEG[3] file. JPEG (pronounced "JAY-peg") is a clever way of compressing an image and storing it in less space than it normally would require. JPEGs come in varying levels of compression: the higher the compression, the smaller the file. But there's no such thing as a free lunch and this is especially true of JPEGs. When you save an image as a JPEG, some of the information in the photo is lost forever. JPEGs are called "lossy" because the decompressed image isn't quite the same as the one you started with. The more you compress a JPEG, the *smaller* the file size and the *greater* the loss of quality in the original image.

Images can be saved at several levels of JPEG quality. Typical choices include something like "low, medium, high" or "basic, normal, fine." Although the intermediate level is fine for most photography, a total eclipse is an exceptional event warranting the highest level setting. This means fewer images can be stored on a memory card but they will be the best quality JPEGs the camera can deliver.

How many images can fit on a given memory card? That depends on two factors: the JPEG quality level and the number of pixels in the image. The storage chart below estimates the capacity for a range of image and memory card sizes assuming a JPEG quality of "medium" or "normal." You can store about twice as many JPEGs at the "low" quality setting and half as many at the "high" quality setting.

Storage Chart—Approximate Number of JPEGs Per Memory Card

	128 MB	*256 MB*	*512 MB*	*1 GB*	*2 GB*	*4 GB*
2 megapixels	134	268	552	1119	2245	4494
3 megapixels	120	240	490	996	2000	4000
5 megapixels	48	95	195	395	800	1595
8 megapixels	36	69	143	290	582	1164
10 megapixels	26	53	109	221	444	887

When you put an empty memory card in your camera, the display shows the estimated number of exposures for the camera's current settings. Change the JPEG quality and watch the numbers change. This is just an estimate, but it is still a useful guide.

The size of a JPEG also depends on the amount of detail in the subject matter. An image of an outdoor landscape (with rocks, trees, grass, and clouds) produces a larger JPEG file than a partially eclipsed Sun (dark sky with a relatively featureless yellow crescent).

A powerful feature of digital cameras is the ability to dial in the ISO sensitivity of your choosing. In bright Sun, you might pick ISO 100, while indoor photography with the available light might call for ISO 800. The downside of higher ISO speeds is the increase in digital "noise." It results from electronic amplification of a weak signal and appears as grainy specks in images shot at higher ISO values. Digital noise usually becomes significant by ISO 1600. You can study the noise in your camera by shooting the same scene using a range of ISO settings and comparing the results.

Fortunately, eclipses are relatively bright so an ISO value of 400 is a good compromise. It's fast enough to minimize blurring from vibrations without sacrificing image quality caused by digital noise.

JPEG vs. RAW File Formats

Each pixel in a JPEG image is composed of three colors: red, green, and blue (RGB). Furthermore, the brightness in each of these colors can vary from black (=0) to completely saturated (=255). This limits the brightness range of an image stored as a JPEG. The exposure latitude of a JPEG is similar to a color slide because the exposure must be nearly perfect for a good result. Many photographers bracket their exposures because it helps to ensure getting a photo with the best exposure.

Some digital cameras also offer a file format called RAW, which stores the image without the "lossy" compression found in JPEGs. Each manufacturer uses its own proprietary format for RAW but all contain the unmodified pixel data from the image sensor in either 12 or 14 bits. This means the RAW image contains information in each color over a brightness range of 4096 to 1 or more instead of the JPEG's more limited range of 256 to 1. The RAW format's greater range makes it ideal for eclipses because the corona's brightness varies by more than 1000 to 1.

There are two big trade-offs in saving images using RAW format. First, RAW files are typically twice as large as the highest quality JPEGs (four times larger than medium quality JPEGs). Second, they take longer to write to memory cards. The write speed of memory cards (rated in MB per second) is an important factor if you plan to shoot a lot of RAW files, but faster cards are more expensive.

Since RAW files take longer to write than JPEGs, you can take fewer shots during totality. Digital cameras have a memory buffer where they temporarily store images before writing them to the memory card. If you shoot a series of pictures in rapid succession, you will eventually fill the buffer, preventing further exposures until the camera catches up by writing more images to the memory card.

Should you shoot eclipses in JPEG or RAW? For most people, the best quality JPEGs are perfectly fine. They allow you to shoot more quickly than RAW and you can store twice as many photos on a memory card. However, if you want the very best quality images with the highest brightness range possible, then shoot in RAW format.

Finally if you plan to edit any of your digital photos, you should first convert all JPEGs to a "lossless" format like TIFF, PSD, or PNG. That's because every time a JPEG image is saved, the image degrades a little more.

Solar Filters

When viewing or photographing the partial phases of any solar eclipse, you must always use a solar filter. A solar filter is also needed for observing all phases of an annular eclipse, because the disk of the Moon does not block the entire face of the Sun. Even if 99% of the Sun is covered, the remaining crescent or ring is dangerously bright. Failure to use a solar filter can result in serious eye damage or permanent blindness. *Do not look directly at the Sun without proper eye protection!*

During totality, however, when the Sun's disk (photosphere) is fully covered by the Moon, it is completely safe to look at the eclipse without any solar filter. In fact, you *must* remove the solar filter during totality or you will not be able to see or photograph the exquisite solar corona and prominences.

Solar filters for telescopes and cameras come in several different kinds of materials. Metal-coated glass filters are the most expensive but they offer excellent resolution and natural-looking color. Handle glass filters with care since they are fragile and break easily when dropped.

Aluminized polyester or Mylar[4] is the shiny silver plastic used in some party balloons and potato chip bags. It was a popular and inexpensive material for solar filters back in the 1970s and 1980s. Unfortunately, aluminized polyester gives the Sun a cold blue color and it is sometimes riddled with tiny pinholes. Black polymer is a safer alternative and is composed of carbon particles suspended in a resin matrix. It is stiffer than polyester and produces a more natural yellow image of the Sun.

In recent years, a new material made especially for solar filters has become available. Balder Planetarium AstroSolar Safety Film is a metal-coated resin with excellent optical quality and high contrast. The company even offers instructions on how to make an inexpensive cardboard cell to mount the filter on your telescope, binoculars, or camera.

Materials and techniques that should *not* be used for solar filters include exposed color film, stacked neutral density filters, smoked glass, crossed polarizer filters, floppy disks, and CDs (compact disks). These materials are ***not safe*** even if the image appears dim and no discomfort is felt while viewing the Sun.

There are three physical configurations of solar filters: eyepiece, off-axis, and full-aperture. Eyepiece filters, furnished with some small telescopes, *are **not safe** and should never be used for viewing the Sun.* The tremendous heat generated at the eyepiece can easily shatter or crack the filter, allowing the full intensity of the Sun's light to be magnified and focused on your retina. Throw eyepiece filters away.

The only safe solar filters for telescopes and cameras are full-aperture and off-axis filters, both of which fit over the objective (front end) of the telescope or camera lens. A full-aperture solar filter is a cap with a solar filter mounted across its entire top. An off-axis solar filter is a cap with a hole off to one side into which the solar filter is mounted. Off-axis filters are cheaper than full-aperture filters because their filters are smaller.

If you use an off-axis solar filter with any catadioptric[5] telephoto lens, the focus of the Sun will change significantly when you remove the filter to photograph totality. You must refocus. The optical field of a catadioptric system is not flat, and focusing with only the edge of the mirror is quite different from focusing with the whole mirror. In a full-aperture solar filter, the filter occupies the entire top of the cap and thus the focus is averaged over the entire surface of the mirror. No refocusing is needed when the filter is removed at the beginning of totality or when the filter is replaced at the end of totality. Full-aperture solar filters are preferred.[6]

If your telescope has a finder scope, be sure to place a small solar filter over its objective lens to protect your eyes and to keep the finder cross hairs from burning. If you don't have filter for the finder, keep its upper lens covered with a cap or small piece of cardboard taped in place.

Most telescope manufacturers and dealers offer some type of solar filter for their products. Advertisements for filters can also be found in all the major popular astronomy magazines (*Astronomy*, *Astronomy Now*, *Sky & Telescope*, *Sky At Night*, *SkyNews*, and *Ciel & Espace* [French]). You can also try a web search for "solar filter telescope" (via AltaVista.com, Google.com, Yahoo.com, etc.).

Single-Lens Reflex Cameras—Film and Digital

Close-ups of the eclipsed crescent Sun and detailed portraits of the solar corona require the use of a single-lens reflex (SLR) camera.[7] SLRs also feature interchangeable lenses, from extreme wide-angle to high-power telephoto. You can also remove the lens and hook the camera body up to a telescope so the telescope provides the optics. The modern film SLR is an electronic marvel that features autofocus, programmed autoexposure, and a built-in motor drive to advance the film. But older SLRs are also fine for eclipse photography. SLRs come in several film formats, but this chapter concentrates on the 35 mm format, which will be referred to simply as the SLR.

The digital single-lens reflex (DSLR) camera is the digital counterpart to the SLR. All DSLRs have motor drives that let you shoot two or more frames per second. Their electronic shutters have sophisticated metering and exposure modes while lenses autofocus on their target in a split second. Current models produce images with 6 to 10 megapixels, but this will increase as new cameras appear.

Although newer lenses usually work on both SLRs and DSLRs made by the same camera manufacturer, the magnification and field of view of a lens is quite different on the two camera types. The imaging area on the 35 mm film of an SLR measures 36 × 24 mm. In comparison, the typical imaging area of most DSLRs is about 24 × 16 mm. The width and height of the DSLR's sensor is two-thirds of the SLR's, while the total area imaged is only 44%. Although a camera lens focuses the same size image of the Sun on both SLRs and DSLRs, it fills a larger fraction of the DSLR's imaging area. This so-called digital magnification factor means an image shot with a DSLR appears to be made with a

Baily's beads at the total solar eclipse of February 26, 1998, aboard the *MS Statendam* off the coast of Curaçao. [600 mm, f/4 Nikon with TC-301 2 × converter, efl = 1200 mm, f/11, 1/250 s with ISO 400 film. © 1998 Johnny Horne]

lens having 1.5 times the focal length compared to the image shot with an SLR using the same lens. The increased image scale on DSLRs has important implications when shooting eclipses.

Super Telephotos and Telescopes

No decision affects your eclipse photography more than the choice of a lens. Wide-angle lenses for SLRs have focal lengths of 35 mm or less, while normal lenses fall within the range of 45 mm to 70 mm. Telephoto (high-magnification) lenses begin at 105 mm and go to 500 mm or 1000 mm. Zoom lenses provide adjustable magnification and new designs cover a wide range of focal lengths.[8] Additional magnification can be obtained with teleconverters. These accessories fit between the camera body and the lens to increase the focal length by 1.4 to 2 times, thus increasing magnification.

The smaller imaging sensors in DSLRs means that wide-angle lenses for this format range from 10 to 20 mm, normal lenses are 35 mm, and telephotos are 70 mm and higher. Zooms work well with DSLRs, and teleconverters are also useful for higher magnification eclipse photography.

The size of the Sun's image is determined solely by the focal length of the lens. A telephoto lens or telescope with a focal length of 500 mm or greater is recommended for shooting eclipse close-ups using an SLR. The 500 mm focal length yields a solar image of 4.5 mm on 35 mm film. Allowing for one solar radius of corona on either side of the Sun, an image of the total eclipse covers about 9 mm on the film. You can calculate the Sun or Moon's image size on your film (in millimeters) for any camera system by dividing the focal length (in millimeters) by 110.[9]

The least expensive 500 mm telephoto lens is a catadioptric telescope. Catadioptric telescopes use a combination of mirrors and lenses to focus the light into your eye or into the camera. The folded light path allows a long focal length to fit within a short portable tube, making small catadioptric telescopes easily portable—ideal for most eclipse photography.[10]

Coupling a 2 × teleconverter with a 500 mm lens will produce a 1000 mm focal length, which doubles the Sun's size to 9 mm on 35 mm film. For the Sun and corona together, their image size increases to 18 mm.

With DSLRs, the 1.5 × magnification factor offers a free ride for eclipse photographers. A 350 mm lens captures the same field of view on a DSLR that a 500 mm does on an SLR. The field imaged with a DSLR and 500 mm lens would require a 750 mm lens on an SLR. You can easily simulate the size of the Sun's disk in a DSLR. Take a coin

The image sizes of the eclipsed Sun and corona are shown for a range of focal lengths on both 35 mm film SLRs and digital DSLRs, which use a sensor two-thirds the size of 35 mm film. Thus, the same lens produces an image 1.5 times larger on a DSLR as compared to film. [Adapted from F. Espenak and J. Anderson: *Total Solar Eclipse of 2008 August 01* (NASA TP-2007–21419, 2007)]

from your pocket and measure its diameter. Now place the coin 110 times its diameter from the lens and shoot a picture of the coin. The full moon also makes a great target for evaluating the magnification of your lens for shooting eclipses.

The ideal focal length for telephoto photography of solar eclipses ranges from 500 to 2000 mm for SLRs (350 to 1350 mm for DSLRs), depending on whether you concentrate on the corona or on the prominences. Again allowing one solar radius of corona on either side of the Sun, you can calculate that lenses with focal lengths longer than about 1400 mm (900 mm for DSLRs) may not contain all the corona, and focal lengths longer than 2500 mm (1700 mm for DSLRs) may not capture the entire solar disk. The longer the lens, the more expensive and the less stable a telescope/telephoto system will be, which means that you will need a heavy-duty tripod or mount, adding to your expense and weight.

Improvements in lens technology have revived refracting telescopes as long lenses for eclipse photography. The new extra-low-dispersion (or apochromatic) lenses eliminate the color halos around images that handicapped older refractors. Apochromatic refractors have wider objective lenses, yet shorter tubes, making them fast cameras. Apos are expensive, they are not as portable as catadioptric telescopes, and they require equatorial mounts. Nevertheless, apos are prized by advanced eclipse photographers for their image quality.[11]

Using a telescope means that you lose the autofocus function (and possibly autoexposure) of your camera. This handicap isn't as serious as it seems because most cameras don't autofocus well with eclipses. The rapidly changing light conditions confuse many autofocus cameras, causing them to hunt for the correct focus. The best solution is to turn autofocus off and focus manually.

Telescope Clock Drives and Polar Alignment

For eclipse photography with focal lengths of 1200 mm (800 mm for DSLRs) or more, a sturdy equatorial mount with clock drive is needed to compensate for the Earth's rotation. This device turns your telescope one revolution per day so that you can track the Sun, Moon, and stars. Without a clock drive, the Sun slowly drifts westward through your camera's field about one Sun diameter every two minutes. Frequent tripod adjustments are required to center the Sun or it will drift out of the frame during the eclipse. A clock drive keeps the Sun centered so you can concentrate on watching and photographing totality. It also permits the use of longer focal lengths and slower shutter speeds without fear of blurring your photos.

Most telescopes are equipped with battery-driven clock drives. This frees you from worrying about AC power, voltages, frequencies, and plug sizes used in other countries.[12] Bring fresh batteries and install them in your telescope clock drive before the eclipse begins.

To track the Sun accurately, the axis of an equatorial mount must be aligned with the celestial pole. This can be accomplished by arriving at your eclipse site the night before the eclipse and aligning on Polaris in the Northern Hemisphere or Sigma Octanis in the Southern Hemisphere. If you are south of the equator, make sure your clock drive has a reversible motor for use in the Southern Hemisphere.

You may need to set up your telescope just hours before the eclipse, with no opportunity for nighttime polar alignment. In this case, a bubble angle finder and a magnetic compass are needed to polar align the telescope. Bubble angle finders let you measure angles with respect to the horizontal and are available in most hardware stores.

First, level your equatorial mount and tripod using the bubble angle finder. Then use the compass to align the azimuth of your polar axis on north. Since magnetic north can differ from celestial north by several degrees, you need to know the offset or magnetic declination for your observing site. You can get this information from topographic maps or from the National Geophysical Data Center web site <http://www.ngdc.noaa.gov/seg/geomag/jsp/Declination.jsp>.

Next, adjust the polar axis of the mount to the latitude of the observing site using the bubble angle finder. The polar axis should be set at an angle from the horizontal equal to your geographic latitude.[13]

If you do not have a clock drive, you must use relatively short exposures because the Earth's rotation causes the Sun to drift westward 1/2° every two minutes. For sharp images, here is a formula for the longest exposure allowable without a clock drive:

$$\text{Exposure (seconds)} = 340 / \text{focal length (millimeters)}$$[14]

For longer exposures, you need a clock drive to compensate for the Earth's rotation.

Camera Tripods

Flimsy tripods are the main reason that eclipse photographs come out fuzzy and blurred. Small portable tripods that are nice for airline travel are not sturdy enough to hold your camera and heavy telephoto lens steady for sharp eclipse pictures. Because you must touch your camera to adjust exposures, your tripod must also dampen any vibrations quickly.[15]

To maximize stability and minimize vibrations, don't extend the tripod's legs more than halfway and do not extend the center column. Adjust the tripod height so that you can reach the camera controls while kneeling or while sitting on a chair. Test your setup by tapping on the camera or tripod while viewing the Sun through your camera. Vibrations should be small and should damp out quickly.

You can help reduce vibrations by suspending some weight under the tripod. Put rocks or sand in a sack or in plastic zip-lock bags and hang them from the center of the tripod using string or duct tape. Setting up the tripod on sand or grass is better than concrete or asphalt because the softer surfaces dissipate vibrations faster.

Before you travel to an eclipse, make sure that your lens on its tripod can be pointed at the Sun's predicted altitude for the eclipse and verify that all controls work smoothly. You don't want to discover that your equipment becomes unstable when you try to view the crucial portion of the sky on eclipse day.

Cable Releases and Right Angle Finders

You cannot take crisp close-ups of a total solar eclipse without using a cable release. Cable releases have a button at one end and the other end connects to the camera. When you press the cable release button, it triggers the camera's shutter without jostling it. Originally, cable releases were spring-loaded devices that screwed into a socket located on the camera's shutter button. Most modern cameras use an electronic cable release that plugs into the camera body.

A right angle finder is a prism/eyepiece device that attaches to your camera's viewfinder. It allows you to look through the viewfinder when your head is above the camera instead of behind it. Some models even have a 2 × zoom feature that lets you magnify the viewfinder image. This is a great focusing aid.

While a right angle finder may not seem essential, it is worth its weight in gold. For most eclipses, your camera will be pointed upward.

Partial phases of the November 3, 1994 total solar eclipse from central Bolivia. [3.5-inch Questar, Thousand Oaks Type 3 solar filter, fl = 1400 mm, f/14, 1/250 s, ISO 400 film. © 1994 Ken Willcox]

Without a right angle finder, you might have to get down on your hands and knees or even lie on your back to look through the camera. A right angle finder allows you to comfortably center and focus the Sun's image. You will also be facing the top of your camera so you can see all the controls during totality.

Photographing the Partial Eclipse

A solar filter must be used with your lens to photograph the partial phases of the eclipse. This requires determining the proper shutter speed and f-ratio for your particular equipment well in advance of the eclipse.

If your camera has a built-in spot meter that covers a smaller area than the Sun's image, you can simply meter on the Sun's disk through your solar filter and use that exposure throughout the partial phases.

Quick and Easy Eclipse Photography

by Patricia Totten Espenak

Is your idea of the perfect eclipse experience—"I just want to watch!"? Do your eyes glaze over when serious eclipse photographers delve into the minutiae of exposure times, f-stops, and cameras that require you to carry the manual with you? *But* do you still have just a tiny desire for some photos of your very own? Then read on.

After ten total solar eclipses, my methods and gear have evolved to accommodate both. I can watch almost the entire eclipse while taking great photos using the following equipment: a comfortable chair, a sturdy tripod, a cable release, a right angle finder, and a solar filter for my telescope or telephoto lens.

I use a Bogen 3001 tripod with a Bogen 3275 (410) compact geared head. The geared head has two large knobs that allow me to make small, precise adjustments in altitude and azimuth every minute or so to easily track the Sun. With the right angle finder, I can check the position of the image and then quickly lean back to view the eclipse. I focus manually on a sunspot and the focus ring is then taped down to secure it. After setting the camera in program mode and matrix metering, I'm ready to go. I just sit back with the cable release in one hand and my eclipse glasses in the other.

Thirty seconds before totality begins, I check to see that the Sun is centered in my viewfinder and I remove the solar filter. A dozen or more shots are quickly taken in the seconds before second contact and again after third contact. This results in nice diamond ring sequences while the camera automatically adjusts the exposure.

During totality, I might take another dozen shots. I also check the Sun's position in the viewfinder, but mostly I'm gazing up at the spectacle while I press the cable release button.

If your camera does not have a spot meter, you need to perform a simple exposure test. Set your equipment up on a sunny day. For SLRs, load the camera with the same kind of film you will use for the eclipse. Carefully center the Sun in your lens or telescope using a solar filter. For telephoto lenses, open the aperture to its widest setting. Shoot one exposure with every shutter speed from 1/15 through 1/2000. Take notes so that you can identify the best exposure after the film is developed. For handy reference, write down the best exposure and tape it to your tripod or solar filter. It should include the ISO, f-number, and shutter speed—for example, "Sun: ISO 400, f/8, 1/125." The exposure doesn't change during the partial phases because the Sun's surface brightness remains the same throughout the eclipse.[16]

Eclipse photography doesn't have to be difficult or interfere with your view of totality. Just use a 500 mm lens and set your camera on "program." This image was shot in Libya during the 2006 total solar eclipse. [Nikon D200 DSLR, Sigma 170–500mm zoom at 500 mm, f/5.0, 0.8 s, ISO 400. © 2006 Patricia Totten Espenak]

After my third contact sequence, I replace the solar filter and return to the leisurely pace of the partial phases.

Capture the moment with your eyes and your camera. You *can* get the best of both!

Patricia Totten Espenak is a retired chemistry teacher who dotes on her granddaughter Valerie and chases eclipses around the world with her husband.

With DSLRs, the exposure test is similar but you can see the results immediately by previewing the images on your camera's LCD screen. Use the histogram function to choose the best exposure. Your camera manual has more information on using the histogram function.

Your best exposure was determined on a sunny day. If the eclipse day has haze or clouds, a longer exposure will be needed to compensate. A thin haze may require an exposure one or two shutter speeds slower than normal, while thicker clouds could call for three or more shutter speeds slower. Use your planned exposure and several longer ones. Film and memory cards are cheap and eclipses don't happen often.

Photographing the Total Eclipse

The brightness of the solar corona changes tremendously as you move away from the Sun's disk. The inner corona shines as brightly as the full moon, but the outer corona is thousands of times fainter. The challenge is to capture both the brightest and faintest parts of the corona. Unfortunately, this variation in brightness is impossible to record in any one exposure because film and digital photography just don't have the dynamic range of human vision. Only your eyes can see the exquisite detail of this celestial event in all its glory. That's why you should view totality with your eyes and not just with your camera.

The good news is that you can photograph some aspect of the corona with almost any exposure you make. There is no one "correct" exposure. Nevertheless, here are some guidelines.

Several factors determine the best shutter speed for a good exposure in your final photograph. A table accompanying this chapter provides recommended shutter speeds for various eclipse phenomena over a range of ISO speeds and lens f-numbers. Each eclipse phenomenon (diamond ring, prominences, corona) has a different brightness value for determining the proper exposure for that aspect of the eclipse.[17] The exposures in the table were determined by photographing more than a dozen solar eclipses, but they are only suggestions. Each eclipse is different and the corona's brightness varies. Weather conditions (haze or clouds) may require longer exposure times.

Bracket your shots on both sides of the ideal exposure and take several photographs at the same setting to assure success. If you use an ISO speed too high, you may discover that your camera does not have a fast enough shutter speed for proper exposure. Determine all of your camera settings before the eclipse so that you know in advance what ISO works best with your equipment.[18]

Many eclipse photographers plan their exposures using the following strategy. Using the ISO and f-number, determine the shortest shutter speed for the prominences (bright) and longest shutter speed for the

Exposure Guide for Solar Eclipse Photography

SUN—partial phases of eclipse and prominences
Use full aperture solar filter for partial phases
Use no filter for prominences

(Film Speed—ISO)

f/	*50*	*100*	*200*	*400*	*800*
2.8	1/2000	1/4000	–	–	–
4	1/1000	1/2000	1/4000	–	–
5.6	1/500	1/1000	1/2000	1/4000	–
8	1/250	1/500	1/1000	1/2000	1/4000
11	1/125	1/250	1/500	1/1000	1/2000
16	1/60	1/125	1/250	1/500	1/1000
22	1/30	1/60	1/125	1/250	1/500
32	1/15	1/30	1/60	1/125	1/250

SUN—total eclipse: inner corona (3° field)
No filter

(Film Speed—ISO)

f/	*50*	*100*	*200*	*400*	*800*
2.8	1/30	1/60	1/125	1/250	1/500
4	1/15	1/30	1/60	1/125	1/250
5.6	1/8	1/15	1/30	1/60	1/125
8	1/4	1/8	1/15	1/30	1/60
11	1/2	1/4	1/8	1/15	1/30
16	1 sec	1/2	1/4	1/8	1/15
22	2 sec	1 sec	1/2	1/4	1/8
32	4 sec	2 sec	1 sec	1/2	1/4

SUN—total eclipse: Inner corona (1° field)
No filter

(Film Speed—ISO)

f/	*50*	*100*	*200*	*400*	*800*
2.8	1/500	1/1000	1/2000	1/4000	-
4	1/250	1/500	1/1000	1/2000	1/4000
5.6	1/125	1/250	1/500	1/1000	1/2000
8	1/60	1/125	1/250	1/500	1/1000
11	1/30	1/60	1/125	1/250	1/500
16	1/15	1/30	1/60	1/125	1/250
22	1/8	1/15	1/30	1/60	1/125
32	1/4	1/8	1/15	1/30	1/60

SUN—total eclipse: outer corona (5° field)
No filter

(Film Speed—ISO)

f/	*50*	*100*	*200*	*400*	*800*
2.8	1/2	1/4	1/8	1/15	1/30
4	1 sec	1/2	1/4	1/8	1/15
5.6	2 sec	1 sec	1/2	1/4	1/8
8	4 sec	2 sec	1 sec	1/2	1/4
11	8 sec	4 sec	2 sec	1 sec	1/2
16	15 sec	8 sec	4 sec	2 sec	1 sec
22	30 sec	15 sec	8 sec	4 sec	2 sec
32	60 sec	30 sec	15 sec	8 sec	4 sec

Note: These exposure tables are given as guidelines only. The brightness of prominences and the corona can vary considerably. You should bracket your exposures to be safe.

outer corona (dim). After totality begins, shoot a sequence of exposures using every shutter speed, starting with the one for prominences and ending with the one for the outer corona. For instance, at ISO 400 and f/11, the recommended exposure for prominences is 1/1000 second and for the outer corona 1 second. The shutter speed sequence would then be: 1/1000, 1/500, 1/250, 1/125, 1/60, 1/30, 1/15, 1/8, 1/4, 1/2 and 1. This is a total of eleven exposures. For additional insurance, repeat the sequence in reverse, ending with 1/1000. With a 36-exposure roll of film, you would still have 14 exposures for the diamond ring and prominences at the beginning and end of totality. A big advantage of digital with a large memory card is the ability to shoot many more images during totality.

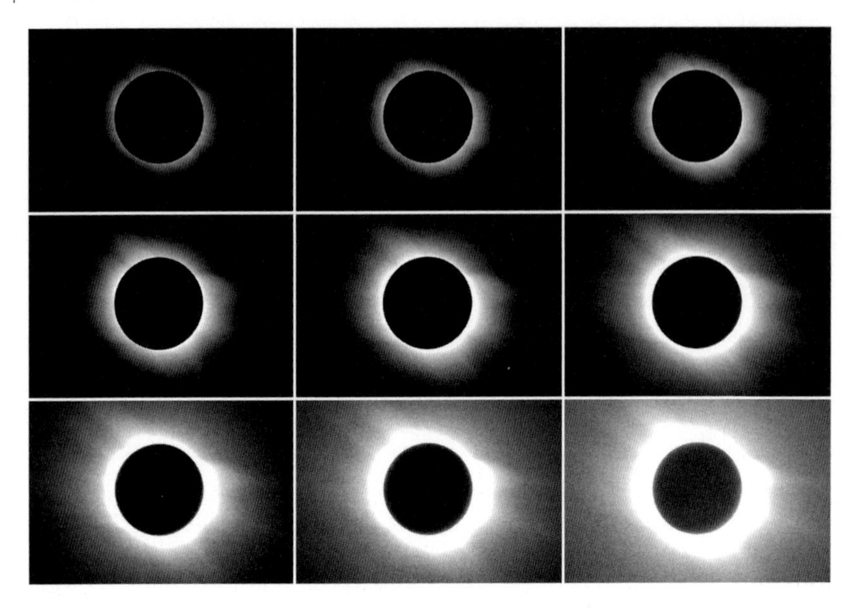

Some photographers shoot a series of exposures to capture the wide range of brightness in the corona. This sequence was made with a DSLR during the March 29, 2006 total solar eclipse from Libya. [Nikon D200 DSLR, Vixen 90 mm fluorite refractor, fl = 810 mm, f/9, 9 exposures: 1/125 s to 2 s, ISO 400. © 2006 Fred Espenak]

Even if your camera's exposure is completely automatic, or if you simply don't want to hassle with exposure settings, you can still get good pictures of the diamond ring and totality using automatic exposure. Load your camera with ISO 400 color negative film (or set ISO 400 on DSLRs) and grab some shots on autoexposure. Although the inner corona and prominences may be overexposed, you will still have some fine souvenirs of the event. Best of all, you can devote most of your time to watching the eclipse rather than fiddling with camera settings. Simplicity is especially recommended if you are a novice photographer or have never seen a total solar eclipse (see vignette on Quick and Easy Eclipse Photography).

The Global Positioning System and Time Signals

The times for key phases of each eclipse are listed for any location along the eclipse path in the NASA eclipse bulletins.[19] To anticipate these events, you should set your watch for the exact time before the eclipse begins.

Coordinated Universal Time[20] is broadcast 24 hours a day over short-wave radio station WWV, Fort Collins, Colorado, and from Hawaii on

WWVH at 5, 10, and 15 MHz. Radio station CHU in Ottawa, Canada also broadcasts time signals at 3.330, 7.335, and 14.670 MHz. The Chinese radio station BPM broadcasts time signals on the same frequencies as WWV. For time signals in other countries, see <http://www.dxinfocentre.com/time.htm>.

If you leave your radio on throughout the eclipse, you will always know the correct time without checking your watch. Unfortunately, there are many locations around the world where radio time signals cannot be picked up. A GPS (global positioning system) receiver can fill this void by not only providing accurate time but also your precise location.

The GPS program was developed by the United States Department of Defense as an accurate way of determining geographic coordinates worldwide. The system uses a set of 24 satellites in orbits 12,545 miles (20,183 kilometers) high and with periods of 12 hours. At any time, 8 to 12 of the GPS satellites are visible from any spot on Earth. A GPS receiver uses signals broadcast from the satellites to determine its three-dimensional position with respect to the satellites. This position is displayed as latitude, longitude, and elevation above sea level, along with the exact Universal Time. Military GPS receivers can determine positions to within inches. GPS receivers for civilian use have an accuracy of about 50 feet (15 meters). You can use them to assure yourself that you are standing within the path of totality.[21]

Keep in mind that the time shown in many consumer GPS receivers can be off by several seconds. This is a cost-saving measure for the GPS manufacturers since it is good enough for most consumers. If you need a GPS receiver that displays accurate time, check the specifications to see that it uses the one-pulse-per-second signal.

Digital Voice Recorders

Digital voice recorders (DVRs) are tiny, battery-operated devices designed for recording meetings, interviews, or classroom lectures for later review. They typically have a built-in microphone, a speaker, and an earphone jack. These pocket-size gizmos can store two or more hours of sound in flash memory. Some even have a USB interface to transfer sound files directly to your Computer's hard drive. Olympus and Sony each offer a range of models with prices starting under $50.

During an eclipse, DVRs are great for recording your observations and reactions.[22] Put in a fresh set of batteries before the eclipse begins. Fifteen minutes before totality, begin recording continuously. You will not only have a helpful and permanent record of your comments, but you will also capture the excitement of your companions.

Photographing Shadow Bands

by Laurence A. Marschall

Visual observations of shadow bands are interesting and fun, but the most useful observations from a scientific standpoint are those using photography or videotape. The blotchy patterns of the bands can be predicted from mathematical models of how sunlight travels through the Earth's atmosphere. Yet few of the many attempts to catch shadow bands on film or tape have been successful. Perhaps a dozen fuzzy photographs and one flickering film are all that are available. The bands move so swiftly and have such low contrast that it is hard to get a good picture.

To photograph the bands, use a high shutter speed (1/250 second or faster) and a high ISO setting (ISO 800 or higher). Digital cameras are preferred because the faint bands can be enhanced by using image processing programs like Photoshop®.

Place a large piece of white cardboard or plywood (5 × 5 feet or larger) on the ground to act as a screen for the shadow bands (see the vignette "Catching Shadow Bands" in chapter 10). Mount your camera on a tripod to avoid vibrations, point it at the screen, and focus manually (autofocus can be fooled in low light).

Select an autoexposure program mode on the camera that favors fast shutter speeds to freeze the quick-moving shadow bands. Many cameras have an exposure program called sports mode. It forces the camera to use high shutter speeds for capturing action while maintaining proper exposure. If your camera does not have sports mode, it will still probably have an autoexposure program that sets the camera to adjust the aperture while holding the shutter speed fixed at a value you set beforehand (my camera calls this the S mode). Using autoexposure mode ensures that you get correct exposures as the ambient light changes, which is of course exactly what happens just before and just after totality, when shadow bands are visible. Put a yardstick on the screen to establish a scale and shoot as many shots as possible.

For videotaping, many hand-held camcorders have a fast-shutter option (1/1000 to 1/4000 second) that should be used to stop the motion of the shadow bands. If you get a distinct record of the shadow bands and their motion,

Some advanced eclipse observers bring a second DVR on which they have pre-recorded timed instructions that they play back through an earphone during the hectic minutes surrounding totality. The DVR playback is synchronized with the time to give audio cues for the minutes remaining before second or third contact, times to remove or replace solar filters, camera settings, and reminders to watch for various eclipse phenomena.

If you've never seen a total eclipse, an audio recording of instructions will help you keep to your plan.

Elusive shadow bands were captured with a video camcorder just before second contact during the 2001 total solar eclipse from Zimbabwe. This photo is an average of 25 video frames and has been processed to enhance the contrast. [© 2001 Dr. Wolfgang Strickling]

it is not just something to admire: it is a remarkable scientific resource. Shadow band researchers (there are perhaps a half dozen worldwide) will beat a path to your door.

Dr. Laurence A. Marschall is W. K. T. Sahm Professor of Physics at Gettysburg College in Pennsylvania

Photographing Pinhole Crescents

Eclipses provide other phenomena that make interesting pictures, such as the crescent images of the partially eclipsed Sun produced by tree foliage. The narrow gaps between leaves act as "pinhole cameras" and each projects its own tiny (and inverted) image of the crescent Sun on the ground. This pinhole camera effect becomes more pronounced as the eclipse progresses.

The eerie twilight of totality is easy to photograph. A DSLR with a wide-angle lens captured both the corona and spectators during the total solar eclipse of March 29, 2006 in Libya. [Can on EOS 20D DSLR, 12 mm lens, f/5.0, 0.8 s, ISO 100. © 2006 Alan Dyer]

You can make your own pinhole camera to project the crescent Sun with pinholes punched in cardboard, or with a wide-brimmed straw hat. You can even produce the effect with the shadow of your hands by loosely lacing your fingers together. Watch the crescents form from the light passing through the gaps between your fingers. The profusion of crescents on the ground or on a person's face makes a nice photographic memento. Almost any kind of camera will work. Just be sure to disengage the automatic flash.

Can a kitchen utensil double as an astronomical instrument? An eclipse chaser used a strainer spoon to project the partial phases onto a T-shirt during the 2006 total eclipse in Turkey. [© 2006 Tunç Tezel]

Landscape Eclipse Photography

If you have an extra camera and tripod, take a few wide-angle shots of the sky, horizon, and landscape before, during, and after totality. This sequence might show the Moon's dark, fast-moving shadow approaching from the west or racing away over the horizon to the east. Place yourself or some other interesting subjects in the foreground to give the

Field of View for Various Photographic Focal Lengths

Focal length	*Field of view (35mm)*	*Field of view (digital)*	*Size of sun*
14 mm	98° × 147°	65° × 98°	0.2 mm
20 mm	69° × 103°	46° × 69°	0.2 mm
28 mm	49° × 74°	33° × 49°	0.2 mm
35 mm	39° × 59°	26° × 39°	0.3 mm
50 mm	27° × 40°	18° × 28°	0.5 mm
105 mm	13° × 19°	9° × 13°	1.0 mm
200 mm	7° × 10°	5° × 7°	1.8 mm
400 mm	3.4° × 5.1°	2.3° × 3.4°	3.7 mm
500 mm	2.7° × 4.1°	1.8° × 2.8°	4.6 mm
1000 mm	1.4° × 2.1°	0.9° × 1.4°	9.2 mm
1500 mm	0.9° × 1.4°	0.6° × 0.9°	13.8 mm
2000 mm	0.7° × 1.0°	0.5° × 0.7°	18.4 mm

Image Size of Sun (mm) = Focal Length (mm) / 110

photos some scale. Full-frame fisheye lenses are especially useful and can produce dramatic images. Use your camera's light meter and bracket exposures if possible. If you know the Sun's altitude during totality, you can use the following table to choose a lens focal length that will include the Sun in the top of the frame. You'll also capture any bright planets near the Sun during totality. You might even discover a new comet.

A multiple-exposure eclipse portrait shot on one piece of film was a success in spite of changing weather conditions. The image was taken near Cedona, Australia on December 4, 2002. [Nikon F-70 SLR, 50 mm, f/5.6, partials: 1/500 s to 1/30 s, totality: 2 s, Baadar Solar Filter for partials, Fuji Reala 100 negatives. © 2002 Geoff Sims]

Multiple Exposure Sequences

Some of the most dramatic eclipse photographs are multiple exposures that show the totally eclipsed Sun accompanied by a sequence of partial phases on either side. This can be accomplished with digital cameras by shooting a series of photos and then combining them into a single image using a photo editing program such as Adobe Photoshop.

With film cameras, the process is more complicated and requires a camera capable of taking multiple exposures without advancing the film. Most modern electronic SLRs do not have this feature. A salesperson at your camera store should be able to show you camera models with multiple-exposure capabilities.

Whether you shoot with a digital or film camera, you also need a sturdy tripod. It will hold the camera fixed to one point in the sky without moving when you recock the shutter between exposures. A cable release helps to further reduce vibrations.

To record all the phases of a total eclipse in one scene, you need to know the field of view of your camera lens. The Sun moves 15° per hour—its own diameter every two minutes. A 50 mm lens (35 mm lens for digital) covers 49° along the diagonal. It should take the Sun about three hours to traverse this distance. Try to orient your camera so that the Sun moves along the diagonal. From first to last contact, a total eclipse lasts about 2¼ hours. Taking an exposure every 5 to 10 minutes provides adequate separation between images on the final photograph.

If you want the total phase to be at the center of the frame, you must calculate times from that point in order to have all of the images equally spaced on both sides of the totally eclipsed Sun. For example, if mid-totality is at 10:32 A.M., then you would add or subtract increments of, say, 5 minutes to that figure to get the times to make your exposures. You may want to allow more of an interval just before and after totality to avoid trampling on the corona.

The altitude and azimuth of the Sun or Moon for a specific eclipse are easily obtained using astronomy computer programs available for your home computer. Other sources of information for upcoming eclipses include popular astronomy magazines, NASA's solar eclipse bulletins (Appendix D) and eclipse web sites.[23]

Eclipse Photography at Sea

For eclipse photography at sea, the pitching and rolling of the ship place certain limits on the focal length and shutter speeds that can be used. In most cases, telescopes with focal lengths of 750 mm or longer can be ruled out because the ship would need to be virtually motionless during totality. You must also contend with vibration from the engines, wind across the deck, and hundreds of people stomping around, so try to locate yourself to minimize these problems.

Checklist for Solar Eclipse Photography

Camera Items

Camera (two, if possible, in case one fails)
Lenses (16 mm to 1500 mm)
Cable release (spare advisable)
Heavy-duty tripod
Extra batteries
Film photography: several rolls of 36-exposure film, ISO400
Digital photography: several 1 GB or larger memory cards
Sack or zip-lock bags and heavy string (fill with sand or rocks and hang from tripod for stability)
Exposure list for partial phases and totality

Telescope Items

Telescope (consult owner's manual for photographic equipment) Mount, wedge, cords, eyepieces, photo adapters (T-ring, etc., to attach camera to telescope), counterweights, drive corrector (and spare fuse), lens-cleaning supplies
Solar filter (full aperture preferred)
Inexpensive solar filter for finder scope
Equatorial mount and clock drive (if available)
Magnetic compass (for polar alignment)
Bubble angle finder (to level mount and align polar axis)
Battery for telescope clock drive (if necessary)
For AC power (if needed): AC converter plug, voltage corrector, and extension cord

General Items

Hand-held solar filter (for viewing partial phases)
Penlight flashlight
Digital voice recorder or cassette recorder and batteries (to record comments during eclipse)
Pocket knife
Tools for minor repairs: screwdrivers (regular, Phillips, and jeweller's), needle-nose pliers, tweezers, small adjustable wrench, Allen wrench set, etc.
Plastic trash bags (to protect equipment from dust or rain)
Roll of duct tape or masking tape for emergency repairs
Optional: Extra digital voice recorder to play prerecorded instructions via earphone
Optional: Shortwave receiver for WWV time signal or GPS (global positioning system)

The film or digital ISO choice can be determined on eclipse day by viewing the Sun through the camera lens (with a solar filter) and noting the image motion caused by the ship. In most cases, an ISO of 400 or higher will be needed. As the ship rocks, notice the range of motion and try to snap each picture at one of the extremes.

Some manufacturers have introduced image stabilization (or vibration reduction) in their longer zoom and telephoto lenses. This helps steady the image and lets you use shutter speeds 2 to 3 steps slower than possible without this feature. Many photographers relied on image-stabilized lenses to shoot the 2005 hybrid eclipse from ships in the South Pacific because no islands were in the path of totality.

Some Final Words

If there is one key to successful eclipse photography, it is *preparation*. Set up and test all your equipment at home to ensure that everything works perfectly together. Design a photographic plan or schedule and stick to it. Keep things simple. Don't try to do too much. Practice for the eclipse with a full dress rehearsal.

Bring extra film or memory cards, batteries, cable releases, and other crucial items.

Finally, don't get so overwhelmed with taking photographs that you deprive yourself of time to actually *look* at the eclipse.

Notes and References

Epigraph: Anonymous. Bartlett's *Familiar Quotations* explains that this saying is the creation of Fred R. Barnard, writing in *Printer's Ink*, March 10, 1927. Barnard called it "a Chinese proverb so that people would take it seriously."

[1] Negatives can be made into slides and slides can be made into prints, but some quality may be lost in this conversion. It is best to select slide or print film according to what you plan to do with your photography.

[2] For the latest information on film products, check photography magazines such as *Popular Photography*, *Outdoor Photography*, and *Petersen's PhotoGraphic*. In addition, film manufacturers and good photo stores provide technical bulletins on the latest films. You can also get film information on the manufacturers' web sites (<www.agfa.com>, <www.fujifilm.com>, <www.kodak.com>, and <www.konica.com>).

[3] JPEG stands for Joint Photographic Experts Group. This committee created the JPEG image format in 1992.

[4] Mylar is a registered trademark of DuPont, which does not manufacture this material for use as a solar filter. Other aluminized polyester products such as "space blankets" should *not* be used for solar filters.

[5] Catadioptric optical systems contain both lenses and mirrors. This lightweight and compact design is commonly used in long-focus telephoto lenses and telescopes.

[6] For more information on safe solar filters, see B. Ralph Chou: "Solar Filter Safety," *Sky & Telescope*, volume 95, February 1998, pages 36–40.

[7] The viewfinder of a single lens reflex camera uses a prism and mirror system that allows you to look directly through the same lens that takes the actual photograph. When you press the shutter button, the mirror flips up and the shutter curtain opens, permitting the image to reach the film. An instant later, the shutter closes and the mirror returns to its original position.

[8] Although zoom lenses usually have smaller apertures and are slower than fixed lenses, they work fine for photographing eclipses if they are of good optical quality.

[9] Michael Covington: *Astrophotography for the Amateur*, 2nd edition (Cambridge: Cambridge University Press, 1999), page 35.

[10] Catadioptric "mirror lenses" are made by all the major camera manufacturers (Canon, Minolta, Nikon, Pentax, etc.), as well as independent lens makers (Phoenix, Sigma, Tamron, Tokina, Vivitar, etc.).

[11] Manufacturers of apochromatic, fluorite, and extra-low-dispersion refractors include AstroPhysics, Meade, Takahashi, Tele-Vue, and Vixen.

[12] If you need AC electricity, check on a source of electricity at your observing site or use a drive corrector with a built-in converter that changes 12-volt DC to 60-cycle, 110-volt AC. Many countries use 220-volt AC electricity. If you want to plug into their power, bring along an AC converter, available in many hardware stores.

[13] This quick-and-dirty polar alignment should be good enough for the eclipse. If you need a better alignment, you might try a technique called drift polar alignment <http://www.MrEclipse.com/Special/alignment.html>.

[14] Michael Covington: *Astrophotography for the Amateur*, 2nd edition (Cambridge: Cambridge University Press, 1999), page 131.

[15] Bogen/Manfrotto, Gitzo, and Slik all make heavy-duty tripods.

[16] Some photographers like to expose one extra stop during the thin crescent phases because the Sun's limb is a little darker than disk center. This step is important only if you shoot slide film because it has a smaller exposure latitude.

[17] For ISOs or f-numbers not listed, use the following formula to determine your exposure.

$$S = f^2/(I \times B)$$

S = exposure or shutter speed (seconds)

f = focal ratio

I = ISO

B = brightness value (see table below)

Brightness value (approx.)	*Eclipse feature*
256	Partial and annular phases (using solar filter)
256	Prominences
32–128	Diamond ring effect
64	Inner corona
4	Middle corona
0.25	Outer corona

For example, taking a picture of prominences on the Sun during totality (B = 256) using a telephoto lens at f/11 and an ISO of 400 would require an exposure of:

$$S = 11^2/(400 \times 256)$$

$$S = 121/102{,}400$$

$$S = 0.00118 \text{ second or } 1/846 \text{ second}$$

Normal SLRs don't have an exposure of 1/846 second, so use the closest shutter speeds, which would be 1/500 and 1/1000 second.

[18] Weather conditions (haze or thin clouds) may require longer exposure times. Use the recommended exposures as a starting point and then bracket during the eclipse, especially if the weather is a factor.

[19] See the NASA Eclipse Website <http://eclipse.gsfc.nasa.gov/eclipse.html>.

[20] Coordinated Universal Time is equal to Greenwich Mean Time to within a second.

[21] The price of GPS receivers has dropped dramatically since the 1990s. They are now available for under $100.

[22] Cassette tape recorders can also be used but they are not as small and lightweight as DVRs. Furthermore, they record and play at variable speeds while DVRs run at a constant rate.

[23] See especially the NASA Eclipse Website <http://eclipse.gsfc.nasa.gov/eclipse.html>.

13
Shadow, Camera, Action!—Capturing an Eclipse on Video

> [The total eclipse of February 26, 1979] began with no ado. It was odd that such a well-advertised public event should have no starting gun, no overture, no introductory speaker. I should have known right then that I was out of my depth. Without pause or preamble, silent as orbits, a piece of the sun went away.
>
> Annie Dillard (1979)

Everyone wants to bring home some piece of their incredible eclipse experience to share with friends and family. The simplest and most effective way to accomplish this is with video. It's really the perfect choice because it records both the sights and sounds of the spectacle. Words fail to convey the beauty of the diamond ring effect, the sudden darkness, or the screams of awe and delight, but a video camera can capture it all. Moreover, you can set up the camera to tape the event completely unattended, leaving you free to focus all your attention on totality.

Video cameras have changed dramatically over the past two decades. Advanced technology and miniaturization have produced compact, portable cameras that shoot great eclipse video and are quite affordable. All you need is a simple tripod, a solar filter, and the video camera itself.

DV Camcorders

Most camcorders on the market today shoot digital video (DV) using miniDV cassettes.[1] These convenient tapes can store 30 to 60 minutes of high quality video with a resolution more than twice that of the old 1980s VHS tapes. Although some DV camcorders store video directly onto mini DVD disks, miniDV tape is less expensive and more readily available should you need more while on an eclipse trip to the far corners of the globe. Some of the newer DV camcorders forgo tape and DVDs altogether by recording onto a built-in hard drive or to flash memory. It remains to be seen whether either of these technologies will replace the miniDV cassette in the next decade.

A Video Primer

Although video comes in a veritable alphabet soup of formats, a quick crash course should sort it all out. The first consumer video cameras were heavy, the VHS tapes were large, and the resolution was low. When 8mm was introduced in the late 1980s, it was a major leap forward. The smaller camcorders and tapes made video truly portable while offering a modest boost in picture quality. By the mid 1990s, the Hi-8 format took center stage with a sizable step up in image resolution.

A serious handicap of all these formats is that they store the video signal as analog data on the tape. This means that the picture and sound quality suffer when a copy is made. And if a copy of a copy is produced (third generation), it gets worse. This is a problem if you want to edit the video or make an archival copy because the original tape may only last 10 or 20 years before it deteriorates and can't be played.

The late 1990s brought a technological revolution with digital video (DV) that completely solved these issues. In DV, the video signal is digitally recorded as a series of 0s and 1s. This means that you can make exact copies of the original video without any loss in quality.

Computers also record all their data as 0s and 1s, so it's no surprise that DV and computers are a perfect match. Video can be transferred directly from the camcorder to your computer using a special camcorder port. It is known technically as an IEEE-1394 port, but some camcorder manufacturers use their own names. DV In/Out, DV terminal, i.Link (Sony), and Firewire (Apple) all mean the same as thing as IEEE-1394. Make sure your camcorder has this port so that you can transfer your video to your computer.

You'll also need a special IEEE-1394 cable for the camcorder-to-computer connection. Check your camcorder manual for details. Most computers have software that allows you to download your video to your computer where it can be edited and saved or burned to a DVD for archiving or sharing with others.

Comparison of Video Formats

Format	*Resolution (lines)*
VHS	240
8mm	280
Hi-8	400
MiniDV[a]	500
HDTV[a]	720 or 1080

[a] with CD-quality sound

Many new features have been added to camcorders throughout their numerous format changes. Bigger zooms, autoexposure with various program modes, autofocus, image stabilization, and wide-screen mode have all made it easier to shoot great video. Most DV camcorders can also shoot stills, but they're typically of much lower resolution than a dedicated still camera. On the other hand, many digital still cameras can also shoot short video clips with sound.

At the moment, it is still best to buy a dedicated DV camcorder if you plan on shooting lots of video. In the coming years, the distinction between still and video cameras will narrow, especially as the price of solid-state memory continues to drop. The day is not too far off when you will be able to buy one camera that shoots both still and video without major compromises.

Zoom Lenses

Camcorder manufacturers describe their zoom lenses in two different ways. Modern zoom lenses are a marvel in optical design, using multiple elements, exotic materials, and complex surfaces. When zooming the lens, the physical spacing of the individual components changes so that the combined focal length and magnification of the lens system varies. This is what is known as the *optical* zoom and is usually represented by a multiplication factor. For instance, a 10 × optical zoom means that the focal length or field of view changes by a factor of ten from wide angle to the maximum telephoto setting.

Many camcorders also have a feature called *digital* zoom. It takes over at the end of the optical zoom's telephoto setting to increase the overall zoom range of the lens. This is accomplished by electronically cropping the image and enlarging it to fill the frame. Since the pixels in the image are also enlarged, the resulting image takes on a "blocky" or mosaic-tile appearance. Digital zooms boast ranges of 100× or more but they produce video with a significant loss in picture quality, especially at the high end of their range. Check the specifications for the optical and digital zoom ranges on the package or owner's manual. For instance, it might say *20 × optical / 200 × digital.* When shopping for a DV camcorder, pay close attention to the optical zoom but ignore the digital zoom. It is a marketing gimmick that you might want to disable or turn off (check the manual). If you *must* use the digital zoom, restrict yourself to the lower end of its range where the pixel size is still acceptably small.

Most people want the Sun to appear as large as possible when videotaping a solar eclipse. The Sun's apparent size depends on two factors: the maximum focal length of the zoom lens and the physical dimensions of the camcorder's CCD imaging chip. Although the optical zoom range is an important consideration, it can't be used to determine which camcorder gives the highest magnification.

Just how big will the Sun appear in a particular camcorder? Take a coin from your pocket and measure its diameter. Place the coin 110 times its diameter from the front of the zoom lens. The coin now appears the same size as the Sun's disk as seen through the camcorder. For instance, a U.S. quarter has a diameter of 15/16 inches (24 mm) so it needs to be placed 103 inches (2.62 meters) from the camcorder. When shopping for camcorders at the local camera or electronics store, bring a coin and a measuring tape or piece of string cut to the proper length. You can now compare the expected size of the Sun through a number of camcorders. Estimate the coin's apparent diameter relative to the vertical height of each camcorder's field of view and take notes. This will be a great help in choosing a new camcorder with a optical zoom lens long enough to match your needs.

Once you have purchased a camcorder, find some time to sit down and *read the manual.* Keep the camcorder nearby so that you can test each feature as you read about it. It might help to spread this homework

This video of the diamond ring effect was shot using an inexpensive Sony Digital 8 camcorder and a 4 × converter lens. A solar filter was removed from the lens 30 seconds before second contact during the 2006 total eclipse from Libya. [Sony TR-320 digital camcorder, Tokina 4 × converter lens, exposure on manual. © 2006 Fred Espenak]

assignment out over one or two weeks. Shoot lots of video to get thoroughly familiar with the camcorder and its operation. Finally, bring the camcorder manual with you on your eclipse trip in case any unexpected questions come up.

Camcorder Accessories

At the maximum telephoto setting, the typical camcorder produces an image of the Sun measuring 1/4 to 1/6 the height of the field of view. This is large enough to produce a pleasing eclipse video. If an even higher magnification is desired, consider one of the after-market video lens converters. These add-on lenses simply screw into the threads of a camcorder's zoom lens and extend its telephoto range by a fixed amount. The most common converters increase the telephoto focal length (and Sun's apparent size) by 2× or 3×. Although there are converters with larger magnification factors (4×, 5×, 8×), they are heavier and more expensive.

A number of manufacturers produce these converters (such as Canon, Raynox, Sony, Sunpak, and Tokina), but few retailers stock them. Try a Google search on the web for "video lens converter" and order one from a company with a good return policy just in case there's a problem. If the camcorder zoom and the converter have different thread diameters, as is often the case, an adapter ring will also be needed.

One downside of using a converter is some loss in image quality but this loss is small compared to using a digital zoom. Another converter issue is that they vignette the field and make the corners black when zooming back to wider angles. This isn't a problem for eclipses since the optical zoom is used at or near maximum telephoto.

Even during normal use it's hard to hold a camcorder steady. The added stress and excitement of totality almost guarantees that a handheld video will be a disaster. This is one time when a sturdy tripod is worth its weight in gold. It not only keeps the camcorder stable and pointed towards the Sun, but also frees you to concentrate on the eclipse.

Since many DV camcorders weigh less than a pound, a heavy-duty tripod is not really needed. Try setting up the camcorder and tripod one evening and shoot video of the Moon. It will give a realistic idea of the tripod's stability and is an excellent exercise in tracking and keeping the Moon in the camcorder's viewfinder.

Many people are surprised to find that they can actually see the Moon slowly drift across the camcorder's field of view. This motion is caused by the Earth's rotation. The same thing happens on eclipse day, so it is useful to remember that the Sun (and Moon) moves its own diameter in about two minutes.

Finally, buy a second camcorder battery as a backup on eclipse day. Charge both batteries the night before the eclipse and put in a new miniDV tape. Bring twice as much tape as you think you'll need. These small cassettes take up little room in luggage and it may be difficult to buy more during your eclipse trip.

Solar Filters

The final item needed to videotape an eclipse is a solar filter. The same kinds of solar filters and techniques work for both still photography and video.

A solar filter is needed during the partial phases of an eclipse but it must be removed smoothly and quickly just before totality begins. Practice to ensure that you can do it without bumping the camcorder or tripod. It would be a catastrophe if the eclipse were knocked out of the camcorder's viewfinder just as totality begins.

The CCD and CMOS imaging chips used in today's video cameras can be exposed to direct sunlight for short periods of time with no ill effects. This means that the solar filter can be removed as much as a minute before totality begins without damaging the camcorder. The diamond ring effect is best seen in the ten seconds before and after totality, so plan to have the solar filter off to capture this stunning spectacle.

A simple and inexpensive solar filter can be made for any camcorder using one half of a pair of the cardboard-mounted eclipse glasses. Just cut one filter (and the cardboard surrounding it) from the eclipse glasses and tape it over the end of a short cardboard tube or box selected to fit loosely over the camcorder's zoom lens. Use extra cardboard to block any holes between the filter and its cardboard holder. A few short pieces of sticky tape (masking, duct, etc.) can hold the filter securely over the zoom while allowing the filter to be removed easily when needed.

With the solar filter in place, try shooting some video of the Sun on a bright clear day. Pay particular attention to the camcorder's autofocus and autoexposure features to make sure they work well with your solar filter (and converter lens, if you use one). This is a problem for many camcorders and may require switching the focus and/or exposure to manual.

Showtime!

It's eclipse day and you've traveled half way around the world to be in the path of totality. Your camcorder, tripod, and filter are ready, but are you? It's critical to have a game plan so you don't get flustered. What exactly should you videotape?

You'll want to record all of totality, but what about the partial phases? They last an hour or more on either side of total phase. Don't use up all your battery power and miniDV tape on the partial eclipse, only to miss totality. Besides, your audience back home certainly doesn't want to watch that much video.

Shooting a 10-second clip of the partial phases every 5 to 10 minutes should be plenty. It will also offer a good opportunity to track the Sun and make sure that everything is working properly. It's also wise to turn off the autofocus because eclipses are notorious for confusing this feature. Set the camcorder for manual focus and adjust it to give the sharpest image in the viewfinder. (Aren't you glad you read the manual?)

Depending on the camcorder, it may or may not do a good job on autoexposure. If the image looks grossly overexposed, it might be necessary to switch to manual exposure and adjust as required. As the partial phases progress, the shrinking crescent provides the chance to check the manual focus again.

It's important to know exactly when totality begins so that you'll be ready. You can get times from the NASA Solar Eclipse Bulletins (see Appendix D). If you are traveling with an organized eclipse tour, the trip leader should also have this information. Synchronize your watch with the correct time. If it has an alarm, set it to ring a minute or two before second contact. This will give you ample warning to check the pointing and focus.

As the partial eclipse progresses, be sure to notice which way the Sun drifts across your viewfinder and recall that it shifts one diameter in two minutes. Several minutes before totality begins, move the Sun to a position so that it will pass through the center of the viewfinder at

Coming Soon: High Definition Digital Video

As good as digital video (DV) is, its days are already numbered. The new kid on the block is High Definition Digital Video (HDV). While DV has excellent resolution with 500 horizontal lines, HDV comes in resolutions of 720 and 1080 lines, bringing it close to the quality of projected film. HDV has a wide-screen aspect ratio of 16:9, which is identical to big LCD and plasma screen HDTVs as well as HD television broadcasts.

As of mid 2007, several major video camera manufacturers have already introduced the first consumer-level HDV camcorders. However, they are still expensive ($1,000 to $2,000) compared to DV camcorders ($200 and up). The new HDV camcorders also have shorter zoom ranges—an important consideration when choosing a camcorder for eclipses. Although the price of HDV camcorders will continue to drop, DV will remain a more affordable choice for several years.

mid totality. This will eliminate additional adjustments until after third contact.

About twenty seconds before second contact, carefully remove the solar filter. As you watch the LCD screen of your camcorder, you will see the corona emerge during the diamond ring stage.

Totality! Make a fast check to verify that the eclipse is well placed in the camcorder. Now spend the rest of the time watching with the naked eye and binoculars.

Before you know it, third contact will mark the end of totality. Watch the diamond ring on your camcorder and replace the solar filter about twenty seconds after totality ends.

Congratulations! You've just captured a total eclipse on video!

Final Thoughts

So much happens on eclipse day that a written script is a great asset. A simple list of shots and the approximate time of each will help you capture the video you want. The script can also remind you to check focus, battery level, and remaining tape. It might sound funny but some people get so excited they forget to remove the solar filter. Make sure that's on your script.

Checklist for Solar Eclipse Video

Camcorder Items

Camcorder (DV or HD)
Video Converter Lens—2× to 5× (optional)
Solar filter (must be easily removable before totality)
Sturdy tripod
Extra batteries
Battery charger
AC adapter plugs for international use (for battery charger)
Extra mini-DV tapes (or whatever tapes your camcorder uses)
Sack or zip-lock bags and heavy string (fill with sand, suspend from tripod for stability)
Exposure list for partial phases and totality

General Items

Pocket knife
Tools for minor repairs: screwdrivers (regular, phillips, and jeweller's), needle-nose pliers, tweezers, small adjustable wrench, allen wrench set, etc.
Plastic trash bags (to protect equipment from rain or dust)
Roll of duct tape or masking tape for emergency repairs

A script also allows you to plan ahead by including time to shoot video of companions and equipment as they set up for the eclipse. You might even tape some brief interviews before and after the eclipse to capture people's reactions. Of course, this is much easier to accomplish if you have a second video camera. Use that second camera to record your observing site and the sky during totality. Place the camcorder on a tripod, set the zoom to wide angle, start recording ten minutes before totality and just let it go. The audio track will also capture all the conversations and excitement during the most majestic celestial event visible from planet Earth.

Notes and References

Epigraph: Annie Dillard: "Total Eclipse" in *An Annie Dillard Reader* (New York: Harper Perennial, 1995).

[1] Sony has a hybrid format called Digital 8. It uses Hi-8 video cassettes but records with full DV quality. These camcorders and tape are less expensive than miniDV, but they are also bigger and less convenient.

14
Getting the Most From Your Eclipse Photos

> Then out upon the darkness, gruesome but sublime, flashes the glory of the incomparable corona, a silvery, soft, unearthly light...
>
> Mabel Loomis Todd (1900)

You've just returned from a fantastic eclipse trip and you're eager to see your photos. If you shot digital, you've already peeked at some images on the camera's built-in LCD display. Now it's time to download the photos to your computer and view them on a big screen.

If you shot the eclipse on film you must be patient a while longer. Bring that precious film to a reliable photofinisher and explain that these are priceless eclipse photos. Good rapport with the technician behind the counter will help ensure that the film is handled with care.

Whether film or digital, you'll want some good prints of the eclipse. But that's just the beginning. Personal computers and image editing software can take you far beyond simple snapshots and fix lots of problems. Film shooters must first convert their negatives or slides into digital files before they can be imported into the computer. Even if you're not interested in editing images on a computer, digital images are easy to archive on CD and to share with friends and family via e-mail.

Photofinishers, Eclipse Photos, and Photo CDs

Most photofinishers do a poor job of printing eclipse photos since they don't know what the prints should look like. Fortunately, many one-hour labs are eager to please if you pick a time when they're not busy. Explain that the photo background should be black and the corona should be white with no colorcast. Use photos in this book as examples.

Three separate images of totality and the diamond rings are combined into one using Photoshop Layers. First, put the total eclipse shot in the bottom layer. The two diamond ring shots in the upper layers have their blending modes set to "lighten" to make them transparent. Total solar eclipse of March 29, 2006 from Libya. [Canon 20Da DSLR, William Optics 66 mm triplet Apo refractor, f/6, diamond ring: 1/500 s, totality: 0.3 s, ISO 100. © 2006 Alan Dyer]

Color slides have one advantage because their development requires no subjective judgment. Nevertheless, the colorcast problem returns when prints are made from the slides.

A trickier issue is how the photofinisher mounts the slides after they're processed. Since the background is completely black, it's hard to determine where one frame ends and the next one begins. Color negatives suffer from the same problem because they are usually cut into strips of four or five exposures. If you shoot a couple of scenic photos at the beginning of each roll, they serve as a guide for the edges of subsequent eclipse images.

Better yet, ask the photofinisher to return the negatives or slides *uncut* and *unmounted*. You will have to perform this chore yourself, but you'll avoid the disaster of having film cut right through the middle of your eclipse images.

Some photofinishers can also scan film and transfer the images to CD. Ask for the highest resolution possible, which should be at least 6 megapixels (2000 × 3000 pixels) for 35 mm film. If your photofinisher doesn't offer this service or the resolution is too low, check the Internet or photography magazines for other scanning services.

The relative sizes of the Sun and Moon are seen in this sandwich of two images. After placing the images in two Photoshop layers, the top layer is made transparent by setting its mode to "lighten." Total solar eclipse of March 29, 2006 from Manavgat, Turkey. [Canon EOS 20D DSLR, Sigma 100–300 mm EX APO HSM at 300 mm with Sigma EX APO 2×, Sun disk: 1/500 s, diamond ring: 1/1000 s, ISO 400. © 2006 Odd Høydalsvik]

If you have negatives or slides scanned on a regular basis, you may want to purchase a dedicated film scanner for your computer. Expect to pay $300 for a basic model with a scanning resolution of at least 2500 dpi (dots per inch).[1]

When printing digital images, ask the photofinisher to turn off the exposure and color correction settings. This will prevent the printing machine from making unwanted corrections to your unusual images. Some digital photographers make their own prints at home. Although you have more control making your own prints, it is more expensive and time consuming than ordering commercial prints.

Image Editing Software

Before making any changes to your digital images, burn the original files onto CD or DVD. Better yet, make two copies and keep them in two different locations (for example, home and work). You will always have copies of the originals if any files are accidentally deleted or the hard disk crashes.

Whether your computer is a Macintosh or Windows PC, there are many image editing programs to choose from. They all have similar tools for the most essential editing tasks. One popular program available for both PCs and Macs is Adobe Photoshop Elements ($100).

Elements contains a subset of the features found in its more expensive big brother Adobe Photoshop. Considering the higher cost ($650) and added complexity, Photoshop is overkill for most home users. The techniques described here will use the tool names and nomenclature found in Elements. If you already own different software, stick with it unless it doesn't meet your needs. It's probably similar enough to Elements to understand this chapter.

Photo-editing programs are hogs for memory. You'll need at least 512 MB of RAM but more RAM is always better. The higher the computer's processor speed, the faster it it will open, edit, and save your images.

You will need a textbook specifically written for your program.[2] A good manual is essential for answering questions or learning about various features. On the other hand, a tutorial text will serve as an invaluable guide by leading you through a set of exercises to master the most important image editing techniques. Check with local bookstores or Internet booksellers to find the books that best suit you.

Cropping, Contrast, and Color

The three Cs of cropping, contrast, and color are the easiest and most effective ways to improve any digital photograph, including eclipse images. So fire up your computer, boot your editing software, and open an image file. Have your program manual or textbook handy to look up instructions on using various features.

In the excitement of photographing totality, it's hard to keep the Sun centered in the camera's viewfinder. Earth's rotation causes the Sun to shift its own diameter every two minutes. Unless you make constant adjustments or use a tracking mount, the Sun will drift. During the partial phases, locating the viewfinder's center is difficult because everything looks black except for the Sun. In most shots, the Sun will not be at the center of the frame. Here is where cropping comes to the rescue. Use your software's cropping tool to trim away enough of the image so that the Sun is at the center.

You can even use the cropping tool to tighten the composition by making the eclipse fill a larger fraction of the frame. Keep in mind that this digital zooming trick will make the pixels appear larger as well. How far can you zoom and still get a decent print without pixelation?[3] Your prints and enlargements should have a resolution with at least 250 pixels per inch. If you crop a photo to 2000 × 2500 pixels, divide these values by 250 pixels per inch to determine that the maximum print size is 8 × 10 inches. Using the same criteria, a 4 × 6 inch print must be a minimum of 1000 × 1500 pixels. Your software should have commands or an information window that gives the dimensions of your image in pixels.

The annular eclipse of October 3, 2005 is seen in its entirety from start to finish. Photoshop's "Layers" feature was used to combine 15 separate images into one. [Canon 10D DSLR, Skywatcher 127 mm Maksutov, fl = 1500 mm, f/11.8 + 0.6× focal reducer, Baader Solar Film Filter. © 2005 Pete Lawrence]

When you're happy with the final crop, save the results by appending "crop" to the original name (use the *Save As...* command in Elements). The original image was probably stored in a lossy format like JPEG, so you'll want to change the new file to a lossless format like TIF or PSD.[4]

Contrast and Color Corrections

Each pixel in a digital image is composed of three colors: red, green, and blue (RGB). And each of these three has a brightness value of 0 to 255 (minimum to maximum). The range of brightness of RGB in each pixel ultimately determines the image contrast and exposure of an image. If your photo appears under- or overexposed, or if it has a flat, lackluster appearance, you can change it by adjusting the contrast. One feature for evaluating the image contrast is called the histogram or levels tool. This tool produces a bar chart with each bar representing a brightness value (0 to 255) and the height of the bar represents the number of pixels with that brightness. A well-exposed photograph will have a bell-shaped histogram with most of its pixels in the middle range of brightness (around 128). If most of the pixels are clustered around the low end of the histogram, the image will appear underexposed. Similarly, when most of the pixels are clustered at the high end of the histogram, the image will appear overexposed. Finally, if the histogram shows that all the pixels fall within a narrow range of brightness, then the image will appear very low in contrast.

The histogram or levels tool can be used to fix all these problems. In Elements the levels tool has three small triangles at the bottom of the histogram. Move the left triangle to the right to make the darkest tones in your image get darker. Move the rightmost triangle to the left to make the lightest tones appear brighter. Move the middle triangle to the left or right to change the midtone brightness in the image.

Experiment with these adjustments while watching how the image changes. With a little trial and error, you can usually improve the contrast and exposure in an image. This is a powerful tool, so check your manual for more information. When satisfied with the result, save it with a new name by appending "contrast" or "levels" to the end of it. Once again, use a lossless file format like TIF or PSD.

Does your eclipse image have an unnatural colorcast to it? You can use the histogram/levels tool to remove it. In Elements, open the *Levels* tool and notice the three little eyedropper buttons to the right of the histogram. Click on the middle one, then move the cursor over to the eclipse image. Click on any point in the image that has the color tint you want to remove, and voilà! The colorcast is instantly neutralized. Don't like the result? Click on another spot. If you change your mind, just hit the cancel button to return to the unaltered image.

Dust and Scratches

Are dust or scratches getting you down? These are especially bad on images shot on film, but even digital camera images can have spots caused by dust lodged on the imaging sensors. There are a number of ways of removing these unwanted defects. The more automatic solutions like the *Dust & Scratch* filter in Elements should be used with care because it can also remove fine details in your eclipse images.[5]

A greater degree of control can be achieved using the *Clone Stamp* tool to individually remove each offending blemish. The tool's magic is performed by copying and pasting a tiny piece of the image that resembles the one with the dust spot or scratch. The art of using this tool involves choosing a good piece to paste over the dust and blend in without showing. You control the brush size and edge sharpness of the *Clone Stamp* tool using a pallet of options. Mastering the tool requires some practice but it is time well spent. Check the software manual for more tips and tutorials on using this tool.

The 2001 total solar eclipse is captured in a sequence of 7 individual images using Photoshop "Layers." The central image of the corona is a composite of several exposures. [Canon Powershot G1 digital camera, Orion 80 mm f/5 refractor, fl = 400 mm, 2× Barlow, 25 mm eyepiece, Orion Solar Filter, f/10, partials: 1/200 s, total: 1/500 s to 1/6 s, ISO 50. © 2001 Fred Bruenjes]

Image Sequences

So far, the techniques covered have involved fixing basic defects in your eclipse images. Now it's time to have some fun. Most people shoot a series of photos over the course of the eclipse. From first to fourth contact, you might have a roll or two of partial phases and another roll for totality. If you shot digitally, you could easily have a hundred or more eclipse images. But most of your viewing audience isn't interested in seeing that many eclipse photos no mater how special they are to you.

With a little creativity, you can take those images and turn them into dramatic time sequences showing various stages of the eclipse. As is often the case, there are many ways to accomplish this with an image editing program. One way to do it in Elements is to use a feature called *Layers*. Imagine that you have several eclipse photos each on a separate negative and you want to combine the negatives into one print. In the old days, you would take the negatives into the darkroom, stack them one on top of the other and print them. *Layers* lets you do the same trick with digital images by taking each one and placing it in its own layer. The stack of image layers is saved as a single file that you can open and make additional changes anytime you like. You can move each image layer completely independently of all other layers.

Let's say you have four images of the partial phases that you'd like to combine into a time sequence. Start by opening the first image in the sequence. Check the pixel dimensions of this image and use it to estimate the size of the final sequence. For instance, if the first image is 2000 × 3000 (width × height), the sequence might be four times as wide or 8000 × 3000. Open a new image file with these dimensions and give it a black background. This will be the new the eclipse sequence file so save it with a name like "Sequence."

Click on the first eclipse image to make it active, select the entire image and copy it. Now click on the "Sequence" file and do a *Paste*.

Elements pastes the first eclipse image into a new layer in the "Sequence" file. Open, copy, and paste each of the remaining eclipse images into "Sequence." You can move each layer independent of the others to create an attractive composition. Trim away any unimportant parts of the upper layer images if they block the Sun in the layers below.

Using the *Layers* feature, the composition possibilities are only limited by your imagination. How about a complete eclipse sequence from first to fourth contact? If you find the length of the sequence growing too long, break it up into a series of rows to make a matrix. Another idea is to design your own poster. It could feature a large image of totality at the center surrounded by smaller images of partial phases, the diamond ring effect, and you with friends and telescopes on eclipse day. Knowing what kind of sequence or poster you'd like to make can even help you plan your photo list for the next eclipse.

Composites of the Corona

Eclipse photographers are often disappointed that their photos of totality look nothing like what they remember. From the striking red prominences to brilliant loops, delicate polar brushes, and long streamers, the Sun's corona is of unsurpassed beauty. Why don't the photos capture this majesty instead of just a fuzzy, colorless halo surrounding a black disk?

It's because the corona covers a wide dynamic range in brightness. The innermost regions just above the Sun's photosphere shine with an intensity thousands of times greater than the tenuous streamers just one solar diameter away. In comparison, the brightness range in a typical daylight landscape or a people portrait is only 150:1. The human eye can take in the corona's enormous brightness extremes but photography, be it film or digital, simply can not.

Capturing the brightest and dimmest parts of the corona can be accomplished by shooting a series of different exposures during totality, as described in Chapter 12. This leaves you with a set of individual pictures that each record a narrow slice of brightness in the corona. Using the digital composite technique, you can take these separate photos and combine them into one image that more closely resembles the corona's visual appearance.

There are many ways to combine multiple images into a corona composite, but the technique described here requires the full power of Adobe Photoshop. The basic idea is to process each photo with an unsharp mask[6] using Photoshop's radial blur filter to separate the brightness information from the fine structure details.[7] The new images are then recombined with the original images in several steps to produce the final composite. Although simple in concept, this technique allows

A computer-processed composite image resembles the naked-eye appearance of the eclipse and shows details in both the inner and outer parts of the corona. The original images were taken during the August 11, 1999 total solar eclipse from Lake Hazar, Turkey. [Nikon N70 SLR, Vixen 90 mm fluorite refractor, fl = 810 mm, f/9, 11 exposures: 1/250 s to 4 s, Kodak Royal Gold 100 negative film. © 1999 Fred Espenak]

you to tease out and reveal subtle details hidden in the individual images (figures 14.1 and 14.2).

If you are an experienced Photoshop user and are eager to squeeze the most from your priceless eclipse images, consider trying the corona compositing technique. But even if you are just a beginner, image editing programs like Adobe Photoshop Elements can help you turn those eclipse snapshots into gorgeous masterpieces worthy of displaying for all to enjoy.

The Corona Composite Technique

Photographs of the corona look nothing like the vivid images etched into your mind. That's because your eyes can take in the corona's huge range of brightness while a photograph only records a narrow slice. The corona composite technique allows you to combine several eclipse photos shot at different exposures into a single image that more closely resembles the corona's appearance to the naked eye. The technique described here is based on recent versions of Adobe Photoshop. While other software may be used to achieve similar results, the tools and terms used below refer to Photoshop.

To begin, select four or more corona images shot over a wide range of exposures.[a] Open each image and carefully remove any dust or scratches using the *Clone Stamp* tool.

Cropping to Center

The composite technique assumes that the Moon's disk lies at the center of each image. This is probably not true of your original eclipse images but you can easily trim away two of the edges to make it so.

Photoshop's Info window displays the x and y coordinates of the cursor as you move it across an image. The origin (x = 0 and y = 0) is in the image's upper left corner. If the coordinates are inches or centimeters, open Photoshop's Preferences window to change the coordinate units to pixels.

Position the cursor along the left edge of the Moon's image and write down its x-coordinate. Repeat for the Moon's right edge. The average of these two values is the x-coordinate of the Moon's center. Now place the cursor along the top and bottom edges of the Moon and write down the y-coordinate in each instance. The average of the two y values is the y-coordinate of the Moon's center.

With these measurements, you can now center the Moon by cropping off the unnecessary parts of the frame. Not surprisingly, this is accomplished using the *Crop* tool. After selecting this tool, position the cursor at the x-y coordinates of the Moon's center as displayed in the *Info* window. Hold down the *Alt* key (*Option* key for Mac users) and drag the mouse diagonally in any direction while holding down the left mouse button.[b] Continue dragging diagonally until the maximum limit is reached (when numbers in the *Info* window stop changing). Release the mouse and keyboard buttons. Double-click in the selected window to execute the crop.

The Moon's disk is now perfectly centered. Save the image in Photoshop format (.PSD) using a new name by appending "-crop" to the old name. Repeat the cropping process with each eclipse image.

The cropped images must all share the same dimensions (in pixels) before they can be combined into the final composite. Measure the width and height of each cropped image by selecting the entire image and noting the dimensions displayed in the *Info* window. Using the smallest width and height values, crop all images to those dimensions using the *Canvas Size* menu command and save.

Radial Blur Filter

The *Radial Blur* filter blurs an image by rotating it through a fixed angle about its central axis. In the process, the brightness values of every pixel at a given distance from center are averaged with all other pixels at the same distance over the chosen angle. An angle of 5° to 10° works well for eclipse composites.

Use the *Radial Blur* filter (under the *Filter* menu) to process each cropped image. This can take a few minutes with large files, so be patient. When complete, save the blurred file with a new name by replacing "-crop" with "-blur." Repeat this step with all the cropped images.

It's interesting to compare the new blurred image to the original image (figures 14.4a and 14.4b). Although the blurred image preserves the brightness and shape of the corona, it is completely devoid of fine details.

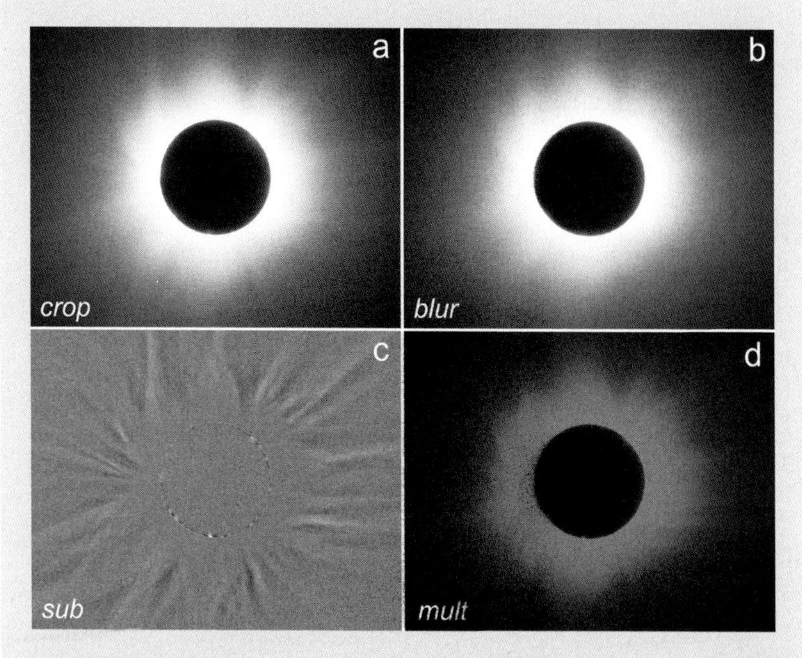

The four primary steps in processing an eclipse image for use in a composite are: (a) crop (cleaned and centered), (b) blur (radial blur filter), (c) sub (subtract *b* from *a*), and (d) mult (multiply *c* by *a*). The original image was shot during the 1999 total solar eclipse. [Nikon N70 SLR, Vixen 90 mm fluorite refractor, fl = 810 mm, f/9, ½ s, Kodak Royal Gold 100 negative film. © 1999 Fred Espenak]

Image Subtraction

If you subtract the blurred image from the cropped image, the resulting image contains all the fine details in the corona without the underlying brightness. Image subtraction can be done using Photoshop's Apply Image command.

Open the cropped and blurred versions of an eclipse image and select the cropped image by clicking on it. Choose the *Apply Image* command (under the *Image* menu). When the dialog box appears, choose the blurred image from the *Source* drop-down menu and select *Subtract* mode from the *Blending* drop-down menu. Set the value in the *Scale* box to 1, the *Offset* box to 128, and click the *OK* button.

When complete, save the newly modified cropped file with a new name by replacing "-crop" with "-sub." Repeat this step with all the cropped/blurred image pairs.

Image Multiplication

The subtracted images appear gray and nearly featureless, but a closer look reveals subtle details in the corona. It's time to combine this information with the cropped images using the *Multiply* mode of the *Apply Image* command.

Open the cropped and subtracted versions of an eclipse image and select (click) the cropped image. Choose the *Apply Image* command and choose the subtracted image from the *Source* drop-down menu. Select *Multiply* mode from the *Blending* drop-down menu and click the *OK* button.

Save the newly modified multiplied file with a new name by replacing "-crop" with "-mult." Repeat this step with all the cropped/subtracted image pairs.

Assembling the Composite

The final composite can now be assembled by adding all the multiplied images together. Open the multiplied images and select the one with the shortest exposure. Use the *Duplicate* command (from *Image* menu) to make a copy of this image and name it "composite." It will form the basis of the corona composite.

Choose the *Apply Image* command and change the *Source* drop-down menu to the image with the second shortest exposure time. Select *Add* mode from the *Blending* drop-down menu and set *Scale* and *Offset* values to 1.4 and 0. Click *OK* to add the images. With the composite image still selected, repeat the *Apply Image* operation on each of the remaining multipled images, working through them from the shortest to longest exposures. The final image contains enhanced structure showing details in both the inner and outer corona. Compare it to the original images.

Variations in the image assembly can fine tune the appearance of the composite image. For instance, a different value for *Scale* can be used in the *Apply Image* step. This will control the contrast and dynamic range of the final composite. While a *Scale* value of 1.4 works well for a four-image composite, a larger value is needed as the number of images increases.

The composite technique presented here is just a starting point. As you learn more about Photoshop and its capabilities, new ideas will present themselves. Don't be afraid to experiment. You will be rewarded with images that match your memories of nature's grandest spectacle.

---

[a]Most digital cameras store the exposure information in the EXIF (Exchangeable Image File Format) record embedded in each digital image. You can read this information with your camera, or with computer software (use Google to search "EXIF reader").

[b]The Macintosh mouse has only one button.

Notes and References

Epigraph: Mabel Loomis Todd: *Total Eclipses of the Sun* (revised) (Boston: Little, Brown, 1900).

[1] Canon, Epson, Konica Minolta, Microtek, Nikon, Polaroid, and Umax all make film scanners in a range of models, features and prices.

[2] Although most image editing programs have on-line manuals or help guides, there's nothing like an old-fashioned book to refer to as you work. The exact titles change frequently to cover new versions of the software. A few recommendations include books from the *Missing Manual* series (O'Reilly Press), the *Visual QuickStart Guides* (Peachpit Press), and the *For Dummies* series (For Dummies Press).

[3] Pixelation means that the individual pixels are large enough to give final detail in the image a blocky appearance.

[4] PSD is the native format for both Elements and Photoshop. JPEGs are lossy because they are compressed files and some image information is lost in the compression process. For more on formats, see "JPEG vs. RAW File Formats" in chapter 12.

[5] If you are also an astrophotographer and shoot photos of the night sky, the *Dust & Scratches* filter in Elements can remove all the stars from your images!

[6] Unsharp masking is an old darkroom technique used to actually increase the apparent sharpness in a photographic negative while lowering its dynamic range. An out-of-focus positive is first made from the negative and developed. This blurred or "unsharp" positive acts like a mask when combined with the original negative. When printed together, the resulting image appears sharper and less contrasty than the original negative. There are digital counterparts to this powerful darkroom technique in programs like Photoshop.

[7] Gerald L. Pellett: "Eclipse Photography in the Digital Age," *Sky & Telescope*, January 1998, pages 117–120.

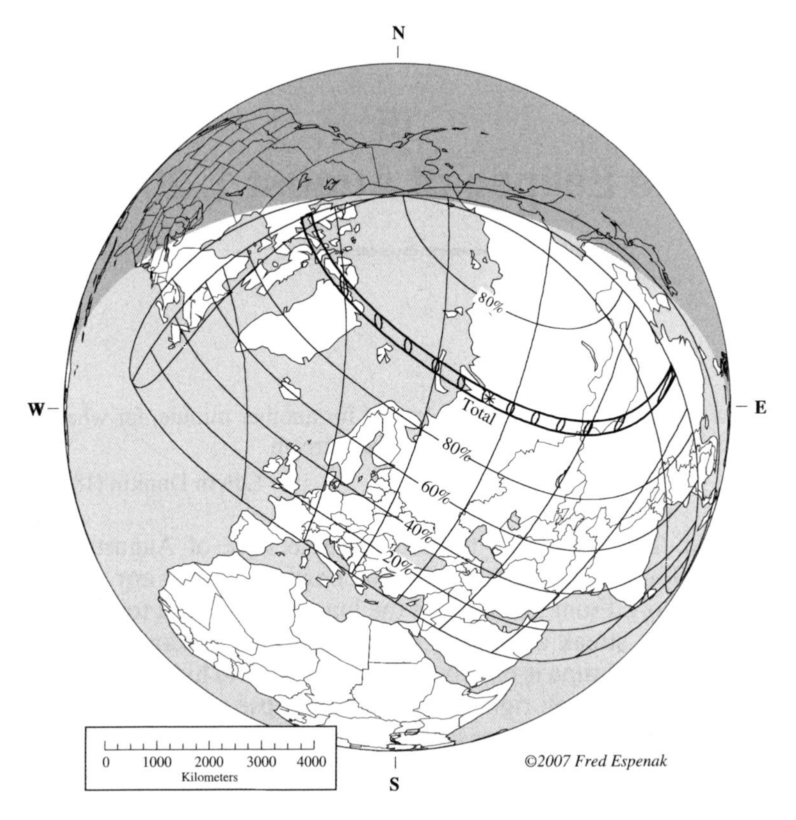

Path of the August 1, 2008 total solar eclipse. For places outside the path of totality, the eclipse magnitude (fraction of the Sun's diameter eclipsed) is given in percent. [Map and eclipse calculations by Fred Espenak]

the total eclipse (about 25%) are woefully against the people of Alert or Grise Ford because of persistent overcast at this time of year, caused primarily by warm air masses that push up from the south and cross icy Arctic Ocean waters, creating low clouds and fog.

The 2008 total eclipse continues its remote itinerary as it clips just a northern bit of Greenland, comes within 350 miles (560 kilometers) of the north pole, and takes aim at Russia where the Ural Mountains mark the divide between Europe and Asia. As the path of the eclipse bends southeastward, it comes ashore in Siberia and soon slides south of the Arctic Circle. It celebrates there with its longest duration of total eclipse—2 minutes 27 seconds—but under ugly weather prospects for the few people on the ground. Weather in Siberia, even in August, is seldom an eclipse observer's paradise.

But now the central line of the eclipse, becoming more sociable, passes close to and between the Russian oil field cities of

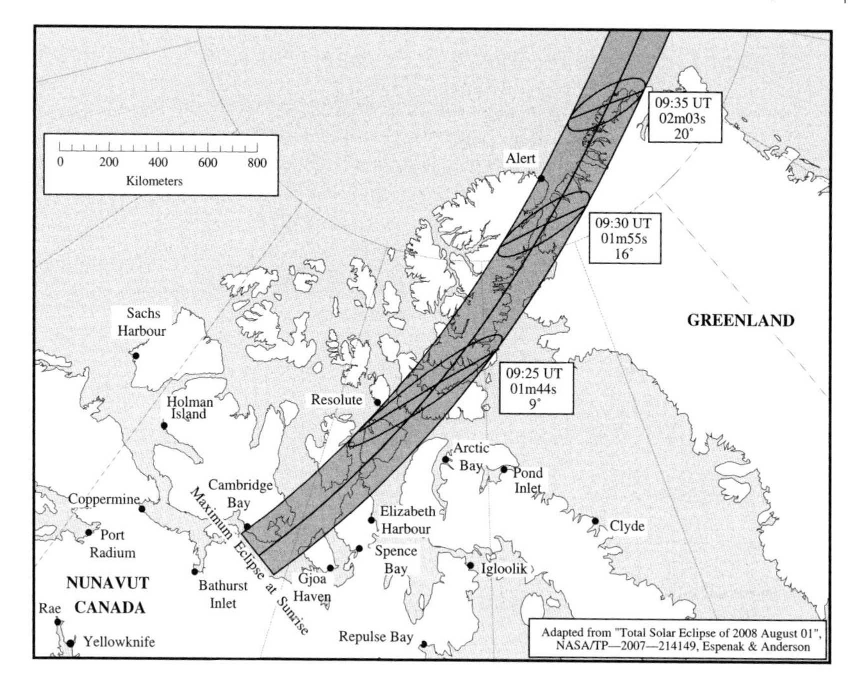

The path of the 2008 total eclipse begins in Canada's Nunavut Territory at sunrise.

Nizhnevartovsk and Strezhevoy, then over Novosibirsk—right across its airport, in fact. Novosibirsk, with a population of 1.4 million, is Russia's third largest city, and the largest city on the itinerary of the 2008 eclipse.

Continuing southeast, the central line of the eclipse passes equally close to Barnaul. Novosibirsk and nearby will have up to 2 minutes 18 seconds of totality. Weather records give the hearty Siberians living between Novosibirsk and the Mongolian border a 50% to 60% chance of seeing totality.

China Bound

As the eclipse leaves Russia, the swath of totality stands briefly in four countries at once: Russia, Kazakhstan, China, and Mongolia. At that boundary, the shadow reaches the Altay Mountains, "Mountains of Gold," a UNESCO World Heritage Site. Glaciers spill off Belukha, the highest peak, which rises almost 15,000 feet (4,506 meters) and is 40 miles (65 kilometers) west of the central line.

Continuing southeasterly through the afternoon, the central line of the eclipse more or less parallels the western side of Altay Range,

Weather Prospects for the 2008 Eclipse

The average cloud amount in August along the eclipse track through Canada is 70% to 80%. The weather prospects gradually improve as the eclipse path curves south through Russia and Mongolia. There, the average cloud amount drops from 70% to 45%. The clearest weather is in northern China, where the average cloud amount of only 30% (probability of clear skies or scattered clouds is 70%).

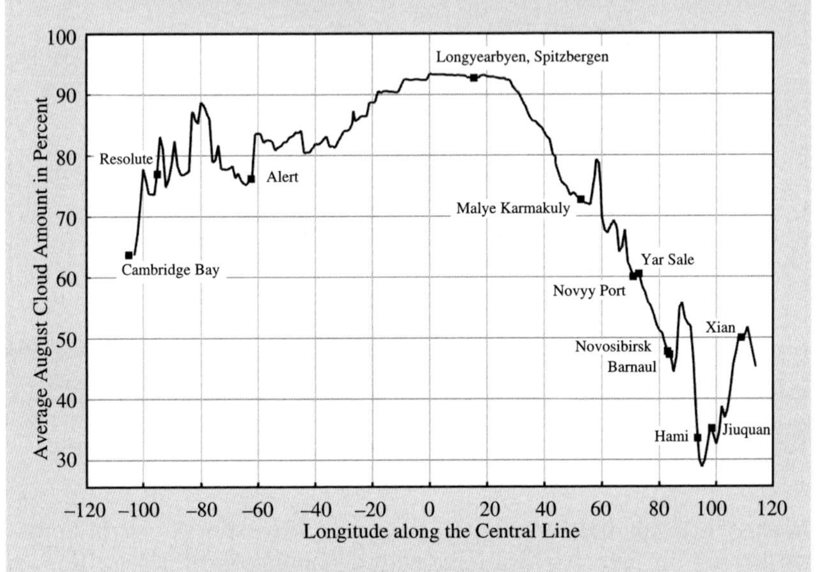

Average cloud amounts along the 2008 eclipse path. [Adapted from *Total Solar Eclipse of 2008 August 01*, NASA TP–2007–214149]

crossing into Mongolia, then into China, then back into Mongolia, then back into China, due to Mongolia's indented border. The Altay Mountains of western Mongolia provide scenic beauty but produce updrafts that all too often (about 50% of the time) cloud the area in August.

The eclipse spends its last minutes in China. It is here, just after the central line of the eclipse leaves Mongolia for the final time and plunges into northwestern China, that prospects for viewing the eclipse of August 1, 2008 are best. This region has deserts to both the east and the west that protect it from the monsoon rains that soak much of Southeast Asia during the summer months. The nearest major town is Hami (about 300,000 people), which just misses seeing totality. Many eclipse chasers will be traveling from Hami about 90 miles (140 kilometers) to the east to the town of Yiwu. Yiwu is well within the zone of totality, just 15 miles (25 kilometers) southwest of the center of the eclipse path. It

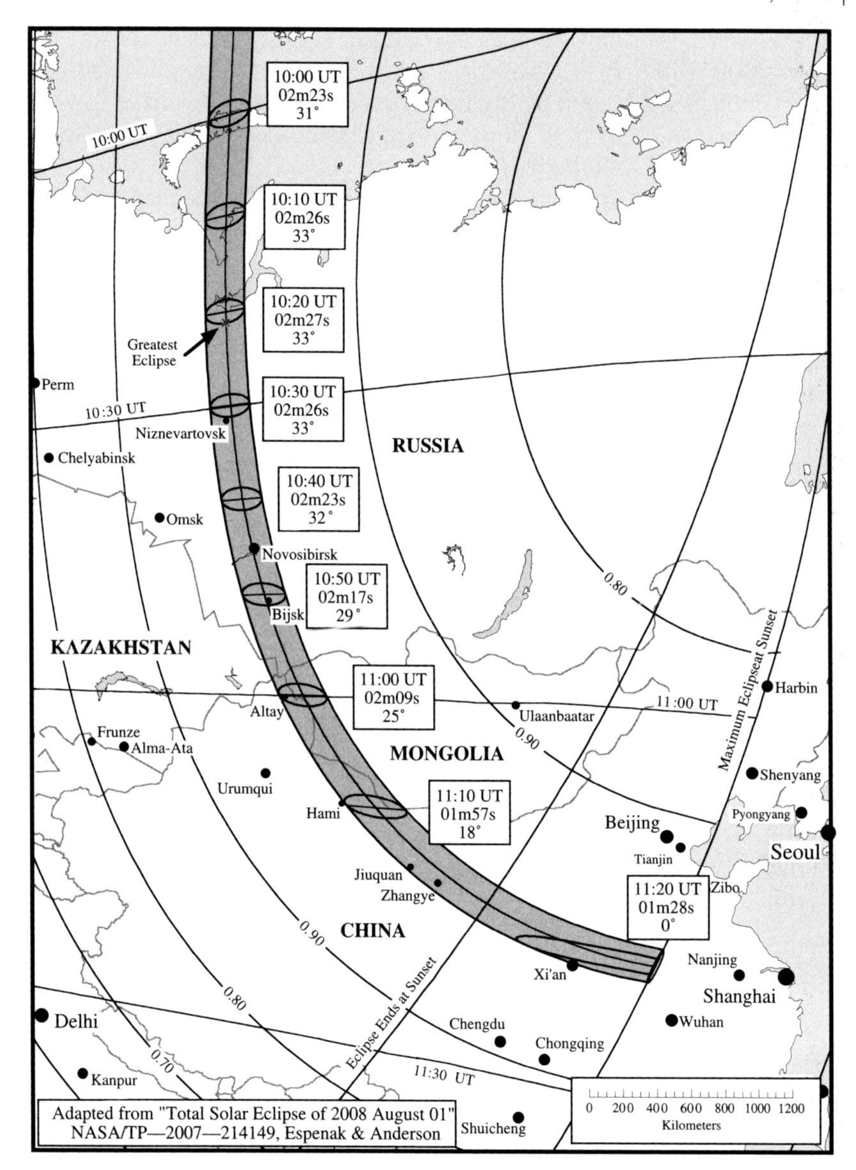

As the 2008 eclipse path swings south through Russia and Mongolia, weather prospects slowly improve. The best chance for clear skies is in northern China.

will experience totality lasting 1 minute 56 seconds with the Sun 19° above the horizon. The Sun at low altitude during the eclipse offers a chance to photograph totality with interesting scenery (landscape, telescopes, buildings, people) in the foreground.

Eastward, ever onward goes the eclipse. Totality ends at sunset 500 miles (800 kilometers) before it reaches sprawling Shanghai and the East China Sea. Cheated by the near miss, Shanghai won't have to wait long for a chance to bask in the Moon's shadow. The very next total eclipse, on July 22, 2009, will pass its way.

But the eclipse of August 1, 2008 is over. The Moon's shadow leaps off the limb of Earth and soars away into space.

Saros Series 126 in Old Age

Last call. Almost. The total eclipse of August 1, 2008 is part of a fading family of eclipses. The twentieth century was its heyday. The twenty-first century is the end of the line. This eclipse family will produce only two more total eclipses before it lapses into partial eclipses to finish out its 1,280-year career of 72 solar eclipses.

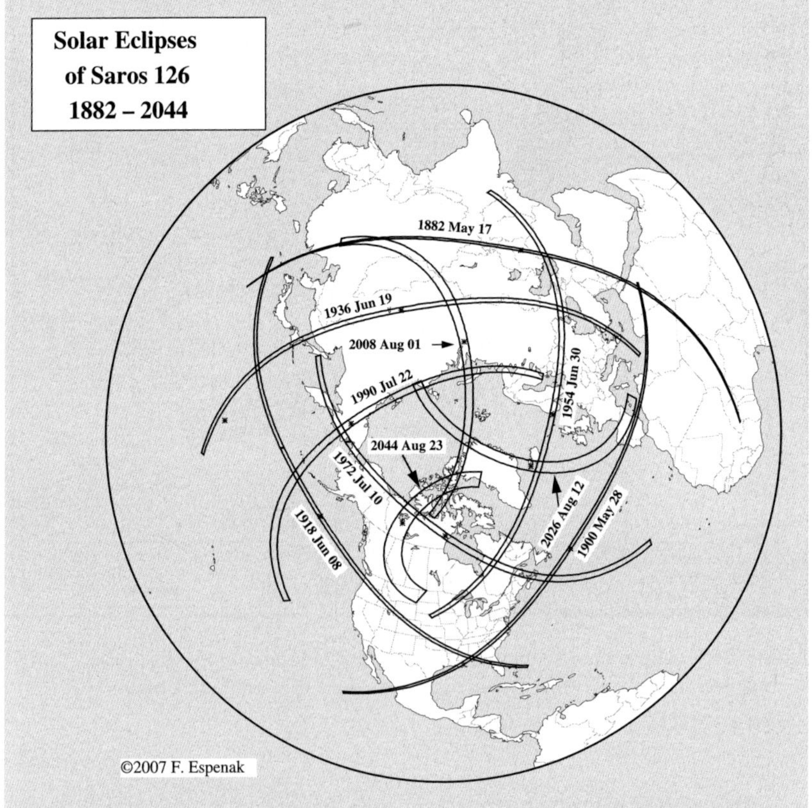

The final ten central eclipses of saros series 126 all take place in the northern hemisphere. The last three total eclipses are 2008, 2026, and 2044. [Map and eclipse calculations by Fred Espenak]

Statistics for Saros Series 126[a]

Number of eclipses: 72
Duration of saros: 1,280 years
Position of 2008 eclipse in series: number 47 of 72

Sequence of eclipses within saros 126:
- 8 partials
  - First eclipse in saros 126 (near south pole): March 10, 1179
  - Saros 126 eclipses occur as Moon crosses descending node
- 28 annulars (exceptionally numerous)
  - First annular eclipse in series: June 4, 1323
- 3 hybrids (annular-totals)
  - First hybrid eclipse in series: April 14, 1828
- 10 totals (stingy both in number and in length of totality)
  - First entirely total eclipse in series: May 17, 1882
  - Last total eclipse in series: August 23, 2044
  - Longest totality: July 10, 1972—2 minutes 36 seconds
- 23 partials
  - Eclipses in series become partial again: September 3, 2062
  - Last eclipse in saros 126 (near north pole): May 3, 2459

[a]Saros data from Fred Espenak and Jay Anderson: ***Total Solar Eclipse of 2008 August 01*** (Greenbelt, Maryland: NASA Goddard Space Flight Center [NASA Technical Publication 2007–214149], 2007), pages 9 and 41.

Each eclipse is part of a family of eclipses called a saros series in which the eclipses occur in sequence at 18-year 11-day 8-hour intervals, each one only slightly different from the one ahead or behind it. But each eclipse track is centered about one-third of the way around the Earth westward of the one before it.

A saros series begins near either the north pole or the south pole and successive eclipses migrate slowly, over 12 or 13 centuries, toward the opposite pole. At the beginning of a saros series, the eclipses are partials. As eclipses reach the middle latitudes, they become annulars, hybrids, or totals. At the end of a saros series, the eclipses revert to partials again. At any one time, about 42 different saros series are bringing eclipses to Earth.

The second-to-last total eclipse in saros series 126 will take place on August 12, 2026, skimming along the east coast of Greenland and visiting westernmost Iceland near eclipse maximum, with totality lasting 2 minutes 19 seconds. Then it will scoot over the Atlantic to end in Spain.

The final total eclipse of saros 126, lasting a maximum of 2 minutes 4 seconds, will occur on August 23, 2044. It will be visible from the west coast of Greenland through the Northwest Territories and Alberta in Canada to, at sunset, portions of Montana and North Dakota in the United States.

And then the spectacle of total eclipses within this series will be over. An uninspiring string of 23 partial eclipses will finish out saros 126, the last in the year 2459.[b]

---

[b]A complete list and maps of all eclipses in saros 126 can be found in the NASA catalog of solar eclipse saros series: <http://eclipse.gsfc.nasa.gov/SEmono/TSE2008/TSE2008.html>

More on the 2008 Total Solar Eclipse

The NASA Eclipse Website has additional details and maps for the 2008 total solar eclipse: <eclipse.gsfc.nasa.gov/SEmono/TSE2008/TSE2008.html>

A special NASA bulletin for the 2008 eclipse contains detailed maps, eclipse times for over 300 cities, and more weather information. Access the NASA Eclipse Website: <http://eclipse.gsfc.nasa.gov/SEpubs/TP214149.html>.

Notes and References

Epigraph: Edwin Dunkin, *Autobiography*, unpublished (compiled by Peter Hingley, Royal Astronomical Society). Dunkin is referring to the total solar eclipse of July 28, 1851 as seen from within the northern edge of the path of totality in Scandinavia.

16
The Eclipse of July 22, 2009

> The effect of totality upon the bystanders was most remarkable. Until the beginning of totality, the murmur of the conversation of many tongues had filled the air; but then in a moment every voice was hushed, and the stillness was so sudden as to be perfectly startling...
>
> Warren De La Rue (1862)

Wednesday, July 22, 2009 marks the return of perhaps the most generous, beneficial, and historic eclipse series in modern times: saros 136. The last time saros 136 visited Earth, it brought the remarkable eclipse of July 11, 1991, which passed over the great telescopes on Mauna Kea in Hawaii and then over Mexico (with a maximum duration of 6 minutes 53 seconds) and, quite graciously, down the slender stem of land that is Central America on into South America.

Before that, saros 136 brought the total eclipse of June 30, 1973 through Africa, with totality lasting as much as 7 minutes 4 seconds. It brought the total eclipse of June 20, 1955, with 7 minutes 8 seconds of totality—the longest of the twentieth century. And it brought the 7-minute-4-second total eclipse of June 8, 1937. Three consecutive eclipses with more than 7 minutes of totality.

Just one cycle earlier, saros 136 produced the eclipse of May 23, 1919, which provided a crucial proof of Einstein's General Theory of Relativity.

It's back.

The total eclipse of 2009 starts just off the west coast of India in the Gulf of Khambhat, part of the Indian Ocean. The eclipse comes ashore at Surat, a city of 2.9 million people 150 miles (240 kilometers) north of Mumbai (Bombay).

This eclipse is a river runner. In India, it pushes upstream on the Narmada River headed northeast, portages to the Ganges River headed east, then swings upstream into the Brahmaputra River and follows it to the eastern Himalayas. After soaring over the mountains into China, the eclipse then cruises down the Yangtze River eastward to Shanghai and the sea.

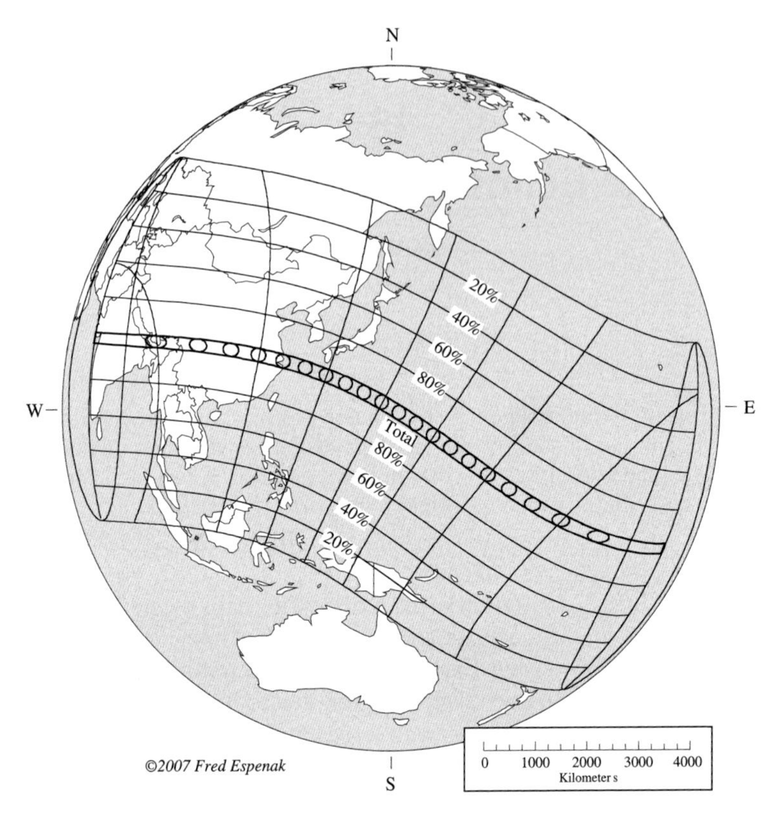

Path of the July 22, 2009 total solar eclipse. The eclipse magnitude (fraction of the Sun's diameter eclipsed) for places outside the path of totality is given in percent.

River Running in India

Headed up the Narmada River in India, the eclipse brings the cities of Indore, Bhopal, and Jabalpur, all with populations over one million, into its shadow. It follows the Narmada River for 450 miles (725 kilometers), about halfway across the Indian subcontinent.

Just south of Allahabad, the zone of totality picks up the Ganges River and traces it for the next 180 miles (290 kilometers), through the large cities of Varanasi and Patna. As the Ganges reaches Bhagalpur, it turns southeast and dips out of the zone of totality to enter Bangladesh. But the eclipse is riverless only briefly, for it picks up the Brahmaputra, flowing from the northeast into the Ganges, and follows the Brahmaputra upstream into the eastern Himalayas.

The path of darkness zips along northeastward across India and cuts across the southeastern corner of Nepal, with the northern edge of totality missing the summit of Mount Everest by 70 miles (110 kilometers).

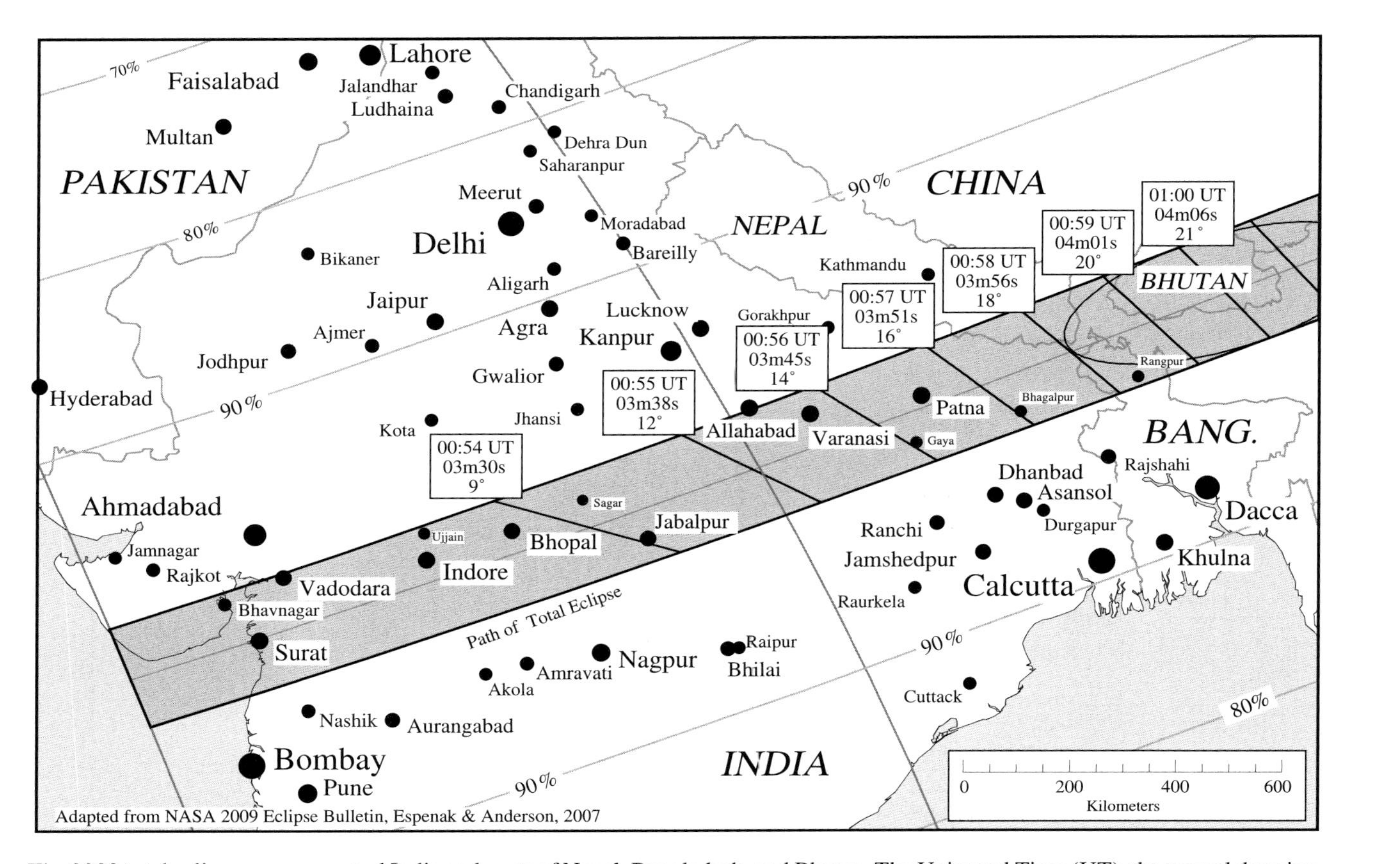

The 2009 total eclipse crosses central India and parts of Nepal, Bangladesh, and Bhutan. The Universal Time (UT), the central duration of totality, and the Sun's altitude are given at one-minute intervals along the eclipse path.

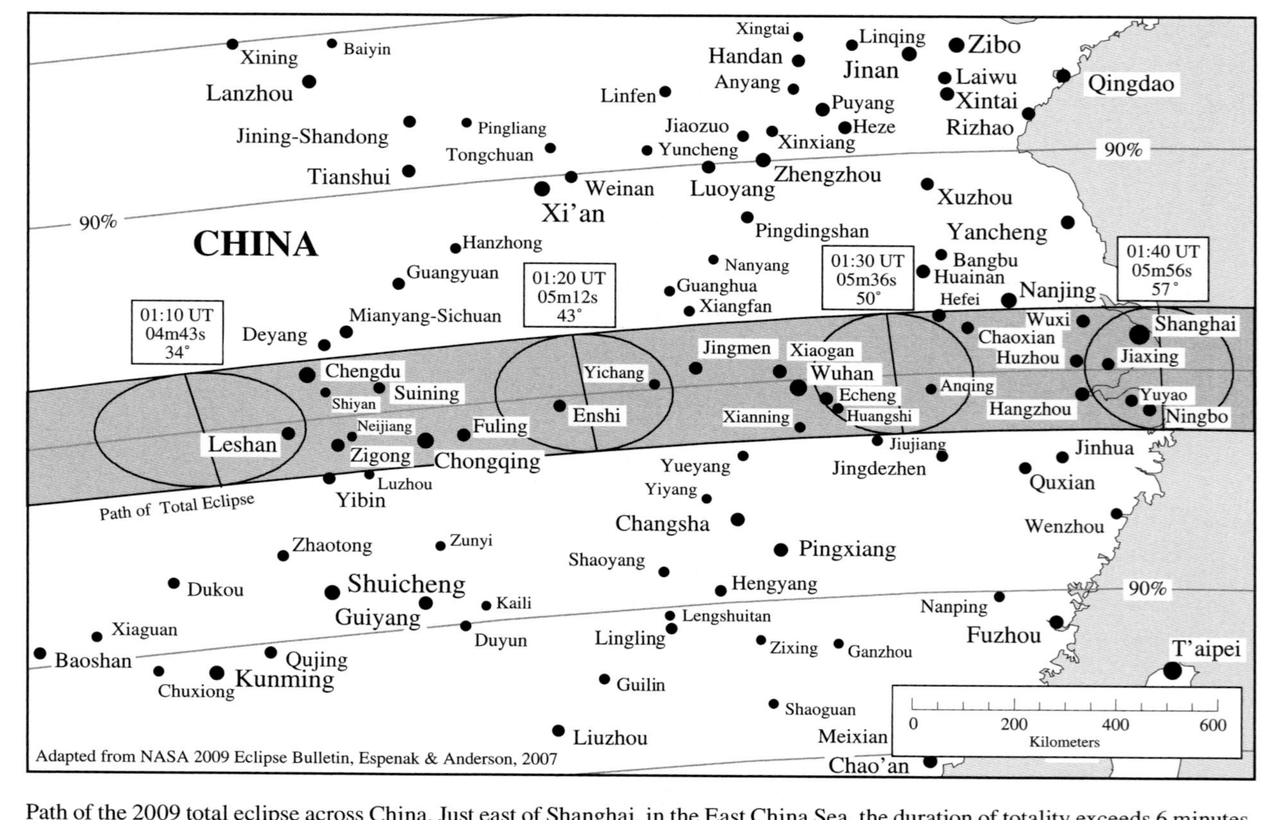

Path of the 2009 total eclipse across China. Just east of Shanghai, in the East China Sea, the duration of totality exceeds 6 minutes.

At that moment, the southern half of totality slices through the northernmost tip of Bangladesh, encountering no major cities. The central line of the eclipse track stays in Bangladesh for only 10 miles (16 kilometers), then is back in the twisted bottleneck of northeastern India for 70 miles (110 kilometers) until it reaches the mountainous nation of Bhutan.

The path of totality, here about 140 miles (225 kilometers) wide, engulfs almost the entire country, missing only the northern bulge. Thimphu, Bhutan's capital and largest city (50,000), gets 2 minutes 52 seconds of totality.

Meanwhile, south of the Himalayas, the southern half of totality glides along the northeasternmost wedge of India, with the Brahmaputra River along the southern edge of eclipse darkness. In places the Brahmaputra is more than 6 miles (10 kilometers) wide. The eclipse has followed the river upstream some 500 miles (800 kilometers). Then the eclipse reaches the eastern Himalayas, where the Brahmaputra comes roaring down from the mountains. The river began its journey on the northern side of the Himalayas in Tibet, flowed east for more than half its course, then made a U-turn down through gorges, including the deepest canyon on Earth, before spilling out onto the Indian plain, flowing west.

As the eclipse parts company with the river and heads into the mountains, it clips the northernmost tip of Burma (Myanmar).

River Running in China

The eclipse now enters China, soaring across the eastern ranges of the Himalayas. The dark shadow flies over the Yangtze River high in the mountains as it flows south from Tibet. Their paths will cross again soon. The central line of totality passes 10 miles (16 kilometers) south of the summit of Gongga Shan (also known as Minya Konka), which rises 24,790 feet (7,558 meters) to touch the shadow. Then, coming down from the mountains, the eclipse picks up the Yangtze River, which has turned east. The Yangtze is the fourth longest river in the world,[1] and the eclipse of 2009 follows it for the next 1,100 miles (1,800 kilometers)—with the river only occasionally and briefly wandering out of the path of totality.

Chengdu, Chongqing, and Wuhan are just three of the multi-million-people cities that will experience a midmorning extinguishing of the Sun. First to enjoy the shadow of the eclipse is Chengdu—maybe. It's one of the cloudiest cities on Earth. Then comes Chongqing, where the Jailing River joins the Yangtze. Halfway between Chongqing and Wuhan, the eclipse looks down upon the Three Gorges Dam, largest hydroelectric dam in the world. It is already producing electricity and controlling floods. A flood in 1954 killed 33,000. But the dam also threatens a number of animals whose habitat is changing from canyon and whitewater river to deep, still lake. Important archeological sites (and perhaps many still unknown) are being drowned. More than a million people have been displaced. The Yangtze once carried Himalayan silt to Shanghai and the sea, where it propped up the city and protected the coast from typhoons. That silt will now build up behind the dam. The Three Gorges Dam gathers 5 minutes 11 seconds of totality.

On down the river flows the eclipse, to Wuhan, where the Han River joins the Yangtze. From Chongqing to Wuhan: 11 minutes by eclipse; 4 days by river.[2] Wuhan is one of China's "Four Furnaces." Chongqing is another. They are centers of heavy manufacturing. But that's not where the title comes from. It's the summer heat and humidity. In Wuhan, the daytime highs in July *average* 99°F (37.2°C).

The eclipse ends its river cruise and transcontinental journey with visits to Hangzhou and Shanghai on the delta of the Yangtze River as it flows into the China Sea. Shanghai, with a population of 18.7 million, is the largest city on the 9,400-mile (15,200-kilometer) itinerary of this eclipse. The central line of the eclipse passes between Shanghai and Hangzhou, with totality lasting 5 minutes 52 seconds.

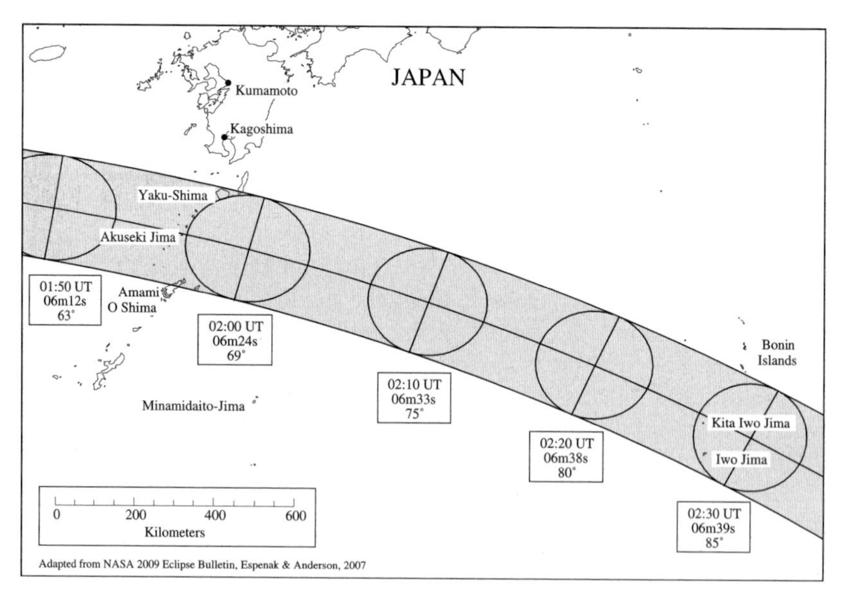

Japan's main islands lie outside the 2009 track of totality, but a few of its small southern islands are more fortunate. Yaku-Shima, Akuseki Jima, Amami O Shima, Kita Iwo Jima, and Iwo Jima enjoy a total eclipse.

Weather Prospects for 2009 Eclipse

The 2009 total eclipse occurs on July 22, which coincides with South Asia's summer monsoon season, bringing cloudy weather to much of the area. The average cloud amount in July along the eclipse track in India is 70% to 80%. It grows worse through the Himalayas (85% chance of clouds), but then quickly improves to 50% to 60% across China. Unfortunately, the weather prospects don't improve significantly as the shadow races across the Pacific.

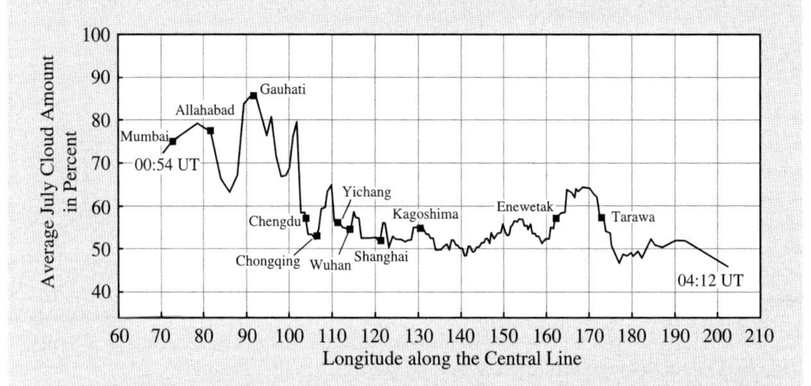

Average cloud amounts along the 2009 eclipse path. [Figure adapted from Fred Espenak and Jay Anderson: "Predictions for the Total Solar Eclipses of 2008, 2009 and 2010," *Proceedings of IAU Symposium 233: Solar Activity and Its Magnetic Origins* (Cambridge University Press, 2006)]

Island Hopping in the Pacific

And then the eclipse sets off across the China Sea, part of the Pacific Ocean, encountering only tiny specks of land for the rest of its time on Earth. It passes over the Ryukyu Islands of Japan, a chain of small islands that arc southwestward from the main islands toward Taiwan.

In the Ryukyus, the zone of totality includes, on the north, the island of Yaku Shima and, on the south, the northern part of the island of Amami O Shima, with the central line almost touching Akuseki Jima. Akuseki Jima is small in size—less than 4 square miles (10 square kilometers)—but receives a large dose of totality: 6 minutes 20 seconds.

For the eclipse of 2009, one last touchdown in Japanese territory remains—just as the eclipse nears its climax. In the Volcano Island group, 750 miles (1,200 kilometers) southeast of the main islands of Japan, the eclipse passes almost centrally over Ishinomura-Kitaio Island (or Kita Iwo Jima), with the face of the Sun disappearing for 6 minutes 34 seconds, just 5 seconds less than the maximum the 2009 eclipse will

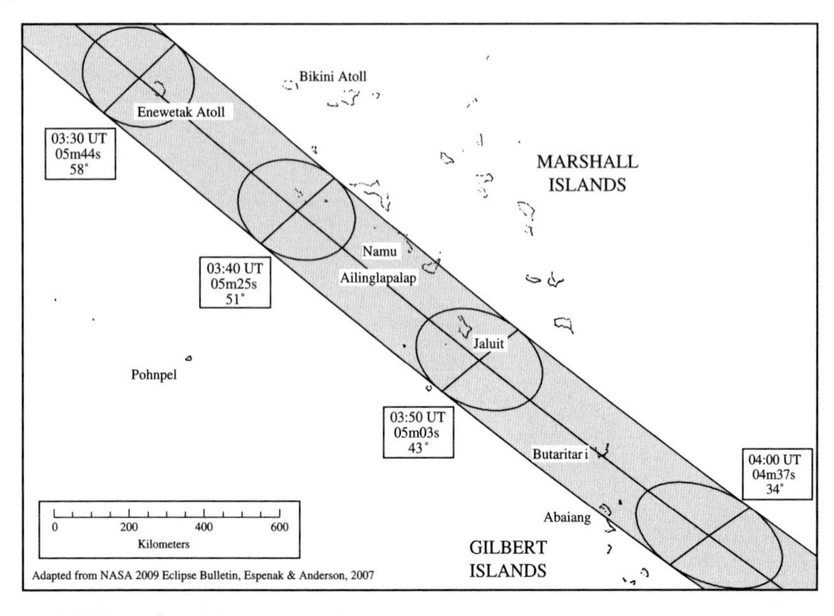

The 2009 total eclipse visits a few small islands and reefs of the Marshall and Gilbert Islands. The duration of totality is dropping but still about 5 minutes.

generate. At greatest eclipse the path of totality has expanded to 161 miles (258 kilometers) across—its widest. Fifty miles (80 kilometers) to the south of the central line, with 5 minutes 13 seconds of totality, lies Iwo Jima.

For the next 1,700 miles (2,800 kilometers), the only people who will see the total eclipse are those on ships or airplanes. Then the eclipse finds Enewetak Atoll in the Marshall Islands, scene of American nuclear weapons testing between 1948 and 1958. The eclipse is now moving across the globe faster and faster as it slides down the curvature of the Earth. Nine hundred miles (1,500 kilometers) farther and the eclipse reaches the Gilbert Islands (Kiribati) and land again—but again not much of it—as it rushes through Butaritari (which lies on the central line and receives 4 minutes 49 seconds of totality) and Marakei in the Gilbert Islands. A more famous island in the Gilberts, Tarawa, lies just south of totality.

More on the 2009 Total Solar Eclipse

The NASA Eclipse Website has additional details and maps for the 2009 total solar eclipse: <http://eclipse.gsfc.nasa.gov/SEmono/TSE2009/TSE2009.html>

A special NASA bulletin for the 2009 eclipse contains detailed maps, more weather information, and eclipse times for hundreds of cities. Access the NASA Eclipse Website: <http://eclipse.gsfc.nasa.gov/SEpubs/TP214169.html>.

Contact Times[a] for the 2009 Total Solar Eclipse

Country	*City*	*Partial Begins*	*Total Begins*	*Total Ends*	*Partial Ends*	*Max. Eclipse*	*Sun Alt.*	*Eclipse Mag.*	*Duration Totality*
India	Ahmadabad	—	—	—	07:20	06:24	3	0.973	
	Bhopal	—	06:22	06:25	07:23	06:24	7	1.063	03m09s
	Bombay	—	—	—	07:19	06:22	2	0.961	
	Calcutta	05:29	—	—	07:31	06:26	17	0.911	
	Delhi	—	—	—	07:25	06:27	9	0.852	
	Indore	—	06:22	06:25	07:22	06:23	6	1.063	03m05s
	Jabalpur	—	06:23	06:26	07:24	06:24	10	1.064	02m58s
	Patna	05:30	06:25	06:28	07:29	06:27	15	1.066	03m44s
	Surat	—	06:21	06:25	07:20	06:23	3	1.062	03m14s
	Varanasi	05:30	06:24	06:27	07:28	06:26	13	1.065	03m01s
Bangladesh	Dacca	05:59	—	—	08:04	06:58	20	0.930	
	Rangpur	06:00	06:57	06:59	08:04	06:58	19	1.067	02m19s
Nepal	Kathmandu	05:31	—	—	07:31	06:28	16	0.962	
Bhutan	Thimphu	05:31	06:28	06:31	07:35	06:29	20	1.068	02m52s
China	Beijing	08:25	—	—	10:45	09:32	49	0.730	
	Chaoxian	08:19	09:31	09:35	10:54	09:33	52	1.076	03m58s
	Chengdu	08:07	09:11	09:14	10:26	09:13	36	1.072	03m15s
	Chongqing	08:08	09:13	09:17	10:31	09:15	39	1.073	04m06s
	Enshi	08:10	09:17	09:22	10:36	09:19	42	1.074	05m05s
	Hangzhou	08:21	09:34	09:40	10:59	09:37	55	1.077	05m19s
	Huzhou	08:22	09:34	09:40	10:59	09:37	55	1.077	05m47s
	Jiaxing	08:22	09:35	09:41	11:00	09:38	56	1.077	05m51s
	Jingmen	08:13	09:21	09:26	10:42	09:23	45	1.074	05m08s
	Leshan	08:06	09:10	09:14	10:26	09:12	36	1.072	04m43s
	Nanjing	08:20	—	—	10:55	09:34	53	0.995	
	Ningbo	08:23	09:37	09:42	11:03	09:40	57	1.077	04m21s
	Shanghai	08:23	09:37	09:42	11:02	09:39	57	1.077	05m00s
	Suining	08:08	09:12	09:17	10:29	09:14	38	1.073	04m29s
	Suzhou	08:22	09:35	09:40	11:00	09:38	55	1.077	04m55s
	Wuhan	08:15	09:24	09:29	10:46	09:27	48	1.075	05m25s
	Wuhu	08:20	09:31	09:36	10:55	09:34	53	1.076	04m57s
	Wuwei	08:19	09:30	09:35	10:54	09:33	52	1.076	05m05s
	Wuxi	08:22	09:35	09:39	10:59	09:37	55	1.077	03m40s
	Xi'an	08:13	—	—	10:35	09:20	42	0.909	
	Xiaogan	08:15	09:23	09:29	10:45	09:26	47	1.075	05m24s
	Yuyao	08:23	09:36	09:41	11:02	09:39	56	1.077	05m02s
	Zigong	08:07	09:11	09:15	10:28	09:13	37	1.072	04m13s
Japan	Akuseki Jima	09:35	10:53	11:00	12:21	10:57	67	1.079	06m20s
	Iwo Jima	10:01	11:26	11:31	12:53	11:28	85	1.080	05m13s
	Kita Iwo Jima	10:00	11:24	11:30	12:52	11:27	84	1.080	06m34s
	Tokyo	09:56	—	—	12:30	11:13	73	0.747	
Marshall Is.	Enewetak	14:09	15:28	15:34	16:43	15:31	57	1.077	05m40s
	Jaluit	14:33	15:46	15:51	16:56	15:49	44	1.074	04m48s
Gilbert Is.	Butaritari	14:42	15:53	15:58	17:00	15:56	38	1.073	04m48s

[a] Contact times are given in Local Standard Time in each country.

Before the eclipse of 2009 can reach Tahiti, it ends its earthly sojourn and takes off into the heavens.

A Legacy

The eclipse of 2009 has the longest duration of totality of any eclipse in the twenty-first century. It will not be surpassed until the year 2132. Such an eclipse, of historic lineage, of great duration, of long over-land travels, deserves better weather than it is likely to receive. Its date coincides with the South Asia's summer monsoon season. In India, viewing prospects are poor. Weather conditions improve once the eclipse crosses the Himalayas into China, but not as much as eclipse viewers would hope. Along the eclipse path through China, the likelihood of clouds is 50% to 60%. Things are no better in the islands of southern Japan or in the Marshall or Gilbert Islands.

There is, alas, no place with highly favorable weather conditions for the July 22, 2009 total eclipse. China or Kita Iwo Jima off Japan, with durations of totality in the 6-minute range, are the most alluring of uncertain destinations.

The next time an eclipse of saros 136 visits Earth will be on August 2, 2027. Totality will be a little shorter, but still as long at 6 minutes 23 seconds. And the weather conditions will be better—but hot—in Spain, North Africa, and the Middle East.

Despite the weather prospects along the 2009 path of totality, this eclipse and the others in its saros series deserve special attention. No eclipse cycle in modern times has been so abundant in length of totality and no eclipse cycle ever has figured so powerfully in the history of science.

The next chapter traces the heritage of the 2009 total eclipse of the Sun—the story of saros series 136.

Notes and References

Epigraph: Warren De La Rue: "On the Total Solar Eclipse of July 18th, 1860, observed at Rivabellosa, near Miranda de Ebro, in Spain," *Philosophical Transactions* of the Royal Society, volume 52, 1862, page 333. De La Rue is referring to the total solar eclipse of July 18, 1860.

[1] The Yangtze is the third longest river on Earth if the combination Mississippi and Missouri Rivers are disqualified.

[2] Chongqing to Wuhan is about 500 miles (800 kilometers) air miles. It is about 800 miles (1,300 kilometers) on the circuitous, canyon-carving Yangtze. The eclipse shadow in central China is traveling about 2,500 mph (4,000 km/hr).

17
The Pedigree of an Eclipse

> [T]he general phaenomenon is perhaps the most awfully grand which man can witness.
>
> George B. Airy (1851)

Every eclipse belongs to a family, a saros series, that begins, evolves, and ends. Each saros series has a character all its own, with distinguishing features, the ability to captivate its viewers, and a chance to participate in human history. Here is the story of one remarkable family of eclipses.

The eclipse family known as saros 136 was born on June 14, 1360, deep in the southern hemisphere, over Antarctica and the southern Indian Ocean. The Moon was at a descending node—crossing the Sun's path on its way south. The Sun just happened to be near that node at the time so that, as viewed from Earth, the Moon grazed the southwestern edge of the Sun. It was a very slight partial eclipse, unnoticeable to the eye without filters, and there was no one there to see.

It was an inauspicious birth, but the firstborn of every saros family is always a slight partial eclipse that brushes the Earth at one of the poles and gives no visual evidence of the splendor that will come as the family matures.

Celestial Clockwork

Time passed. The years rolled by. Forty-one other solar eclipses touched the Earth, but they came from other saros families. Then, after 6,585 days, the Moon had completed 223 lunations and the Sun had passed by the descending node of the Moon 19 times. The two cycles matched almost exactly, forcing the Sun and Moon to meet under very nearly the same circumstances that prevailed 18 years 11⅓ days earlier. Another eclipse was inevitable.

But the new eclipse was not identical to the first. The Moon's node was not exactly where it had been 18 years earlier. It had backed around the Sun's orbit and was now 0.477 degrees east of its previous position, so this time the Moon cut off a slightly larger portion of the Sun's light. It was still a very slight (0.199) partial eclipse, noticeable only with optical filters in the south Atlantic Ocean and Antarctica, and again no one was there to see.

Each 18 years and a few days brought a new solar eclipse of saros family 136, each a little farther north on the whole; each a partial eclipse, but each time covering a little more of the Sun. During 126 years there were eight partial eclipses, until on August 29, 1486, the Moon's diameter covered 98.6% of the Sun's, leaving at maximum eclipse only a thin crescent of the Sun visible as seen from near the south pole.

The next eclipse in the cycle, September 8, 1504, was different from those that preceded it. The Moon passed directly across the disk of the Sun as seen from near the Antarctic coast. It would have been a total eclipse if the angular size of the Moon's disk had been large enough, but the Moon's elliptical orbit had carried it farther from the Earth than average, so it appeared smaller—too small to cover the Sun completely. At maximum eclipse, the Sun's surface still shone around the circumference of the Moon to form a bright ring of light—an annular eclipse—lasting at most 31 seconds.

Throughout the sixteenth century, each of the six eclipses of saros 136 was an annular eclipse, gradually migrating northward. And with each eclipse, the Moon was a little closer to Earth, so that the Moon's

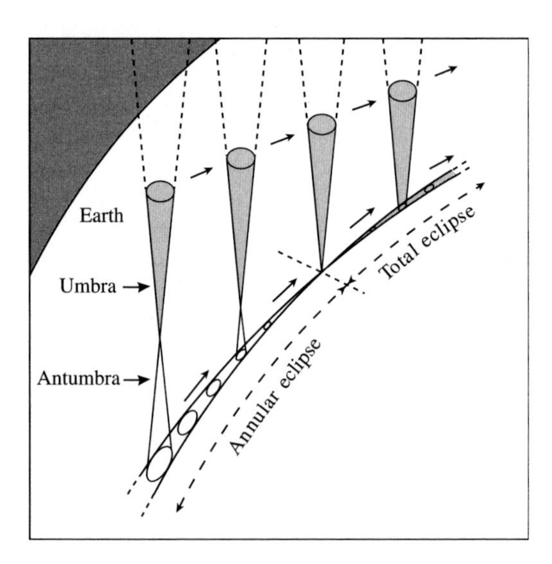

An occasional solar eclipse may start off annular but become total as the roundness of the Earth reaches up to intercept the shadow. The eclipse then returns to annular as the curvature of the Earth causes its surface to fall away from the shadow. Such eclipses are called hybrids or annular-total eclipses.

apparent size grew larger and the duration of the annulus got *shorter.* On November 12, 1594, the annular eclipse lasted 4 seconds at its midpoint. The disk of the Moon was almost large enough to completely hide the Sun.

At the next eclipse, it did—for just *1 second.* On November 22, 1612, in the southeastern Pacific Ocean and Antarctica, birds, fish, and whales saw an eclipse that was annular all along the central path until

The Extinction of Total Solar Eclipses

In 1695, Edmond Halley discovered that eclipses recorded in ancient history did not match calculations for the times or places of those eclipses. Starting with records of eclipses in his day and the observed motion of the Moon and Sun, he used Isaac Newton's new theory of universal gravitation (1687) to calculate when and where ancient eclipses should have occurred and compared them with eclipses actually observed more than 2,000 years earlier. They did not match. Halley had great confidence in the theory of gravitation and resisted the temptation to conclude that the force of gravity changed as time passed. Instead, he proposed that the length of a day on Earth must have increased by a small amount.

If the Earth's rotation had slowed down slightly, the Moon must have gained the angular momentum to conserve the total angular momentum of the Earth–Moon system. This boost in angular momentum for the Moon would have caused it to spiral slowly outward from the Earth to a more distant orbit where it travels more slowly. If, 2,000 years earlier, the Earth had been spinning a little faster and the Moon had been a little closer and orbiting a little faster, then eclipse theory and observation would match. Scientists soon realized that Halley was right.

But what would cause the Earth's spin to slow? Tides. The gravitational attraction of the Moon is the principal cause of the ocean tides on Earth. As the shallow continental shelves (primarily in the Bering Sea) collide with high tides, the Earth's rotation is retarded. The slower spin of the Earth causes the Moon to edge farther from our planet.

From 1969 to 1972, the Apollo astronauts left a series of laser reflectors on the Moon's surface. Since then, scientists on Earth have been bouncing powerful lasers off these reflectors. By timing the round trip of each laser pulse, the Moon's distance can be measured to an accuracy of several inches. The Moon is receding from the Earth at the rate of about 1.5 inches (3.8 centimeters) a year.

As the Moon recedes from Earth, its apparent disk becomes smaller. Total eclipses become rarer; annular eclipses more frequent. Total eclipses are moving toward extinction. When the Moon's mean distance from the Earth has increased by 14,550 miles (23,410 kilometers), the Moon's apparent disk will be too small to cover the entire Sun, even when the Moon's elliptical orbit carries it closest to Earth. Total eclipses will no longer be possible.

How long will that take? With the Moon receding at 1½ inches a year, the last total solar eclipse visible from the surface of the Earth will take place 620 *million* years from now. There is still time to catch one of these majestic events.

the eclipse reached its midpoint, where the surface of the Earth was closest to the Moon and hence the Moon appeared largest. At that location, the Moon's disk for just an instant completely covered the Sun and the eclipse (technically, at least) became total. The dark shadow of the Moon caused by saros 136 for the first time actually touched the Earth. Then, as modestly as a young boy and girl touching hands for the first time, the shadow and the Earth pulled apart as the Earth's surface curved away beneath the extended shadow. Just the slightly greater distance from the Moon that the Earth's curvature provided was enough to make the Moon once again appear too small to completely cover the Sun. The Sun peered out from around the entire circumference of the Moon, and the eclipse was annular once more.

The eclipse of 1612 and all five eclipses of saros 136 in the seventeenth century were hybrids of this kind: beginning as annular, becoming total, then returning to annular again. And with each eclipse, the duration of totality was a little longer. The eclipse of January 17, 1703 was a hybrid too, but of a rare sort. It started total and spent almost all its time on Earth as total, slipping into annular only at the end. Totality now lasted as long as 50 seconds.

The Big Leagues

On the twenty-first eclipse of saros 136, the Earth was near its closest to the Sun, so the Sun's angular size was greatest and it was hardest to eclipse. But the Moon in its orbit was close enough to Earth so that its apparent size was larger than average. Thus, finally, on January 27, 1721, the eclipse was total all along the central path, with totality lasting a maximum of 1 minute 7 seconds.

Each subsequent eclipse of saros 136 was total, and the duration of totality was rapidly increasing. By 1793 the eclipse lasted a maximum of 2 minutes 52 seconds over the Indian Ocean and provided a shorter show for the Aborigines in Australia and the newly arrived British settlers. The eclipse path was shifting northward. As the American Civil War ended, saros 136 crossed the 5-minute plateau into the realm of rare and great eclipses as it stretched from South America to southern Africa. The totality of April 25, 1865 lasted 5 minutes 23 seconds.

Now almost everything was conspiring to make saros 136 one of the greatest in history. The eclipses were occurring later and later in the spring, when the Sun was farther from Earth and hence smaller in apparent size, enabling total eclipses to last longer. At each eclipse, the Moon was getting closer and closer to perigee (its position closest to earth), so that the Moon was near a maximum apparent size, allowing

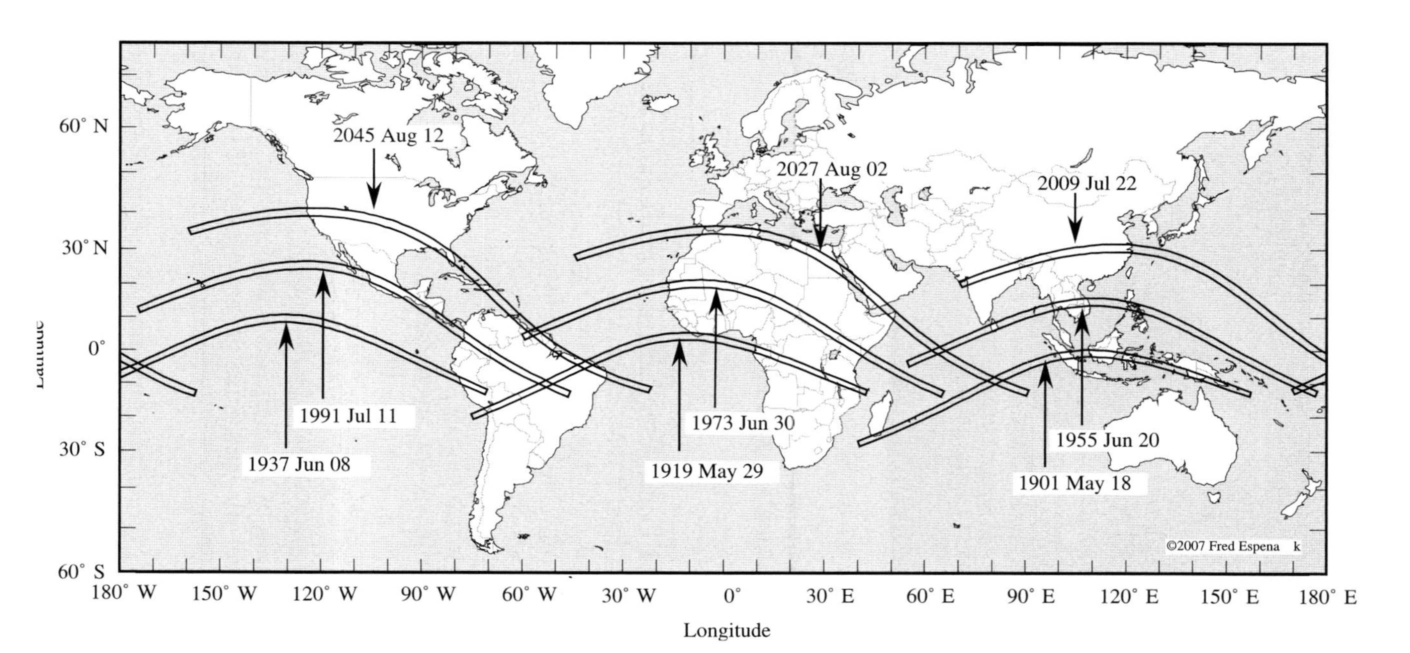

Paths of totality for six past and three future eclipses of saros 136. Successive eclipses shift westward and northward. For odd-numbered saroses, eclipses shift westward and southward. [Map and eclipse calculations by Fred Espenak]

In its most recent visit before 2009, saros 136 produced the total solar eclipse of July 11, 1991. Totality lasted almost 7 minutes in Baja, Mexico. [Nikon FE SLR, Celestron C-90 Maksutov, fl = 1000 mm, f/11, composite image with exposures of 1/125 s to 1 s, Kodachrome 64 slide film. © 1991 Fred Espenak]

total eclipses to last longer. Every eclipse of saros 136 in the twentieth century would last longer than 6 minutes. And the twentieth century was prepared to make good use of this rare gift.

The eclipse of May 18, 1901 brought astronomers from all over the world to Indonesia to continue their analysis of the solar atmosphere, still best seen inside the shadow of the Moon. They had at most 6 minutes 29 seconds within which to work.

The next eclipse in the cycle lasted even longer—6 minutes 51 seconds at its peak, a totality long enough to make it memorable in and of itself. But that was not why eclipse number 32 in saros 136 became the most famous in history. Astronomers used that eclipse to measure star positions around the darkened Sun and concluded that starlight passing by the Sun had been bent, just as Einstein had predicted in his recently completed general theory of relativity. Thus the total eclipse of May 29, 1919 marked a major turning point in the history of science.

Saros 136, however, had its own dramatic schedule to fulfill, independent of the brilliance or depravity of man. It returned on June 8, 1937, as the clouds of war darkened over Europe and the Far East, with a 7 minute 4 second eclipse—the first to last over 7 minutes since 1098. Unfortunately, this amazing sight passed almost entirely across the expanse of the Pacific Ocean, avoiding land where scientists could deploy their instruments.

Saros 136 reached its climax on June 20, 1955, with an eclipse that, at maximum, lasted 7 minutes 8 seconds, the longest since June 20, 1080 (7 minutes 18 seconds), and the longest until June 25, 2150 (7 minutes 14 seconds). The shadow visited Sri Lanka, Thailand, Indochina, and the Philippines. Never again would saros 136 equal that duration of totality. Yet few saroses even come close.

June 30, 1973 brought an eclipse, visible across Africa, that lasted as long as 7 minutes 4 seconds.[1] The last offering of saros 136 in the twentieth century happened on July 11, 1991, bringing up to 6 minutes 54 seconds of totality and tracing a path from Hawaii through Mexico, Central America, Colombia, and Brazil.

The six longest eclipses of the twentieth century were in 1901, 1919, 1937, 1955, 1973, and 1991—every one of them a member of the same eclipse family. Saros 136 is one of the greatest eclipse generation cycles in recorded history.

Closing Out a Career

Now, gradually, the glory is beginning to fade. Saros 136 is declining. The Moon is receding from perigee, making its disk appear smaller; it will not cover the Sun as well. The eclipses are occurring ever later in the year, when the Earth is actually closer to the Sun so that its disk appears larger and harder to eclipse. Eclipses will become steadily shorter, but they will still be total and will still be comparatively long for quite some time. The glory will fade slowly.

Saros 136 did not end in 1991. It returns on July 22, 2009, with a 6-minute-39-second display and a central path across India, China, and the western Pacific. The total solar eclipse of 2009 is the longest of the twenty-first century.

Saros 136 will be back on August 2, 2027, racing over North Africa, Saudi Arabia, and the Indian Ocean, bringing darkness for up to 6 minutes 23 seconds. It will pay its first visit to the continental United States on August 12, 2045, streaking from northern California through Florida, the longest totality for the continental United States in the calculated history of eclipses. Over the Caribbean, it will last for 6 minutes 6 seconds. But that will be the last eclipse of saros 136 to surpass the 6-minute mark.

Three saros cycles (54 years) later, on September 14, 2099, it will return to North America, slicing across southwestern Canada and plunging southeastward across the United States to the mouth of the Chesapeake Bay and off into the Atlantic. At maximum, east of the Windward Islands, the eclipse will last 5 minutes 18 seconds.

Saros 136 is aging, but its eclipses are still total and more than long enough to lure people by the thousands into its shadows. By May

The path of the Moon's shadow on Earth during the total solar eclipse of July 11, 1991, captured in this composite of 8 NASA GOES-7 weather satellite images. [Courtesy of William Emery and Chuck Fowler, University of Colorado]

13, 2496, however, a total eclipse of saros 136 will have dwindled to 1 minute 2 seconds for hardy travelers in the Arctic. It is the last of 44 total eclipses in this remarkable saros family.

At the next visit of this saros, May 25, 2514, the dark shadow of the Moon will miss the Earth, passing above the north polar region. The remaining six eclipses in the sequence will all be partial as well, steadily declining in the Moon's coverage of the face of the Sun. On July 30, 2622, there will be one final partial eclipse, virtually unnoticeable near the north pole. Saros 136 will have died.

Saros number 136 will have performed on Earth for 1,262 years and created 71 solar eclipses. Not an exceptionally long career, but what a record! Of its 71 eclipses, 15 will have been partial, 6 annular, 6 hybrid (combination annular and total), and *44 total*. The typical saros offers an average of only 19 or 20 total eclipses.

Saros 136 brought the people of the twentieth century three eclipses with totality exceeding 7 minutes and all six of the longest eclipses in that century. Its three eclipses in the first half of twenty-first century will all exceed 6 minutes, the longest eclipses to be seen in that span of time. If this saros were an athlete, its shadow-black jersey with its corona-white number 136 would be retired to hang in glory in the Eclipse Hall of Fame.

While not as spectacular as a total eclipse, annular eclipses are still fascinating events. An image of the annular eclipse of May 31, 2003 through thick clouds from Olafsfjördur, Iceland. [Nikon 8008 SLR, Vixen 80 mm fluorite refractor, fl = 640 mm, TeleVue 2× barlow, efl = 1720 mm, no solar filter, f/21, 1/250 s, Kodak Royal Gold 200 negatives, ©2003 Fred Espenak]

A composite image of the partial eclipse of May 31, 2003 was captured on a single piece of film from Bergen, Norway. [Canon EOS Rebel G (500N) DSLR, Canon EF 28–105 mm at 65 mm, f/1.8, 1/60 s, Fuji Sensia 100, ©2003 Odd Høydalsvik]

A multiple exposure captures phases of the annular solar eclipse of October 3, 2005 from Madrid, Spain. [Olympus OM-1 SLR, Zuiko 300 mm f/4.5, Baader solar filter, f/8 filtered, f/22 unfiltered, 1/125 s, Fuji Provia 100F, ©2005 Cees Bassa]

A sunrise in partial eclipse was seen from Les Roches, Guyana during the annular eclipse of September 22, 2006. [35 mm SLR, 100 mm Maksutov telescope, fl = 1000 mm, f/10, no solar filter, Konica/Minolta ISO 100 negative, ©2006 Daniel Fischer]

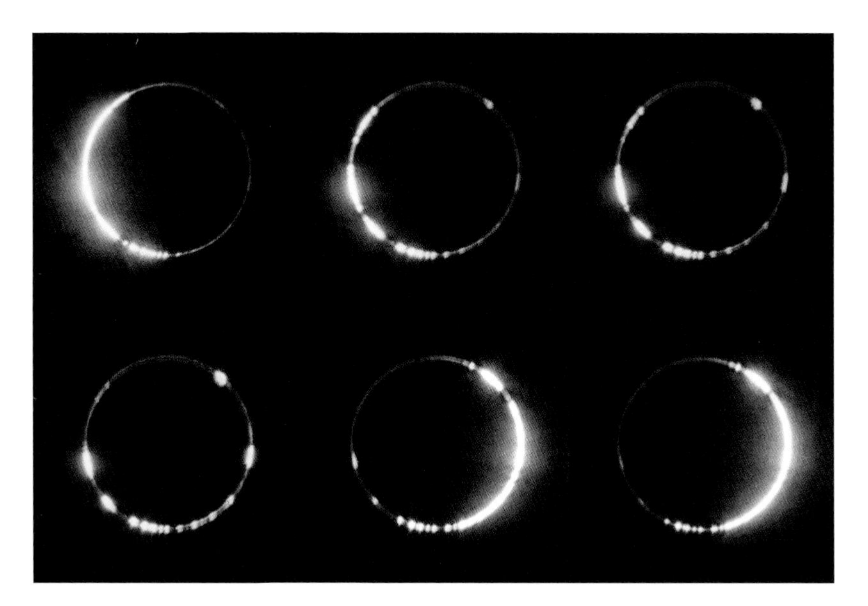

A unique beaded annular eclipse is seen from a small aircraft 44,000 feet above Iceland during the annular eclipse of October 3, 1986. This sequence of six images was shot in rapid succession and covers just 6 seconds. [Nikon FM2 SLR, 400 mm f/6.3, f/22, 1/500 s, Kodacolor 100 negatives, ©1984 Glenn Schneider]

The Moon's disk was surrounded by a diamond necklace during the beaded annular eclipse of May 30, 1984 from Pendleton, South Carolina. At beaded annular eclipses, the Moon is very nearly the same size as the Sun so the Moon's higher mountains break the annular ring into a series of beads and crescent segments. [35 mm SLR, 3-inch f/15 Jaegers refractor, fl = 1140 mm, no solar filter, f/15, 1/500 s, ISO 64 slide film, ©1984 Johnny Horne]

Eclipse chasers aboard the MS *Paul Gauguin* experienced a real cliff-hanger. Hidden behind thick clouds, the Sun emerged into view just as the diamond ring formed at second contact. The total solar eclipse of April 9, 2005 was observed from the South Pacific west of the International Data Line. [Canon EOS 20D DSLR, 10–20 mm lens at 10 mm, f/4, 1/4 s, ISO 100, ©2005 Alan Dyer]

Shooting the April 8, 2005 eclipse from the rolling deck of the MV *Galapagos* was a challenge. An image-stabilized lens made it possible to capture enough frames to produce a composite image of the corona in Photoshop. [Canon EOS 1Ds Mark II DSLR, Canon 500 mm f/4 IS lens, composite of 7 exposures: 1/1000 s to 1/15 s, ISO 400 ©2005 Fred Espenak]

The ship MV *Galapagos* brought 90 passengers into the narrow mid-Pacific path of the total solar eclipse of April 8, 2005. Totality lasted a scant 29 seconds. [Canon EOS 300D DSLR, Canon EF-S 10–22 mm zoom, set at 22 mm, f/4.5, 1/25 s, ISO 800 ©2005 Olivier Staiger]

Earthshine is revealed on the dark side of the Moon in a digital composite image of the March 29, 2006 total eclipse from Side, Turkey. [Canon 10D DSLR, Skywatcher 80ED Pro f/7.5 refractor, fl = 600 mm, ©2006 Pete Lawrence]

Temple of Apollo and the 2006 total eclipse were captured in a multiple exposure sequence from Side, Turkey [Olympus C-5060 digital, 6 mm, f/2.8, partials 1/125 s, totality 1/30 s, Baadar solar filter for partials, ©2006 Geoff Sims]

The diamond ring effect just before second contact was photographed from Libya during the total solar eclipse of March 29, 2006. [Nikon D200 DSLR, 170–500 mm zoom at 500 mm, f/5.0, 0.8 s, ISO 400, ©2006 Patricia Totten Espenak]

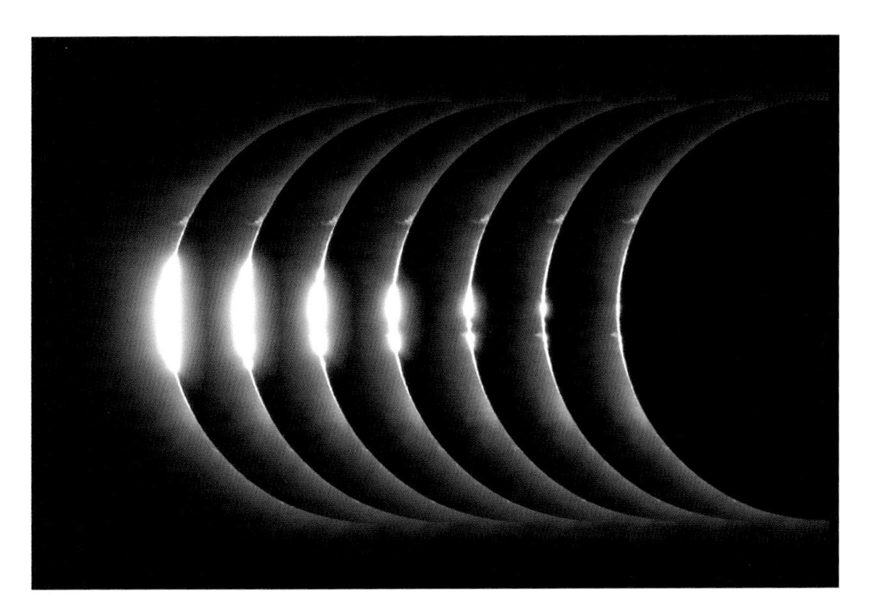

A time sequence of seven images captures the break-up of the crescent Sun into Baily's beads immediately preceding second contact. The total eclipse of March 29, 2006 was photographed from Jalu, Libya. [Nikon D200 DSLR, Vixen 90 mm fluorite refractor, fl = 810 mm, f/9 1/1000 s, ISO 200, ©2006 Fred Espenak]

A long trek across Niger's Sahara Desert was required for this splendid view of prominences and chromosphere which was shot just after second contact during the 2006 total eclipse. [Hasselblad 2 ¼ format, 152 mm Maksutov, fl = 2,900 mm, Astro-Physics 900GTO mount, f/19, 1/4 s, Kodak Royal Gold 25 (ASA 25), ©2006 Jacques Guertin]

Details in the corona are revealed in this Photoshop composite that was processed to emphasize fine structure. The images were obtained during the 2006 eclipse from the Sahara Desert in northern Libya. [Nikon D200 DSLR, Vixen 90 mm fluorite refractor, fl = 820 mm, f/9, composite of 22 exposures: 1/1000 s to 1 s, ISO 200, ©2006 Fred Espenak]

A multiple exposure sequence captures the partial phases and totality during the March 29, 2006 total eclipse from Side, Turkey. [35 mm SLR, 17 mm f/2.8 Peleng fisheye, f/5.6, exposures: 1/500 s for partials, 3 s for totality, Fuji Superia 100, ©2006 Tunç Tezel]

The Eclipse Family of Saros 136

Number		*Date*	*Type*	*Duration*[a]
1	1360	June 14	Partial	[0.050]
2	1378	June 25	Partial	[0.199]
3	1396	July 5	Partial	[0.346]
4	1414	July 17	Partial	[0.489]
5	1432	July 27	Partial	[0.626]
6	1450	August 7	Partial	[0.757]
7	1468	August 18	Partial	[0.876]
8	1486	August 29	Partial	[0.986]
9	1504	September 8	Annular	0:31
10	1522	September 19	Annular	0:23
11	1540	September 30	Annular	0:17
12	1558	October 11	Annular	0:12
13	1576	October 21	Annular	0:08
14	1594	November 12[b]	Annular	0:04
15	1612	November 22	Hybrid	totality: 0:01
16	1630	December 4	Hybrid	totality: 0:07
17	1648	December 14	Hybrid	totality: 0:15
18	1666	December 25	Hybrid	totality: 0:24
19	1685	January 5	Hybrid	totality: 0:35
20	1703	January 17	Hybrid	totality: 0:50
21	1721	January 27	Total	1:07
22	1739	February 8	Total	1:28
23	1757	February 18	Total	1:52
24	1775	March 1	Total	2:20
25	1793	March 12	Total	2:52
26	1811	March 24	Total	3:27
27	1829	April 3	Total	4:05
28	1847	April 15	Total	4:44
29	1865	April 25	Total	5:23
30	1883	May 6	Total	5:58
31	1901	May 18	Total	6:29
32	1919	May 29	Total	6:51
33	1937	June 8	Total	7:04
34	1955	June 20	Total	7:08
35	1973	June 30	Total	7:04
36	1991	July 11	Total	6:53
37	2009	July 22	Total	6:39
38	2027	August 2	Total	6:23
39	2045	August 12	Total	6:06
40	2063	August 24	Total	5:49
41	2081	September 3	Total	5:33
42	2099	September 14	Total	5:18
43	2117	September 26	Total	5:04
44	2135	October 7	Total	4:50
45	2153	October 17	Total	4:36
46	2171	October 29	Total	4:23
47	2189	November 8	Total	4:10
48	2207	November 20	Total	3:56

The Eclipse Family of Saros 136 (*Continued*)

Number		*Date*	*Type*	*Duration*[a]
49	2225	December 1	Total	3:43
50	2243	December 12	Total	3:30
51	2261	December 22	Total	3:17
52	2280	January 3	Total	3:04
53	2298	January 13	Total	2:52
54	2316	January 25	Total	2:42
55	2334	February 5	Total	2:32
56	2352	February 16	Total	2:24
57	2370	February 27	Total	2:17
58	2388	March 9	Total	2:10
59	2406	March 20	Total	2:03
60	2424	March 31	Total	1:55
61	2442	April 11	Total	1:46
62	2460	April 21	Total	1:34
63	2478	May 3	Total	1:21
64	2496	May 13	Total	1:02
65	2514	May 25	Partial	[0.952]
66	2532	June 5	Partial	[0.824]
67	2550	June 16	Partial	[0.685]
68	2568	June 26	Partial	[0.544]
69	2586	July 7	Partial	[0.397]
70	2604	July 19	Partial	[0.252]
71	2622	July 30	Partial	[0.105]

Breakdown of solar eclipses in saros 136:

Total	44
Annular	6
Hybrid (annular/total)	6
Partial	15
Solar eclipses in saros 136:	71
Span of saros 136:	1,262 years

Source: NASA Saros 136 web page: <http://eclipse.gsfc.nasa.gov/SEsaros/SEsaros136.html>.

Note: Saroses with even numbers occur at the descending node; they start near the south pole and progress northward. Saroses with odd numbers occur at the ascending node; they start near the north pole and progress southward.

[a] Duration of totality or annularity is given in minutes:seconds. If eclipse is partial, its magnitude (fraction of the Sun's diameter covered by the Moon) appears in brackets.

[b] Henceforth on the Gregorian calendar, which replaced the Julian calendar in 1582 and dropped 10 days from that year to keep March at the beginning of spring.

Saros Series Statistics

	Range	*Average*
Number of solar eclipses in a saros series	69–87	72
Timespan for a series	1,226–1,551 years	1,280 years

At any time, 42 saros series are running simultaneously.

Notes and References

Epigraph: George B. Airy: "On the Total Solar Eclipse of 1851, July 28," in Bernard Lovell, editor: *Astronomy*, volume 1, The Royal Institution Library of Science (Barking, Essex: Elsevier Publishing, 1970), page 1.

[1] Michael Rogers wrote about the eclipse of June 30, 1973 for the general public in "Totality—A Report," published in *Rolling Stone* magazine on October 11, 1973. Some of the scientific information is wrong, but Rogers does a remarkable job of capturing the excitement of a total solar eclipse. For this article, he won the AAAS-Westinghouse Science-Writing Award.

18
The Eclipse of July 11, 2010

> The appearance of the corona, shining with a cold unearthly light, made an impression on my mind which can never be effaced, and an involuntary feeling of loneliness and disquietude came upon me.
>
> John Couch Adams (1852)

Eclipses almost seem to have personalities of their own. The eclipse of 2009 was an extrovert, seeking audiences in India and China, the two most populous countries in the world. By contrast, the total solar eclipse of July 11, 2010 prefers its privacy. Almost its entire course lies across the South Pacific Ocean, with only a few islands, mostly atolls, along the path of totality.

The most notable exception is Easter Island, a lonely volcanic plot in the middle of the Pacific Ocean that is over 2,000 miles (3,200 kilometers) from Chile and Tahiti, the nearest centers of population.

When the eclipse finally reaches a continent—South America—it comes ashore near the remote southern tip, where the transcontinental distance has narrowed to 300 miles (500 kilometers). There, in the midst of southern hemisphere winter, the eclipse seeks out a national wilderness park in Chile, hops over the tail end of the Andes Mountains into Argentina, admires a spectacular mountain range and three glacier-fed lakes, tires of Earth, and leaves.

Too bad. This is an impressive eclipse if you can manage to get into its path. At its peak—in the Pacific Ocean 1,000 miles (1,600 kilometers) from any island—totality lasts 5 minutes 20 seconds. The path of totality is an ample 161 miles (259 kilometers) in width.

Darkening Polynesian Waters

On July 11, 2010, the eclipse that tries not to be seen begins in the South Pacific Ocean 1,100 miles (1,800 kilometers) northeast of New Zealand and almost 500 miles (800 kilometers) southeast of Tonga. There is no island closer.

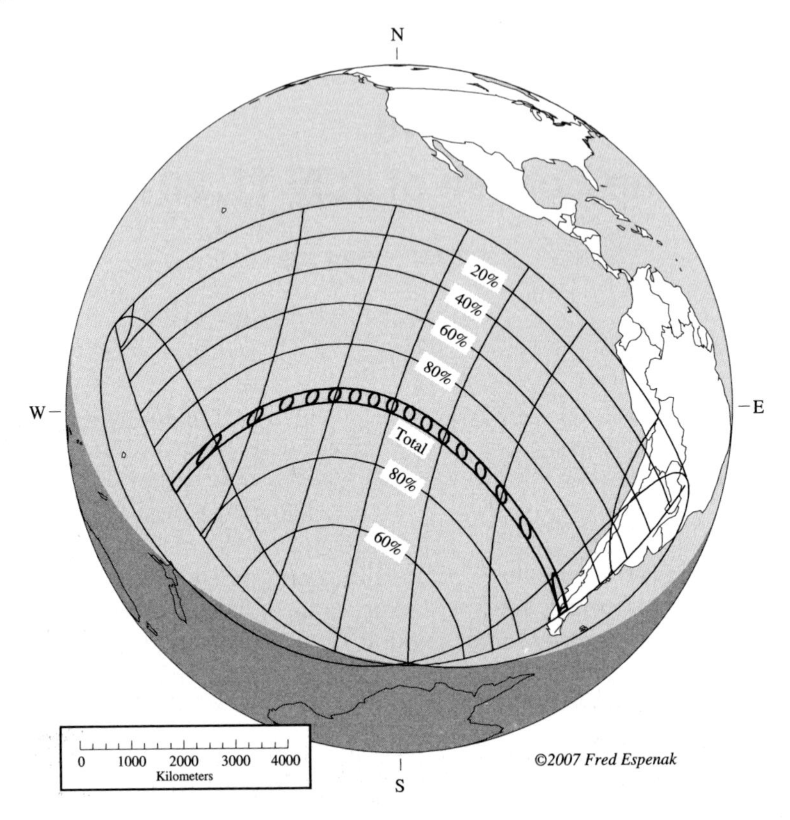

Path of the July 11, 2010 total solar eclipse—mostly over the South Pacific Ocean. The eclipse magnitude (fraction of the Sun's diameter eclipsed) for places outside the path of totality is given in percent. [Map and eclipse predictions by Fred Espenak]

The stealth eclipse rushes northeastward for almost 1,000 miles before it stumbles on an island, Mangaia, southernmost of the Cook Islands, bringing 3 minutes 18 seconds of totality. Mangaia is shaped like an egg, measures about 5 by 6 miles (8 by 10 kilometers) across, and has a population of about 500.

Six hundred miles (1,000 kilometers) farther, the eclipse dodges Tahiti, with the shadow edge passing about 50 miles (80 kilometers) south of its city of Papeete. What? Be seen by the 170,000 people on the island? Not this eclipse.

From Mangaia to Hikueru, the next inhabited island that the eclipse enshadows, the distance is almost 1,100 miles (1,800 miles). Hikueru is an eye-shaped coral atoll in the Tuamotu Islands—9 miles (15 kilometers) across, mostly lagoon. Hikueru once had prosperous pearl oyster beds, famed for their black pearls. Only about 100 people remain on the island today, but they get 4 minutes 20 seconds of totality.

The eclipse flows across several uninhabited atolls in the Tuamotu chain until, 275 miles (450 kilometers) from Hikueru, it encounters

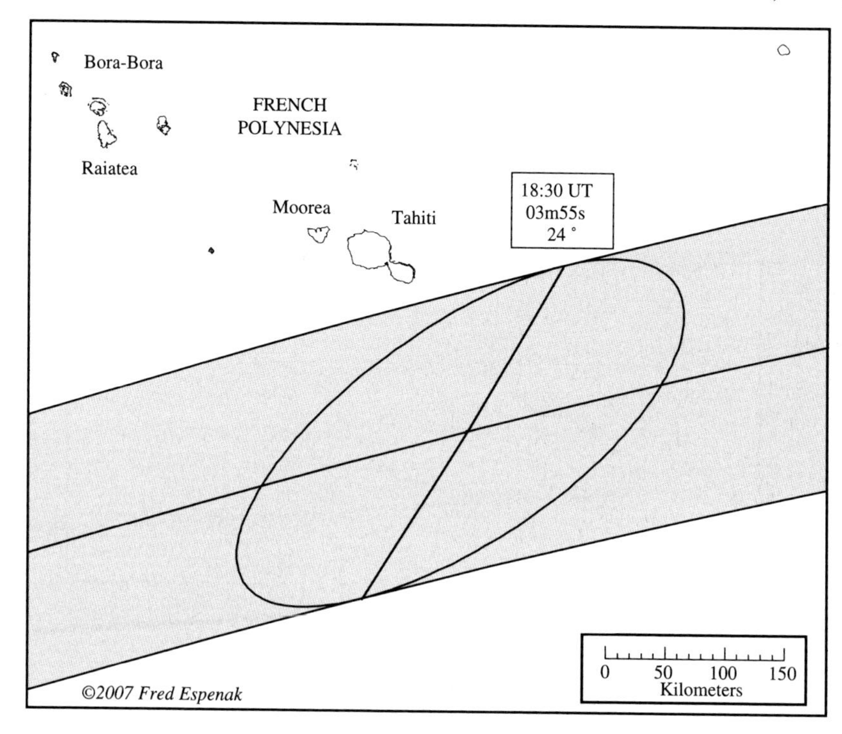

The path of the 2010 total eclipse just misses Tahiti. Cruise ships will provide an opportunity to observe totality. [Map and eclipse predictions by Fred Espenak]

another island with a human population, Tatakoto Atoll. Tatakoto is shaped like a winking eye, about 8 miles (14 kilometers) long by 2 miles (3 kilometers) wide. The atoll has a few dozen residents to greet 4 minutes 22 seconds of total eclipse.

Not for another 2,000 miles (3,200 kilometers) does the 2010 eclipse encounter land, but when it does, it's Easter Island (Isla de Pascua), a kindred soul to the eclipse because it too dwells in solitude. Easter Island is one of the most isolated places on Earth.

On its journey between Takakoto and Easter Island—halfway in between—the eclipse of 2010 celebrates eclipse maximum—by itself, of course—in the middle of the Pacific, shedding 5 minutes 20 seconds of totality on no one, unless they sneak into the shadow track in a ship or plane.

Totality for Easter Island

But what a reception awaits the 2010 eclipse on Easter Island, where eclipse observers from all over the world will converge, likely

Weather for the 2010 Eclipse

Weather prospects for the total solar eclipse of 2010 are only so-so. For the atolls of French Polynesia, the likelihood of clouds is about 45%. For Easter Island, cloudiness averages 56% on July 11. Along the eclipse path in Chile and Argentina, the probability of clouds is 70–80%, with 55% at El Calafate.[a]

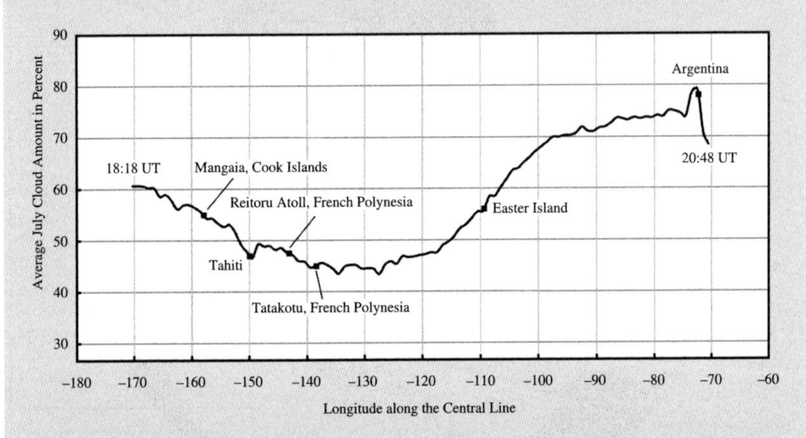

Average cloud amount along the 2010 eclipse path. [Figure adapted from Fred Espenak and Jay Anderson: "Predictions for the Total Solar Eclipses of 2008, 2009 and 2010," *Proceedings of IAU Symposium 233: Solar Activity and Its Magnetic Origins* (Cambridge University Press, 2006)].

Fans of eclipse cruises will find Tahiti an attractive destination with the best weather prospects along the path. For those who wish to observe the 2010 eclipse from land, Easter Island is the favored locale. El Calafate, Argentina is a long-shot second choice because the Sun's altitude is so low.

[a] Weather information provided by Jay Anderson.

outnumbering the Easter Islanders themselves, which is not hard because there are only about 3,800 of them.

Easter Island was created by three major (now extinct) volcanoes. The volcanoes stand at the corners of this triangular island whose dimensions are 14 by 7 miles (23 by 11 kilometers). The largest volcanic craters are a mile (1½ kilometers) wide. Dozens of smaller ones pock the landscape. Almost the entire Easter Island population lives in one town next to the international airstrip, whose runway is nestled between volcanic craters. NASA lengthened the runway so Space Shuttle astronauts can use it in an emergency.

Easter Island is most famous for its statues—the hundreds of giant stone faces and torsos that stand on the hillsides all around the island.

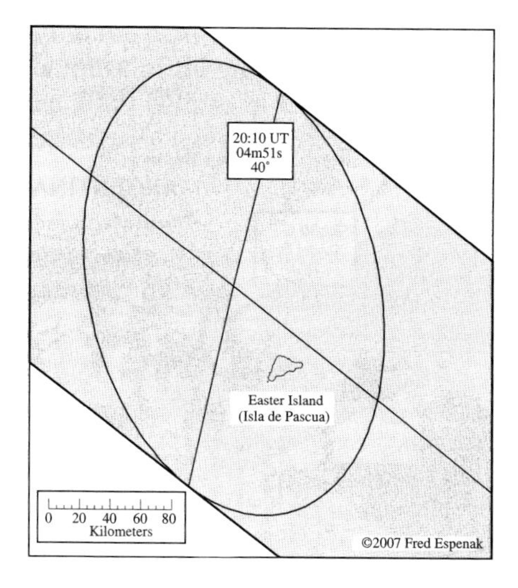

Easter Island is completely enveloped by the 145-mile-wide (233-kilometer-wide) eclipse path. The remote island and its mysterious statues will be a prime destination for 2010 eclipse chasers. [Map and eclipse predictions by Fred Espenak]

They average about 13 feet (4 meters) tall and were probably intended as images of the gods. The largest one erected outside the quarry where the stones were mined is almost 33 feet (10 meters) high and weighs about 82 tons (74 metric tons).[1] The *moai* were built by the advanced seafaring Polynesians who used the Sun and stars and ocean swells to navigate the Pacific and who settled on Easter Island perhaps 1,200 years ago.[2]

Eclipse followers will observe the eclipse standing side by side with the great stone faces. The people on Easter Island will probably be the largest population to see the 2010 total eclipse of the Sun.

Eclipse at Sunset

Then again nothing—no land—for 2,500 miles (4,000 kilometers), until the eclipse, heading southeast, reaches the coast of South America, in far southern Chile. There the Sun, about to set, is darkened prematurely as the eclipse enfolds the vast wilderness of Bernardo O'Higgins National Park. Here lies the South Patagonian Ice Field. No cities. No towns. Accessible only by boat and helicopter.

The eclipse leaps from Chile into Argentina where the Andes Mountains feature a range of peaks of stunning, barren grandeur. Highest among them is Mount Fitz Roy, also called El Chaltén, "smoking mountain," because of frequent clouds streaming across its peak. Mount Fitz Roy is a sheer granite arrowhead rising 11,072 feet (3,405 meters) above sea level and more than a mile (1,800 meters) almost

Contact Times[a] for the 2010 Total Solar Eclipse

Country	*City*	*Partial Begins*	*Total Begins*	*Total Ends*	*Partial Ends*	*Max. Eclipse*	*Sun Alt.*	*Eclipse Mag.*	*Duration Totality*
Cook Is.	Mangaia Island	07:15	08:19	08:23	09:37	08:21	14	1.048	03m18s
	Oneroa	07:15	08:19	08:23	09:37	08:21	14	1.048	03m19s
French	Amanu Island	07:25	08:41	08:45	10:13	08:43	33	1.054	04m11s
Polynesia									
	Anaa Island	07:19	08:33	08:35	10:00	08:34	28	1.053	02m43s
	Hao Island	07:25	08:42	08:45	10:13	08:43	33	1.054	02m57s
	Haraiki Island	07:21	08:36	08:40	10:06	08:38	31	1.054	04m03s
	Hikueru Island	07:22	08:37	08:42	10:08	08:39	31	1.054	04m20s
	Marutea Island	07:21	08:36	08:39	10:06	08:38	31	1.054	03m00s
	Mehetia Island	07:17	08:29	08:31	09:54	08:30	25	1.052	01m45s
	Papeete, Tahiti	07:16	—	—	09:50	08:27	24	0.984	
	Ravahere Island	07:23	08:39	08:43	10:10	08:41	32	1.054	03m33s
	Tatakoto Atoll	07:27	08:45	08:50	10:19	08:48	36	1.055	04m22s
Easter Is.	Mataveri	12:41	14:09	14:13	15:34	14:11	40	1.056	04m44s
Argentina									
	El Calafate	16:44	17:48	17:51	—	17:50	1	1.045	02m47s
	Lago Viedma	16:45	17:49	17:52	—	17:50	2	1.045	02m27s
	Tres Lagos	16:45	17:50	17:51	—	17:51	1	1.045	00m51s

[a] Contract times are given in Local Standard Time in each country.

Notes and References

Epigraph: John Couch Adams: "On the total Eclipse of the Sun, 28 July 1851, as seen at Frederiksvaern," *Memoirs of the Royal Astronomical Society*, Volume 21 (1852). In 1845, Adams had used mathematics to demonstrate that an undiscovered planet lay beyond Uranus and where in the sky it could be found. English astronomers ignored the young man's work, but Urban Leverrier in France subsequently made his own calculation and Johann Galle used that information to discover Neptune in 1846.

[1] Liesl Clark: "Secrets of Easter Island," *Nova* television documentary first aired February 15, 2000: <www.pbs.org/wgbh/nova/easter/civilization/first.html>

[2] There is no certainty about when the Polynesians first settled on Easter Island, but about 800 A.D. seems to be best supported. See Whitney Dangerfield: "The Mystery of Easter Island," Smithsonian.com, April 2007: <http://www.smithsonianmagazine.com/issues/2007/april/easter.php>

[3] Mount Fitz Roy and its neighbors will not block the view from El Calafate. They are 40 miles (65 kilometers) distant. From El Calafate, totality should last 2 minutes 47 seconds. At mid-eclipse, the Sun will be 1.4° above the horizon (if the horizon were flat).

[4] It's the size of Pennsylvania, which ranks thirty-third among states in area.

19
The All-American Eclipse of 2017

> [A] total eclipse of the Sun...is the most sublime and awe-inspiring sight that nature affords.
>
> Isabel Martin Lewis (1924)

Get out the date book: August 21, 2017. That's when a total eclipse of the Sun at last returns to the United States. For a landmass its size—third largest country on Earth—the United States has been shortchanged on totality in recent times. On July 11, 1991, a total eclipse passed over Hawaii. But the last total eclipse to visit the continental United States was February 26, 1979—skimming the northwestern tier of states. From 1979 to 2017—38 years without totality.

By comparison, China, almost the same size as the United States, in that same span of time booked total eclipses in 1980, 1997, 2008, and 2009.

The eclipse of 2017 is an all-or-nothing affair. Most total eclipses visit a succession of countries as their paths cross thousands of miles. Not 2017. It starts 2,400 miles (3,800 kilometers) out in the Pacific Ocean and crosses no land before it reaches the United States. It then traverses the country diagonally from northwest to southeast for 2,400 miles (3,800 kilometers). Once it leaves America and sets out over the Atlantic, its totality will not touch land again. The eclipse will travel 2,800 miles (4,500 kilometers) across the Atlantic but vanish from the surface of the Earth before it can reach Africa.

Coming to America

The total solar eclipse of 2017 first touches Earth in the Pacific Ocean 1,000 miles (1,600 kilometers) south of the western Aleutian Islands of Alaska and 1,200 miles (2,000 kilometers) north-northwest of the Hawaiian Islands. Sailors in the area will see the Sun rise totally eclipsed. Totality reaches the American coast 27 minutes later, having crossed half the Pacific at an average speed of more than 5,200 miles per hour (8,370 kilometers per hour).

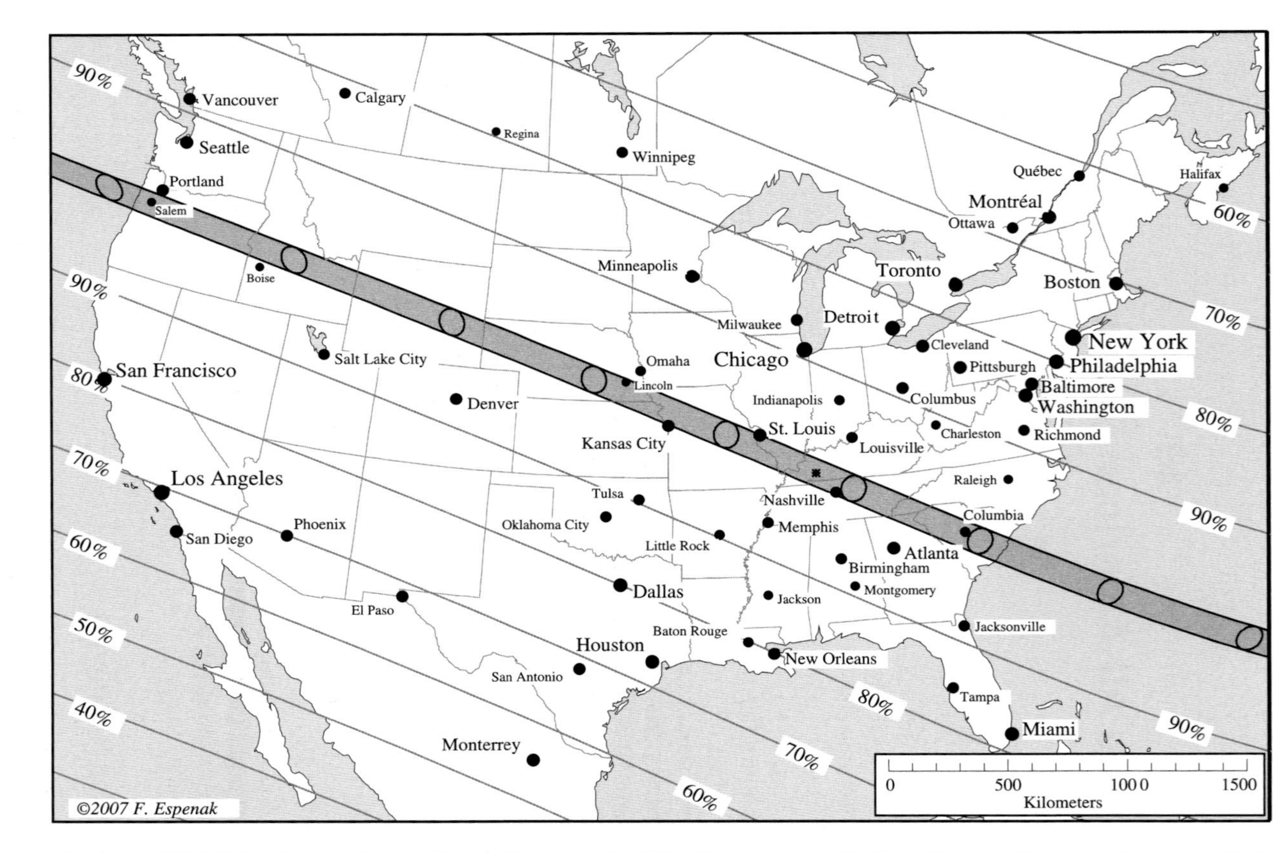

The August 21, 2017 eclipse track runs diagonally across the USA. For places outside the path of totality, the eclipse magnitude (fraction of the Sun's diameter eclipsed) is given in percent. [Map and eclipse predictions by Fred Espenak]

The eclipse comes ashore in Oregon near Cape Foulweather (let's hope not), with the path of totality stretching from Cloverdale to Waldport. The central line passes about 10 miles (16 kilometers) south of Lincoln City, bringing 1 minute 59 seconds of totality. The eclipse bypasses Portland—20 miles (30 kilometers) too far north.

It appears that the eclipse of 2017, a presidential inauguration year, has a hankering for politics, for in its campaign swing across the nation, the eclipse pays a visit to five state capitals and winks at two more in passing. The first is Salem, Oregon, and the eclipse delivers on its promises: 1 minute 55 seconds of totality.

Now the eclipse ascends the Cascade Mountains, bringing totality to Mt. Jefferson, second highest mountain in Oregon, a dormant volcano towering 10,495 feet (3,199 meters) above sea level. But the limits of total eclipse pass a little south of the even higher, volcanic Mt. Hood and a little north of the slightly shorter Three Sisters cluster of volcanoes.

Then down the eastern slope of the Cascades the eclipse slides and over Warm Springs, Mitchell, Prairie City, and Unity, the duration of total eclipse now exceeding two minutes. It crosses Interstate Highway 84 halfway between Baker City, Oregon and Ontario, on the Snake River, which marks the end of Oregon and the beginning of Idaho.

At Baker City, the eclipse begins to trace in reverse much of the Oregon Trail that brought more than a third of a million settlers from Missouri River towns to the Northwest between 1843 and 1868. From Baker City, the immigrants drove their wagons north and west to the Columbia River, then usually floated downstream to the site of Portland and the good farmland of the Willamette Valley to its south.

The eclipse skirts Boise, capital of Idaho, the edge of totality passing 15 miles (25 kilometers) to the north. Sun Valley is more fortunate. How could a solar eclipse resist a name like that? The ski and summer resort, lying just south of the centerline, receives 1 minute 38 seconds of totality.

As much as this eclipse seems to target state capitals, it seems even more to prefer high mountains. Darkness at noon (well, 30 minutes short of noon) descends on Borah Peak, tallest mountain in Idaho (12,667 feet; 3,861 meters).

In its trek across Idaho, the eclipse clips a stubby protrusion of the "beard" of Montana. About 10 square miles (26 square kilometers) of the Beaverhead Mountains and its national forest in southwesternmost Montana receive a few seconds of totality.

In eastern Idaho, Idaho Falls and Rexburg luxuriate in eclipse shadow. Rexburg, almost on the central line, gets 2 minutes 15 seconds of totality.

Then the eclipse enters Wyoming, and it's time for more mountain climbing. In a few seconds, the eclipse bounds up and over the Grand Tetons, with the central line passing just 8 miles (13 kilometers) south of

Deprivation of Totality

The continental United States has been without a total eclipse of the Sun since 1979. Most of the individual states have suffered without a total eclipse for far longer. For each state through which the central line of the 2017 total eclipse passes, here's a breakdown of the last time that state experienced totality—although seldom in the same area. Many states have been waiting a long time for their opportunity. Total eclipses of the Sun rarely come to you. You must go to them.

State	*Most recent total eclipse*	*Location within state*
Oregon	February 26, 1979	Along northern border
Idaho	February 26, 1979	Through the middle of the panhandle
Wyoming	June 8, 1918	Southwestern corner
Grand Teton Peak	January 1, 1889 July 29, 1878	
Nebraska	August 7, 1869	Northeastern corner
Kansas	June 8, 1918	Southwestern corner
Missouri	August 7, 1869	Northeastern corner
St. Joseph	June 16, 1806	[211 years between total eclipses]
Illinois	August 7, 1869	Northwest to southeast through center of state, encompassing over half of it
Kentucky	August 7, 1869	All but western end & northeastern tip
Tennessee	August 7, 1869	Northeastern corner
Nashville	July 29, 1478	[539 years between total eclipses]
North Carolina	March 7, 1970	East coast
South Carolina	March 7, 1970	East coast

spectacular Grand Teton Peak (13,770 feet; 4,199 meters), second highest mountain in Wyoming, 10 miles (16 kilometers) south of Jenny Lake, and just 2 miles (3 kilometers) north of Teton Village. This eclipse does seem to enjoy recreational areas.

The central line of the eclipse, now traveling 1,840 miles per hour (2,960 kilometers per hour), touches down at the Jackson Hole airport, landing eastward on the single north-south runway. The town of Jackson Hole, southern gateway to Grand Teton National Park, is 9 miles (15 kilometers) to the south and thus very close to the center of the eclipse's path.

The total eclipse of 2017 doesn't visit Yellowstone National Park, just north of Grand Teton. The northern limit of totality barely misses the south entrance to America's first national park.

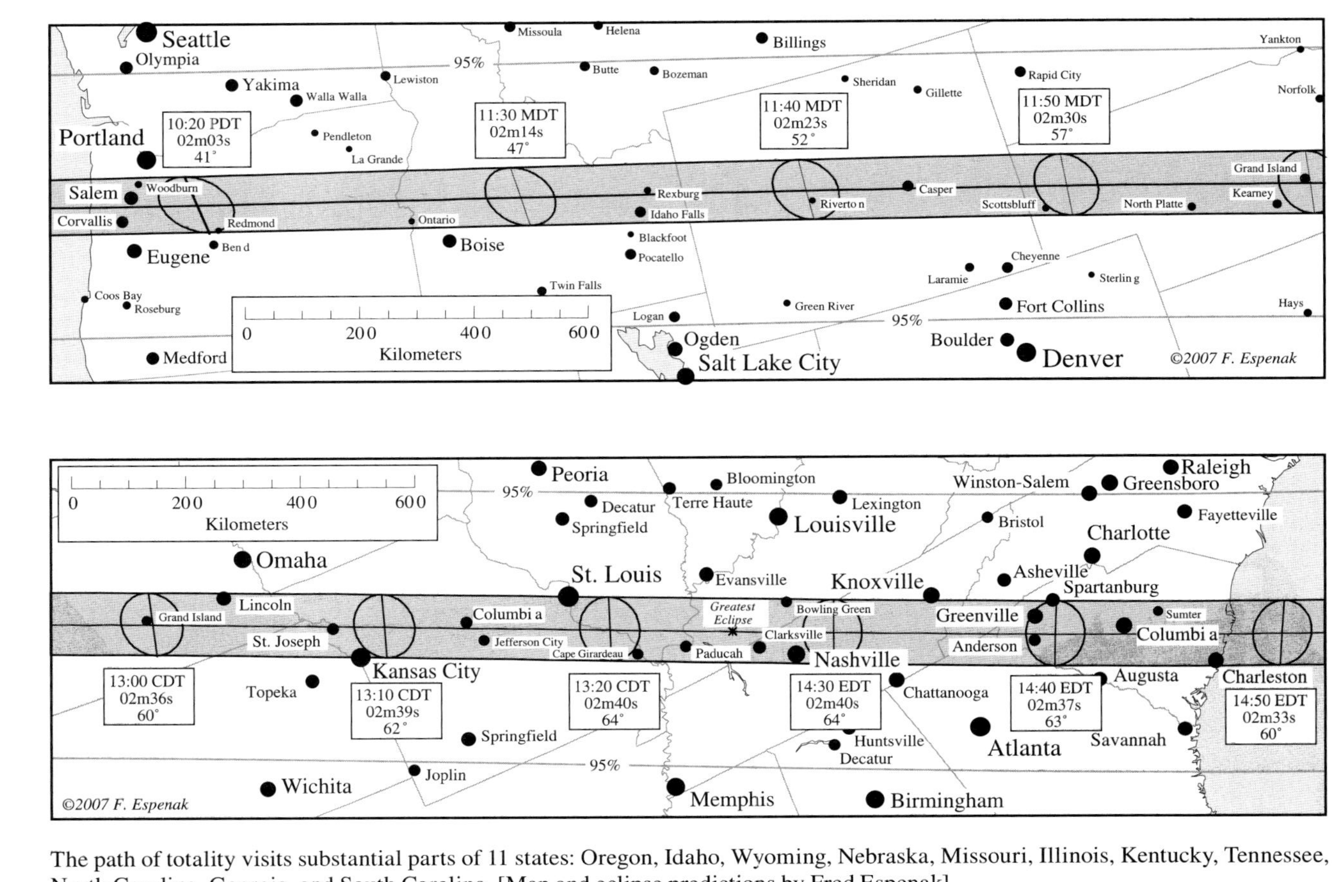

The path of totality visits substantial parts of 11 states: Oregon, Idaho, Wyoming, Nebraska, Missouri, Illinois, Kentucky, Tennessee, North Carolina, Georgia, and South Carolina. [Map and eclipse predictions by Fred Espenak]

Crossing the Divide

On across Wyoming sweeps the eclipse, angling southeast, rushing up and over the Continental Divide in the Wind River Mountain Range and enveloping Gannett Peak, tallest mountain in Wyoming (13,804 feet; 4,207 meters). Then down the Wind River Basin the shadow slides, with the southern edge of the Owl Creek and Bighorn Mountains to the north and the northern edge of the Rattlesnake Hills to the south. It was across this high plateau that the wagon trains rolled westward on the Oregon Trail more than a century and a half ago. Ahead of the settlers, south of the path of totality, lay South Pass, the highest elevation on their journey. South Pass was the halfway point—3 months and 1,000 miles (1,600 kilometers) from Independence, Missouri; 3 months and 1,000 miles still to go to the Oregon coast.

Central Wyoming is a wind-carved, wind-burnished landscape with an arctic climate. Homesteaders there used to say: "If summer falls on a weekend, let's have a picnic."[1]

Straight through Casper, Wyoming goes the center of the eclipse, jay-walking Cy Avenue and South Poplar Street without stopping and looking in all directions. Traffic may halt for 2 minutes 26 seconds and look up.

Southwest of Casper, at the edge of eclipse totality, lies massive, mound-shaped Independence Rock on the Oregon Trail. If the settlers reached this milestone by July 4, they would celebrate. If they didn't, snow would likely catch them in the mountains ahead.

At Casper, the eclipse picks up the North Platte River and follows it through Douglas and across one last range of the Rockies, the northern portion of the Laramie Mountains, bathing Laramie Peak (10,274 feet; 3,131 meters) in totality.

Then the mountains are gone and the eclipse races off out of Wyoming, into the stubby panhandle of Nebraska, and onto the Great Plains. Still it follows the North Platte River, which meanders along the southern half of the path of totality, through Scottsbluff. It was westward along the North Platte River that the Oregon Trail settlers came, admiring the grandeur of Scott's Bluffs and regretting the detour needed around this obstacle. They averaged 1½ miles per hour (2½ km/hr). Fifteen miles was a good day.

Totality then skims Chimney Rock. Travelers on the Oregon Trail considered this pipe of rock, rising from a conical mound and towering 325 feet (100 meters) above the landscape, to be the most spectacular landmark along the Oregon Trail. It was a long day's journey west by wagon from Chimney Rock to Scottsbluff. The eclipse shadow, traveling east, makes the trip in half a minute.

The central line runs 2 miles (3 kilometers) south of Alliance, Nebraska, which is surrounded by hundreds of farm fields that are circular rather than rectangular. This land is arid high plain and needs irrigation. The circular

Contact Times[a] for the 2017 Total Solar Eclipse in the USA

State	*City*	*Partial Begins*	*Total Begins*	*Total Ends*	*Partial Ends*	*Max. Eclipse*	*Sun Alt.*	*Eclipse Mag.*	*Duration Totality*
California	Los Angeles	09:06	—	—	11:45	10:21	48	0.694	
	San Francisco	09:01	—	—	11:37	10:15	43	0.802	
Colorado	Denver	10:23	—	—	13:15	11:47	57	0.933	
District of Columbia	Washington	13:18	—	—	16:02	14:43	56	0.844	
Florida	Miami	13:27	—	—	16:21	14:58	64	0.823	
Georgia	Atlanta	13:06	—	—	16:02	14:37	65	0.971	
Idaho	Boise	10:11	—	—	12:50	11:27	46	0.994	
	Idaho Falls	10:15	11:33	11:35	12:58	11:34	50	1.029	01m49s
Illinois	Carbondale	11:52	13:20	13:23	14:47	13:21	64	1.031	02m37s
	Chicago	11:54	—	—	14:43	13:20	59	0.889	
Kansas	Kansas City	11:41	13:09	13:09	14:36	13:09	63	1.031	00m21s
	Leavenworth	11:41	13:07	13:09	14:35	13:08	62	1.030	01m34s
	Topeka	11:39	—	—	14:34	13:07	62	0.989	
Kentucky	Bowling Green	12:59	14:27	14:28	15:53	14:28	63	1.030	00m57s
	Paducah	12:54	14:22	14:25	15:49	14:23	64	1.031	02m20s
Louisiana	New Orleans	11:58	—	—	14:57	13:30	71	0.799	
Massachusetts	Boston	13:28	—	—	15:59	14:47	50	0.702	
Michigan	Detroit	13:03	—	—	15:48	14:28	57	0.831	
Missouri	Cape Girardeau	11:52	13:20	13:22	14:48	13:21	64	1.031	01m46s
	Columbia	11:46	13:12	13:15	14:40	13:14	63	1.031	02m37s
	Independence	11:42	13:09	13:10	14:36	13:09	63	1.031	01m14s
	Jefferson City	11:46	13:13	13:16	14:41	13:14	63	1.031	02m28s
	St. Louis	11:50	—	—	14:44	13:18	63	1.000	
Nebraska	Grand Island	11:34	12:59	13:01	14:27	13:00	60	1.030	02m34s
	Lincoln	11:37	13:03	13:04	14:30	13:03	60	1.030	01m19s
New York	New York	13:23	—	—	16:01	14:45	53	0.769	
Oregon	Corvallis	09:05	10:17	10:19	11:37	10:18	40	1.027	01m40s
	Portland	09:06	—	—	11:39	10:19	40	0.991	
	Salem	09:05	10:17	10:19	11:38	10:18	40	1.027	01m55s
Pennsylvania	Philadelphia	13:21	—	—	16:01	14:44	54	0.800	
South Carolina	Anderson	13:09	14:38	14:40	16:03	14:39	63	1.030	02m34s
	Charleston	13:17	14:46	14:48	16:10	14:47	62	1.030	01m27s
	Columbia	13:13	14:42	14:44	16:06	14:43	62	1.030	02m30s
	Greenville	13:09	14:38	14:40	16:03	14:39	63	1.030	02m11s
	Sumter	13:15	14:44	14:45	16:07	14:45	61	1.030	01m49s
Tennessee	Clarksville	12:57	14:26	14:28	15:52	14:27	64	1.031	02m19s
	Cleveland	13:03	14:33	14:34	15:59	14:34	64	1.031	01m04s
	Knoxville	13:05	—	—	15:59	14:34	63	0.997	
	Nashville	12:58	14:27	14:29	15:54	14:28	64	1.031	01m55s
	Oak Ridge	13:04	14:33	14:34	15:58	14:34	63	1.030	00m24s
Texas	Dallas	11:40	—	—	14:39	13:10	69	0.800	
	Houston	11:47	—	—	14:46	13:17	72	0.730	
Utah	Salt Lake City	10:14	—	—	13:00	11:34	51	0.925	
Washington	Seattle	09:09	—	—	11:39	10:21	39	0.931	
Wyoming	Casper	10:22	11:43	11:45	13:09	11:44	54	1.029	02m26s
	Cheyenne	10:24	—	—	13:14	11:47	56	0.975	

Eclipse calculations by Fred Espenak

[a] Local Times are in Daylight Saving Time.

fields have a center pivot system with a sprinkler arm that sweeps around the field like the minute hand on a clock. From the Sun's and Moon's perspective looking down, the farmers have drawn more than a thousand giant round images, some light, some dark, as if to honor the creators of eclipses. Alliance is rewarded with 2 minutes 30 seconds of totality.

The North Platte River meanders out of the eclipse path as it approaches Ogallala, but then veers back inside the path of totality as it approaches the town of North Platte. Here too, just west of North Platte, the eclipse picks up Interstate Highway 80, which follows the river.

I-80 lies within the shadow of totality for the next 250 miles (400 kilometers), although the highway usually lies toward one or the other edge of the path of eclipse darkness. For the longest duration of totality, which in central Nebraska is about 2 minutes 35 seconds, observers

Weather Prospects for the 2017 Eclipse

by Jay Anderson

The eclipse of August 21, 2017 appears at a propitious time: the summer thunderstorm season is winding down and retreating southward; the Arizona monsoon is breaking; and the storm-carrying jet stream has not yet begun its journey southward from Canada. The dry and generally sunny fall season is about to begin. After a 38-year totality drought in the continental United States, this eclipse arrives to the open arms of a friendly August climatology.

The westerly winds that bring weather systems onto the North American continent first have to cross the mountain barriers that protect the Pacific coast. The moisture-laden air is forced to rise up the windward slopes, giving seaward-facing Washington and Oregon well-deserved reputations for clouds and precipitation. However, the reputation is for the coast only. The air descends and dries as it settles into successive interior valleys, and eventually out onto the Great Plains. For eclipse-chasers in the Northwest, the mountain valleys offer the best chances of sunny weather on eclipse day. The Willamette Valley of Oregon is especially favored, very nearly matching the best cloud prospects along the entire track.

Beyond the Willamette Valley, cloudiness increases again as the air rises across the Blue Mountains, and then repeats its drying as it descends into the Snake River Valley. Past Boise, Idaho, the by now semi-dry Pacific air rises once again to cross the Rocky Mountains, descends briefly into the eastern branch of the Snake River Valley, and finally descends for good past Casper, Wyoming onto the Great Plains. Each rise is accompanied by an increase in cloudiness; each descent brings a Chinook-like drying and an increase in sunshine and eclipse prospects. As many eclipse-chasers learned in the February 1979 eclipse, the valleys are effective cloud-eaters even when large weather systems cross the Northwest.

Across the Great Plains, from eastern Wyoming to the Mississippi River and on into central Tennessee, the cloud-cover prospects are similar. This huge region, in the late stages of summer, has a reputation for generous sunshine, and it is

want to be on or close to the central line. Reaching the central line is easy using north-south roads off I-80. What I-80 provides is the quickest way of chasing a break in the clouds if local weather disappoints.

Next to receive the eclipse is Kearney. Nearby once stood Fort Kearny,[2] an unfortified outpost built to protect the pioneers headed west on the Oregon Trail. It was beloved because it was a place where letters could be mailed to those left behind.

Between Kearney, near the southern limit of totality, and Lincoln, near the northern limit, I-80 cuts diagonally across the southeast-headed eclipse path. The centerline of the eclipse passes just a little south of downtown Grand Island, straight over the island in the North Platte River that gave the city its name, and right across I-80 5 miles (8 kilometers) southeast of the island.

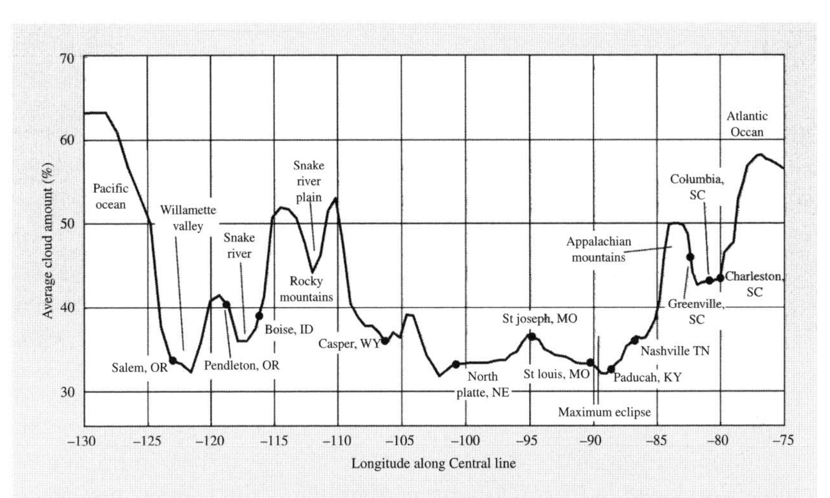

Average cloud amounts along the 2017 eclipse path. [Weather data courtesy of Jay Anderson]

there that the best prospects for a view of the eclipse are found. Average cloud amounts for this part of the track range (in percentage) from the low to mid 30s. The lowest cloud amount along the central line is found to the west of North Platte, Nebraska, but is closely challenged by locations along the eclipse path at the Kentucky-Tennessee border, where totality is at its maximum duration.

As the shadow track reaches the eastern half of Tennessee, it moves into rougher terrain and crosses the Appalachian Mountains beyond Knoxville. Average cloudiness rises significantly through this part of the track, peaking as high as 50% as it crosses the Appalachians at the Tennessee-North Carolina border. The final mountain descent to the Atlantic Ocean brings a small 7% or 8% improvement before the track heads out over the water.

Jay Anderson is a meteorologist at the University of Manitoba in Winnipeg.

On to Lincoln, Nebraska and another state (capital) visit. Lincoln lies near the northern edge of the eclipse. From the steps of its unique, high-rise capitol building, the total phase of the eclipse will last 1 minute 26 seconds. The eclipse bypasses Omaha, Nebraska's largest city, which is 35 miles (55 kilometers) too far north.

Cutting Corners

On its way out of Nebraska into Missouri, the total phase of the eclipse clips the tiniest corner of southwesternmost Iowa, with perhaps 2 square miles (5 square kilometers) of farm fields and one farmhouse receiving a few seconds of totality.

The eclipse then skims the northeast corner of Kansas and leaps the Missouri River into St. Joseph, Missouri, one of the great embarkation points for the Oregon Trail. This eclipse takes an interest in modern embarkation points as well, racing through the airport without a security check. It then parades through downtown St. Joe, bringing 2 minutes 38 seconds of totality.

Kansas City, Missouri, 40 miles (65 kilometers) south of St. Joseph, straddles the southern edge of the eclipse path. Just east of Kansas City, within the path of totality, is Independence, where the Oregon Trail began. Two thousand miles (3,200 kilometers) lay ahead of those travelers. Three hundred fifty thousand people began the journey between 1843 and 1868 in search of a new life in a country they had never seen. As many as 35,000 died along the way, most of disease and accidents.

At Independence, the interstate-highway-friendly eclipse picks up I-70 and brings it Moon-shadow shade all the way to St. Louis. North of I-70, Carrollton and Marshall, Missouri lie very nearly on the central line of totality. Just south of Carrollton, the central line of the eclipse crosses the meandering Missouri River three more times in 4 miles (6 kilometers).

The central line passes right between Columbia and Jefferson City in the center of the state, with plenty of totality for both. On the steps of the state capitol in Jefferson City, totality lasts 2 minutes 28 seconds.

The central line passes 9 miles (14 kilometers) southwest of Fulton, Missouri, where in 1946, ten months after the end of World War II in Europe, Winston Churchill spoke at Westminster College and introduced a new phrase into the world's vocabulary, saying that an iron curtain had descended upon central and eastern Europe. The eclipse provides a dark curtain of a much less ominous and far more aesthetic kind. It lasts 2 minutes 34 seconds, not 44 years.

As it proceeds to and past Columbia and Jeff City and Fulton, the center track crosses the Missouri River three more times.

Like Kansas City, half of St. Louis enjoys the total eclipse. The northern limit of totality passes 2 miles (3 kilometers) south of the Gateway

Future American Eclipses

If the Moon has been stingy about using the Sun to cast its shadow on the United States in recent decades, the Sun and Moon will be more generous in the next 35 years after 2017. On April 8, 2024, not quite seven years after the eclipse of 2017, totality will return to the United States, again racing diagonally across the country, this time from southwest to northeast, from Texas to Maine, bringing as much as 4 minutes 28 seconds of total eclipse. The paths of the 2017 and 2024 eclipses will intersect where Missouri, Kentucky, and Illinois meet and the Ohio River merges into the Mississippi.

Twenty years will then elapse until the United States has totality again, but it will return in a burst of three total eclipses over a period of 7½ years:

August 23, 2044	Montana and North Dakota
August 12, 2045	California through Florida
March 30, 2052	Florida and Alabama through Georgia and South Carolina

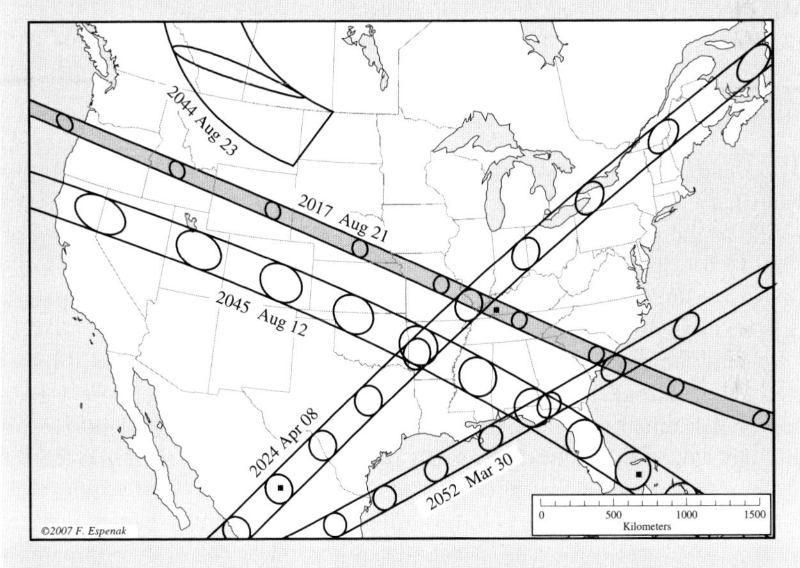

Beyond 2017, the next total solar eclipses through the continental USA are in 2024, 2044, 2045, and 2052. [Map and eclipse predictions by Fred Espenak]

Arch on banks of the Mississippi River. For the western and southern suburbs though, there is a precious minute or so of eclipse darkness. But there are up to 2 minutes 40 seconds of eclipse darkness some 35 miles (60 kilometers) to the southwest or south, near Union or St. Clair or close to Hillsboro, De Soto, and Festus.

The eclipse reaches the Mississippi River town of Ste. Genevieve, Missouri and angles through downtown. The central line then actually enters Illinois before it crosses the Mississippi River because the boundary between Missouri and Illinois was set before the river changed its course during a spring flood more than a century ago. The central line passes diagonally through Kaskaskia, population 9, the first capital of Illinois. For 4 miles (6 kilometers), the central line, still west of the Mississippi, finds itself back in Missouri, then returns to Illinois at another abandoned oxbow in the river and finally crosses the Mississippi.

Caught in the swath of totality, just north of the central line, are Carbondale and Marion, Illinois. Carbondale owes its name to the coal mined all around it in southern Illinois. Marion is home to a federal maximum-security penitentiary where the prisoners are serving a good deal more than 2 minutes 28 seconds in eclipse.

Both cities hold the enviable distinction of also lying within the path of America's next total solar eclipse in 2024. For any city, the gap between visits by a total eclipse of the Sun averages 375 years. Carbondale and Marion will experience the magic twice in only 6⅔ years.

Into the South

The shadow of the Moon leaves Illinois by crossing the Ohio River into Kentucky, bringing totality to Paducah, about 18 miles (29 kilometers) south of the central line. Interstate Highway 24 lies within the band of totality for the next 200 miles (300 kilometers) from Goreville, Illinois to Nashville, Tennessee, for those who may need to do some speedy eclipse chasing in case of summer afternoon cloud build-up.

As the eclipse enters Kentucky, the southern half of the path of totality sweeps across the Kentucky portion of the Land Between The Lakes National Recreation Area, where the Tennessee and Cumberland Rivers are dammed side by side a few miles before they flow into the Ohio River.

At Hopkinsville, Kentucky, the eclipse reaches its maximum—its longest duration of totality—2 minutes 40 seconds. At greatest eclipse, the path of totality of this All-American is 71 miles (115 kilometers) wide. The Sun is 64° above the horizon.[3]

On into Tennessee the eclipse glides, with Clarksville, Springfield, and Westmoreland spread out west to east across the state line and across the eclipse path to welcome it. The eclipse envelops Nashville, yet another state capital. Nashville also boasts the largest metropolitan population of any city fully in the path of totality—1.5 million. The central line passes 25 miles (40 kilometers) north of downtown.

Closer to the centerline is Lebanon, Tennessee, home of the Nashville Superspeedway. The stock cars that race there sometimes turn laps at 200 mph (320 km/hr). The Moon's shadow as it passes over the track will be traveling 1,450 mph (2,330 km/hr).

At Nashville, the eclipse exits Interstate Highway 24 and transfers to I-40. I-40 courteously stays within the lane of totality for 170 miles

(275 kilometers), up and over the Cumberland Mountains, from Nashville to the outskirts of Knoxville.

On its climb over the Cumberland Mountains, westernmost chain of the Appalachian Range, the eclipse visits Cookeville and Crossville on I-40, each 5 miles (8 kilometers) north of the central line. Sparta, Tennessee is not on I-40 but it is on the central line, and receives 2 minutes 39 seconds of totality. The duration of total eclipse has begun to decline, measurably but slowly.

In Kentucky, totality covered the Tennessee River as it emptied into the Ohio River. Now the eclipse crosses the giant-U-shaped Tennessee River again in eastern Tennessee. Spring City, on Watts Bar Lake, and Watts Bar Dam, creator of that lake in that portion of the Tennessee River, are just south of the central line. Also in the path of totality on the Tennessee River and its lakes are Kingston, Loudon, and Lenoir City.

Downtown Knoxville lies outside the path of total eclipse, but some western and southern suburbs, including Oak Ridge and its national laboratory, catch a few seconds of totality. The eclipse is even less kind to Chattanooga, chugging by to the north, venturing no closer than outlying suburb Soddy-Daisy.

Suddenly this eclipse seems like even more of a world traveler than it is—and with an exceptionally large footprint—as it visits Lebanon, Sparta, Athens, Dayton, Philadelphia, Cleveland, and Louisville, all in the same few moments—all wistfully named towns in the Tennessee Valley. Dayton is where John Scopes was tried in 1925 for teaching evolution. Closer to the central line are Athens, Sweetwater, and Madisonville.

Then it's up and over the main branch of the Appalachian Mountains with more than half of Great Smoky Mountains National Park lying in the zone of totality. Clingmans Dome (6,643 feet; 2,025 meters) is near the northern limit of the eclipse. Cade's Cove is closer to the central line. Gatlinburg and Pigeon Forge, gateways to the Park on the Tennessee side, are excluded from totality. Bryson City, North Carolina, at the east entrance to the Smokies, gets just under two minutes of totality. Franklin gets half a minute more, and then the eclipse leaves North Carolina behind. It catches the lightly populated northeastern corner of Georgia and plunges centrally across much smaller South Carolina from Greenville through Charleston, enveloping half the land area of the state.

Last Stops

When the eclipse came ashore in Oregon, it visited the capital of that state. Since then, the eclipse has passed across substantial areas of nine states—and has visited five state capitals: Salem, Oregon; Lincoln, Nebraska; Jefferson City, Missouri; Nashville, Tennessee; and now Columbia, South Carolina, bestowing 2 minutes 30 seconds of totality on the capitol building. The eclipse narrowly missed Boise, Idaho and Topeka, Kansas.

Now the eclipse reaches the South Carolina coast, with totality stretching from Pawley's Island in the north (but not Myrtle Beach) to Sullivan's Island (but not Kiawah Island) in the south. Totally within totality is Charleston, South Carolina, where the American Civil War began, but the city lies toward the southern edge of total eclipse. Fort Sumter in Charleston harbor gets 1 minute 27 seconds of corona time. Charleston's northern suburbs, North Charleston and Mt. Pleasant, are closer to the central line and earn nearly 2 minutes of totality.

And then the eclipse of August 21, 2017 leaves the United States behind and plunges out across the Atlantic Ocean, never to touch land again. Bermuda is about 500 miles (800 kilometers) too far north. The Bahamas are about that same distance too far south. Totality misses Antigua and Barbuda, at the northeast corner of the Caribbean Sea, by about 350 miles (600 kilometers). The path of totality is narrowing. The shadow of the Moon lifts off the waters of the Atlantic Ocean at sunset

The Saga of Saros 145

The eclipse of 2017 cuts diagonally across the United States, with eclipse maximum in Kentucky at latitude 37° north. A total eclipse that peaks at middle latitudes belongs either to a young saros whose central eclipses are just beginning or to an old saros whose flashiest eclipses are dwindling to an end. An eclipse whose greatest occultation occurs near the equator is a saros in its prime.

This All-American eclipse of 2017 belongs to saros series 145. Its odd number tells us that this saros is migrating from the north polar regions to the south. So this saros, peaking in mid-northern latitudes, is still young and strengthening. The 2017 eclipse is number 22 in a sequence that will total 77. It is the sixth total eclipse for saros 145. There will be 35 more.

Saros 145 began in 1639 with a tiny partial eclipse. Then followed 13 more partial eclipses, each obscuring a little more of the Sun. Finally, in 1891, the center of the Moon passed across the center of the Sun. But the Moon was a little too far from Earth at that crossing, so it wasn't quite big enough to fully cover the bright disk of the Sun completely. The eclipse was annular. It would be the only annular eclipse that saros 145 would produce.

In 1909, the next eclipse in the series was a hybrid—starting annular, becoming total, then returning to annular again. After that, saros 145 stopped performing hybrid eclipses as well. One and done.

Instead, saros 145 specializes in total eclipses. It premiered its first on June 29, 1927 for England, Scandinavia, and eastern Russia. Crowds were pleased. It performed encores, with the duration of totality increasing each time:

July 9, 1945	1 min 15 sec	Idaho, Montana, Canada, Greenland, Scandinavia, Russia
July 20, 1963	1 min 40 sec	Alaska, Canada, Maine
July 31, 1981	2 min 02 sec	Russia, Pacific Ocean

about 200 miles (300 kilometers) southwest of the coast of Africa, near the equator. The eclipse is over.

But the shadow will continue to sweep through space and will find the Earth again, bringing to a most fortunate few a total eclipse of the Sun. In the meantime, you can almost hear it singing as it goes:

I've been everywhere, man
I've been everywhere, man
'Cross the deserts bare, man
I've breathed the mountain air, man
Of travel, I've had my share, man
I've been everywhere

Been to Broken Bow, Arapahoe, St. Joe, Missouri
Albany, Easley, Kearney, Shoshoni
Greenhorn, Pine Grove, Sweet Home, Antelope
Midvale, Homedale, photograph with telescope

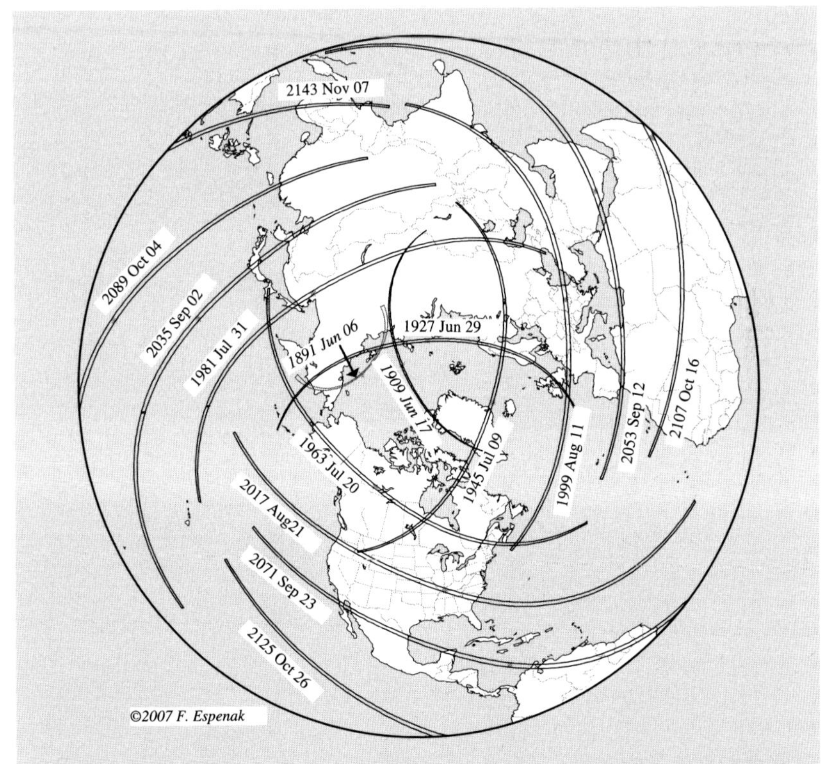

First 15 central eclipses of saros 145. The tracks shift west and south with each succeeding eclipse. The first eclipse (1891) is annular; the second (1909) is hybrid. All the rest from 1927 to 2143 are total eclipses. Saros 145 brings the 2017 eclipse across the United States and provided the 1999 eclipse across Europe and southwestern Asia. [Map and eclipse calculations by Fred Espenak]

Casper, Culver, Cascade, Idaho
Plattsburg, Orangeburg; no pause, off I go

I've been everywhere, man
I've been everywhere, man
'Cross the deserts bare, man
I've breathed the mountain air, man
Of travel, I've had my share, man
I've been everywhere

To Seneca, Etowah, Paducah, Sun Valley
Columbia, Aurora, I bring totality
Toledo, Scio, Shaniko, Oregon
Callaway, Norway, Lemay, Jefferson

August 11, 1999	2 min 23 sec	England, France, Germany, Austria, Hungary, Romania, Bulgaria, Turkey, Iraq, Iran, Pakistan, India

On August 21, 2017, totality will last as long as 2 minutes 40 seconds. The duration of totality will increase to 2 minutes 54 seconds in 2035 for audiences in China, Korea, and Japan; then 3 minutes 4 seconds in 2053 for Portugal, Spain, North Africa, and Saudi Arabia.

But this eclipse series is just hinting at spectacles to come. Saros 145 is one of those rare eclipse families that creates 7-minute total eclipses, approaching the 7-minute-32-second limit for duration of totality.

June 14, 2504	7 min 10 sec	Australia, Micronesia
June 25, 2522	7 min 12 sec	Southern Africa
July 5, 2540	7 min 04 sec	South America

After these, as saros 145 slides toward the south pole, eclipse times decline. The final total eclipse for saros 145 takes place in 2648.

For the next 361 years, the saros will fade away with weaker and weaker partial eclipses, until the last one on April 17, 3009.[a]

Summary for Saros 145

Total eclipses	41
Annular eclipses	1
Annular-total (hybrid) eclipses	1
Partial eclipses	34
Number of eclipses	77
Duration of saros 145	1,370.3 years

[a]Saros data from Fred Espenak and Jay Anderson: ***Total Solar Eclipse of 1999 August 11*** (Greenbelt, Maryland: National Aeronautics and Space Administration Goddard Space Flight Center [NASA Reference Publication 1398], 1997), pages 12, 104.

A complete list and maps of all eclipses in saros 145 can be found in the NASA catalog of solar eclipse saros series: <http://eclipse.gsfc.nasa.gov/SEsaros/SEsaros145.html>.

More on the 2017 and 2024 Total Solar Eclipses

The NASA Eclipse Website features interactive maps of the 2017 and 2024 eclipses using Google maps. You can zoom in on any part of each eclipse path to reveal towns, rivers, lakes, and roads. The websites for the two eclipses are:

<http://eclipse.gsfc.nasa.gov/SEgoogle/SEgoogle2001/TSE2017google.html>

<http://eclipse.gsfc.nasa.gov/SEgoogle/SEgoogle2001/TSE2024google.html>

Shawneetown, Pawnee, Crowheart, Tecumseh
Gasconade, I bring shade; for Merna, corona

I've been everywhere, man
I've been everywhere, man
'Cross the deserts bare, man
I've breathed the mountain air, man
Of travel, I've had my share, man
I go places you haven't been
I've been everywhere

Been to Clarksville, Nashville, Crossville, Tennessee
Knoxville, Maryville, Greenville, Liberty
St. Louis, Metolius, Madras, Depoe Bay
Corvallis, Thermopolis: making night of day
Jackson, Atchison, Riverton, Lebanon
Silverton, Torrington, Charleston, I'm gone.[4]

Notes and References

Epigraph: Isabel Martin Lewis: *A Handbook of Solar Eclipses* (New York: Duffield, 1924), page 3.

[1] John McPhee: *Rising from the Plains* (New York: Farrar Straus Giroux, page 15.

[2] The fort was named for General Stephen Watts Kearny. A later mistake by the post office added an "e" to the name of the town that has never been removed.

[3] During the total eclipse of August 21, 2017, the axis of the shadow cone passes 1,731miles (2,785 kilometers) north of the center of the Earth.

[4] With gratitude for the song "I've Been Everywhere," which was written by Australian composer Geoff Mack in 1959. It gained fame when recorded by Australian singer Lucky Starr in 1962. That same year, American singer Hank Snow adapted the lyrics for American place names and made the song popular in the United States. Johnny Cash recorded "I've Been Everywhere" in 1996 and it became one of his signature pieces.

20
Coming Attractions

Each eclipse has at least one phenomenon
that makes it special.

Stephen J. Edberg (1990)

Maps for Every Eclipse 2008–2030

Between the years 2008 and 2030, a 22-year period, the Moon will eclipse the Sun 52 times. This interval samples at least one eclipse from every saros series currently producing eclipses. The eclipses during this period fall into the following categories:

	2008–2030	*Compared to –1999 to +3000*[1]
Total	15 = 28.8%	26.7%
Annular	17 = 32.7%	33.2%
Hybrid	2 = 3.8%	4.8%
Partial	18 = 34.6%	35.3%

The following pages offer 52 global maps, one for each eclipse. The odd saddle-shaped zone in each map shows the region where the partial eclipse is visible. The magnitude of each eclipse (maximum fraction of the Sun's diameter covered) is shown in increments of 25%, 50%, and 75%. This allows you to quickly estimate the magnitude for any location within the eclipse path. For central eclipses, the path of either totality or annularity is plotted.

The Moon's penumbral shadow typically produces a zone of partial eclipse (during both partial and central eclipses) that covers 25% to 50% of the daylight hemisphere of the Earth. In comparison, the Moon's

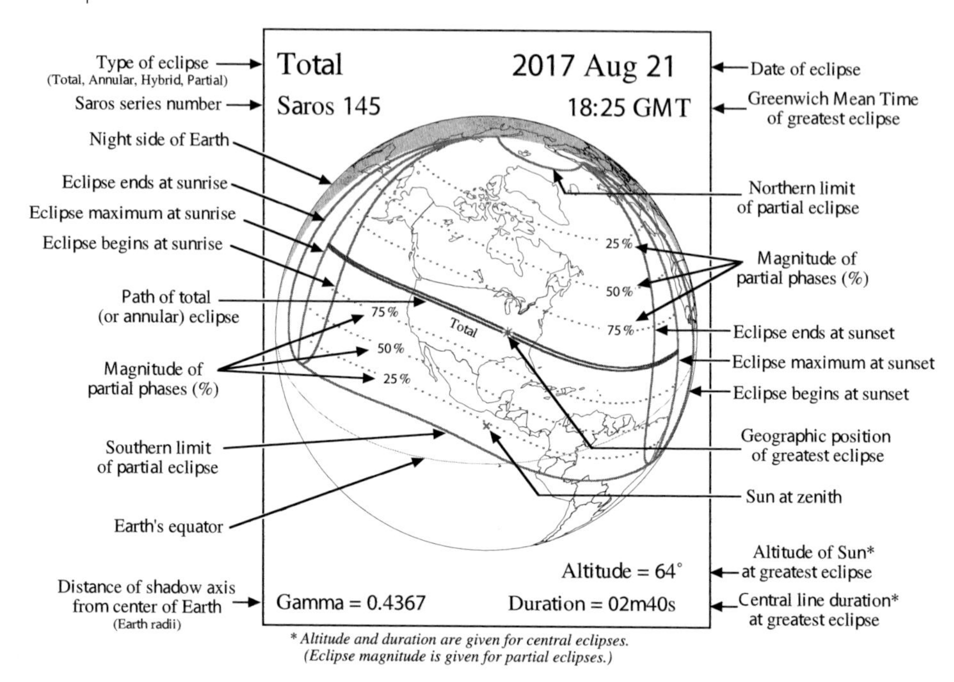

** Altitude and duration are given for central eclipses.*
(Eclipse magnitude is given for partial eclipses.)

umbral shadow (total eclipse) or antumbral shadow (annular eclipse) is much smaller: its path covers less than 1% of the Earth's surface.

Additional information on each map can be identified using the key above.

On pages 286–288 are three maps: North and South America, Europe and Africa, and Asia and Australia. These detailed maps show the paths of every central solar eclipse (total, annular, and hybrid) from 2008 through 2030. The maximum duration of totality or annularity as well as a list of all countries within each central eclipse path can be found in Appendix A, which covers eclipses from 2008 through 2060.

Use *Totality: Eclipses of the Sun* to plan your own voyage into the Moon's shadow. Remember, words and pictures can never fully convey the wonder of a total eclipse of the Sun: to stand in the path of totality, in the light of the corona. You must see one for yourself—or two, or...

Hope to meet you there.

Notes and Reference

Epigraph: Stephen J. Edberg, personal communication, March 1990.

[1] Fred Espenak and Jean Meeus: *Five Millennium Canon of Solar Eclipses: –1999 to +3000* (Greenbelt, Maryland: NASA Goddard Space Flight Center [NASA/TP–2006–214141], 2006.

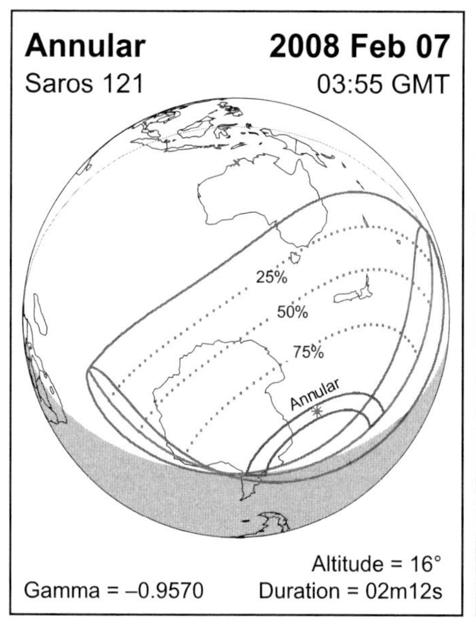
Annular
2008 Feb 07
Saros 121
03:55 GMT
25%
50%
75%
Annular
Altitude = 16°
Gamma = –0.9570
Duration = 02m12s

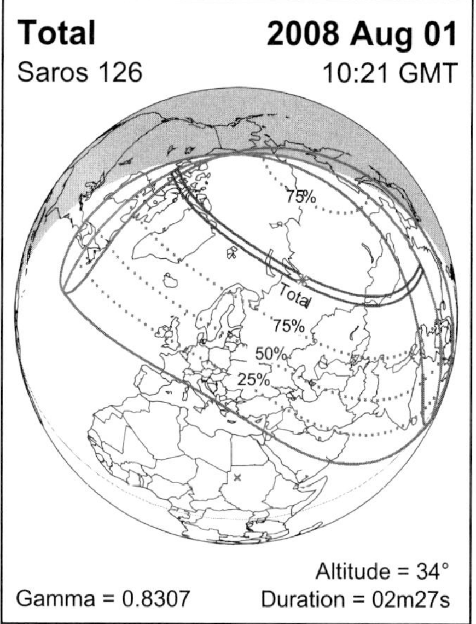
Total
2008 Aug 01
Saros 126
10:21 GMT
75%
Total
75%
50%
25%
Altitude = 34°
Gamma = 0.8307
Duration = 02m27s

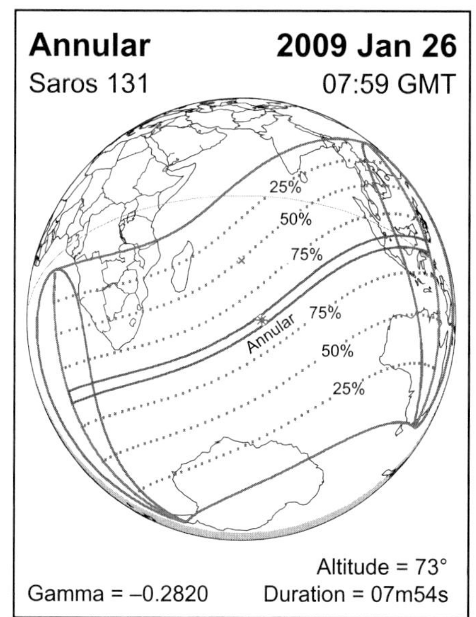
Annular
2009 Jan 26
Saros 131
07:59 GMT
25%
50%
75%
Annular
75%
50%
25%
Altitude = 73°
Gamma = –0.2820
Duration = 07m54s

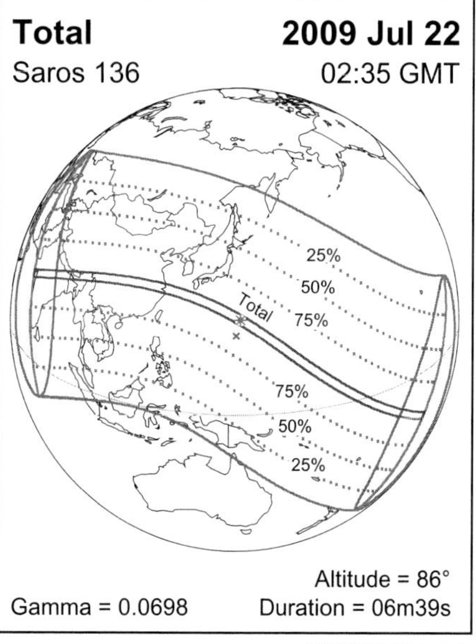
Total
2009 Jul 22
Saros 136
02:35 GMT
25%
50%
75%
Total
75%
50%
25%
Altitude = 86°
Gamma = 0.0698
Duration = 06m39s

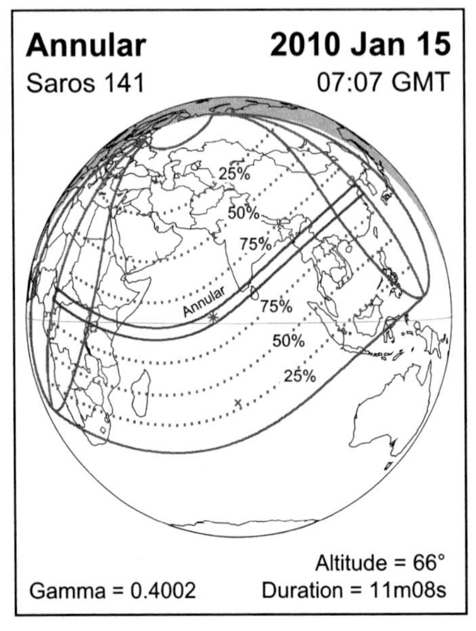
Annular
2010 Jan 15
Saros 141
07:07 GMT
25%
50%
75%
Annular
75%
50%
25%
Altitude = 66°
Gamma = 0.4002
Duration = 11m08s

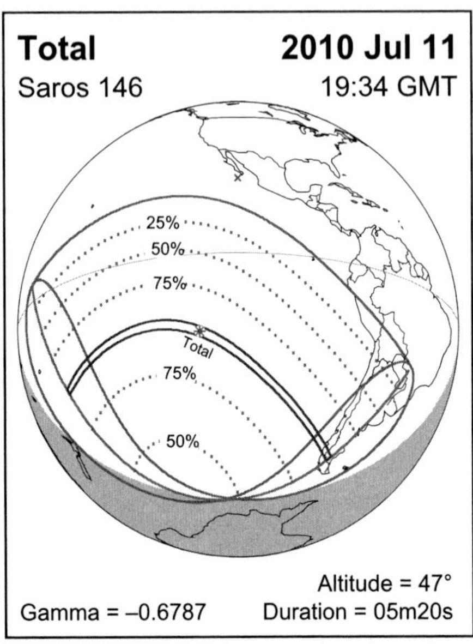
Total
2010 Jul 11
Saros 146
19:34 GMT
25%
50%
75%
Total
75%
50%
Altitude = 47°
Gamma = –0.6787
Duration = 05m20s

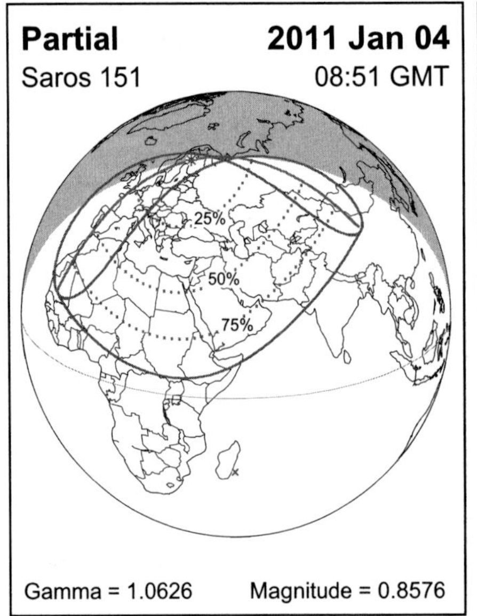
Partial
2011 Jan 04
Saros 151
08:51 GMT
25%
50%
75%
Gamma = 1.0626
Magnitude = 0.8576

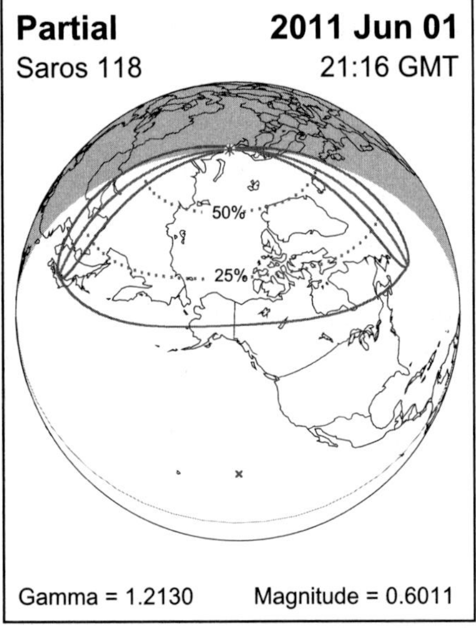
Partial
2011 Jun 01
Saros 118
21:16 GMT
50%
25%
Gamma = 1.2130
Magnitude = 0.6011

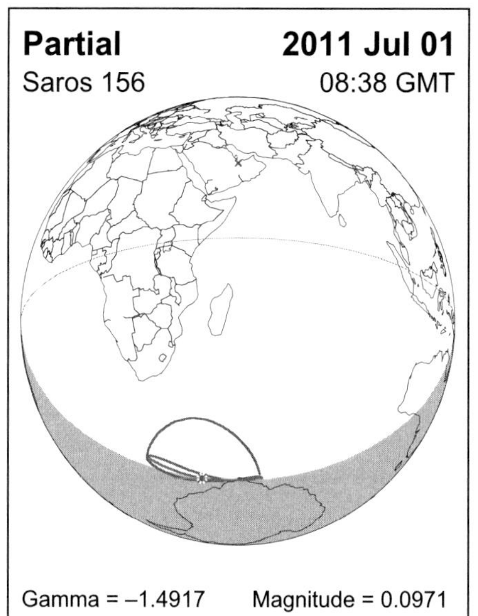

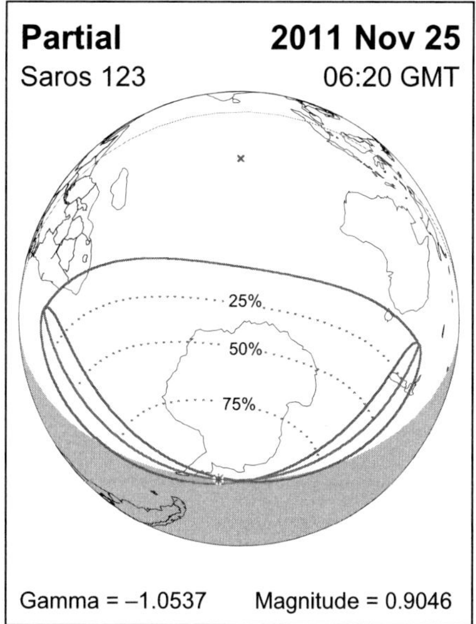

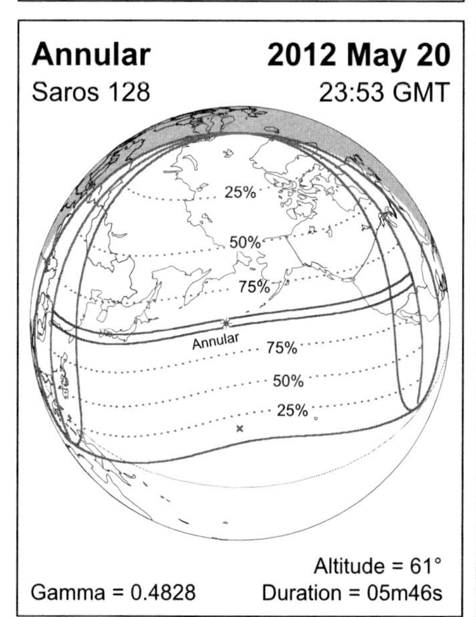

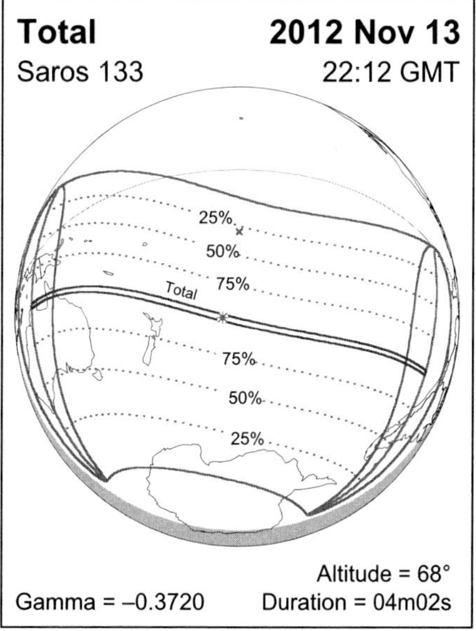

©2007 F. Espenak

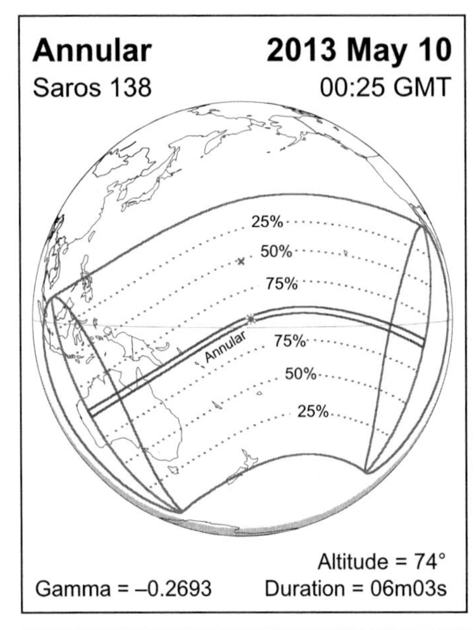
Annular
2013 May 10
Saros 138
00:25 GMT
25%
50%
75%
Annular
75%
50%
25%
Altitude = 74°
Gamma = –0.2693
Duration = 06m03s

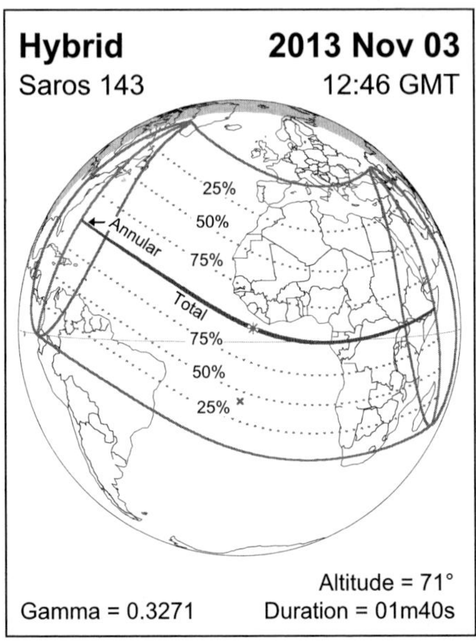
Hybrid
2013 Nov 03
Saros 143
12:46 GMT
25%
50%
Annular
75%
Total
75%
50%
25%
Altitude = 71°
Gamma = 0.3271
Duration = 01m40s

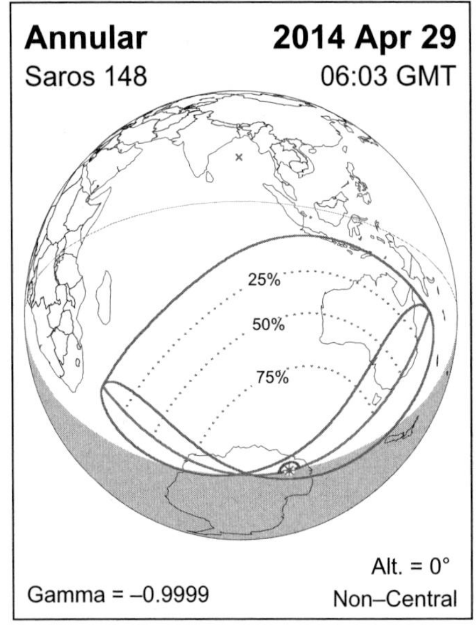
Annular
2014 Apr 29
Saros 148
06:03 GMT
25%
50%
75%
Alt. = 0°
Gamma = –0.9999
Non–Central

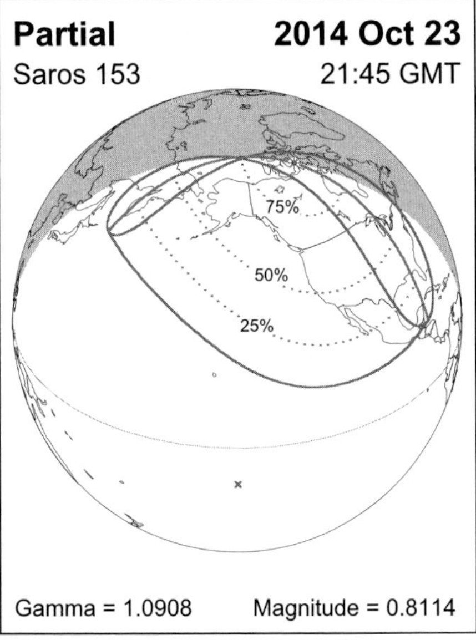
Partial
2014 Oct 23
Saros 153
21:45 GMT
75%
50%
25%
Gamma = 1.0908
Magnitude = 0.8114

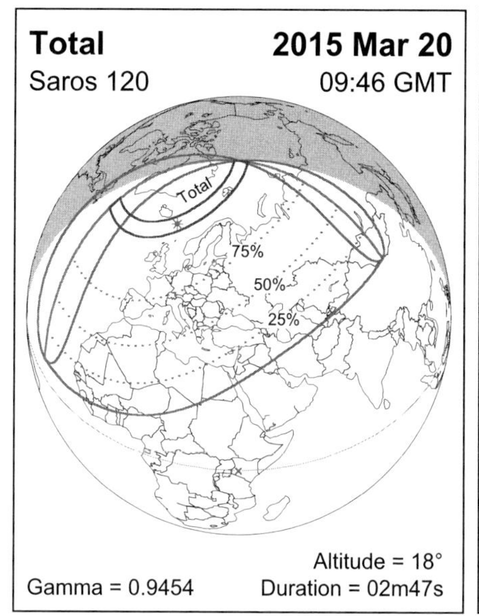
Total
2015 Mar 20
Saros 120
09:46 GMT
Total
75%
50%
25%
Altitude = 18°
Gamma = 0.9454
Duration = 02m47s

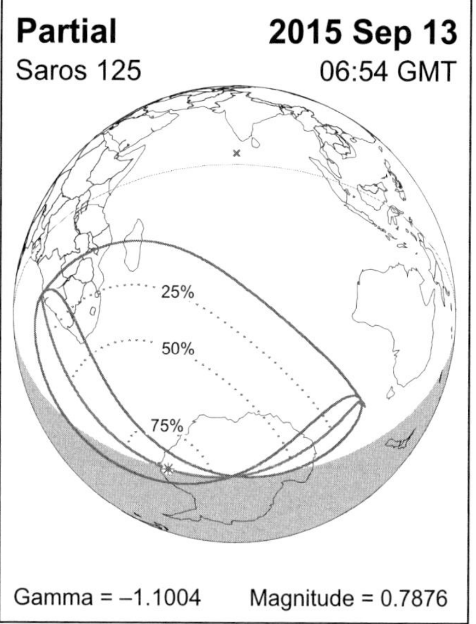
Partial
2015 Sep 13
Saros 125
06:54 GMT
25%
50%
75%
Gamma = −1.1004
Magnitude = 0.7876

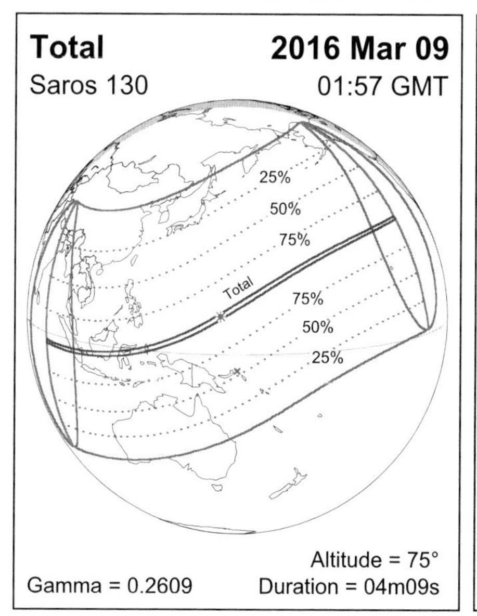
Total
2016 Mar 09
Saros 130
01:57 GMT
25%
50%
75%
Total
75%
50%
25%
Altitude = 75°
Gamma = 0.2609
Duration = 04m09s

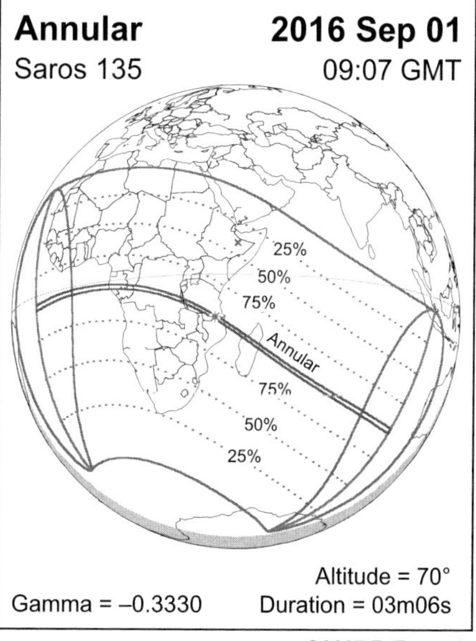
Annular
2016 Sep 01
Saros 135
09:07 GMT
25%
50%
75%
Annular
75%
50%
25%
Altitude = 70°
Gamma = −0.3330
Duration = 03m06s

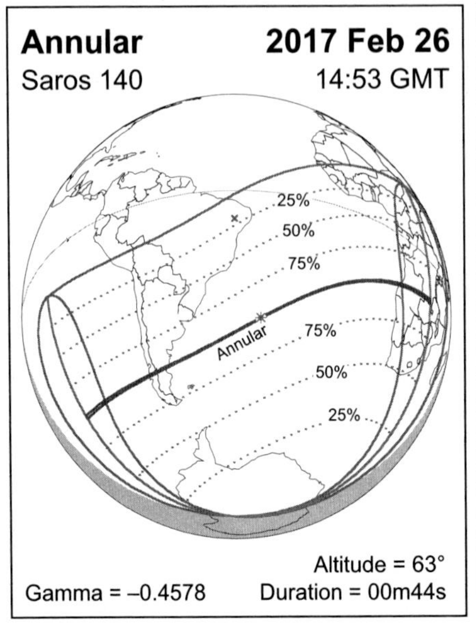
Annular
2017 Feb 26
Saros 140
14:53 GMT
25%
50%
75%
Annular
75%
50%
25%
Altitude = 63°
Gamma = –0.4578
Duration = 00m44s

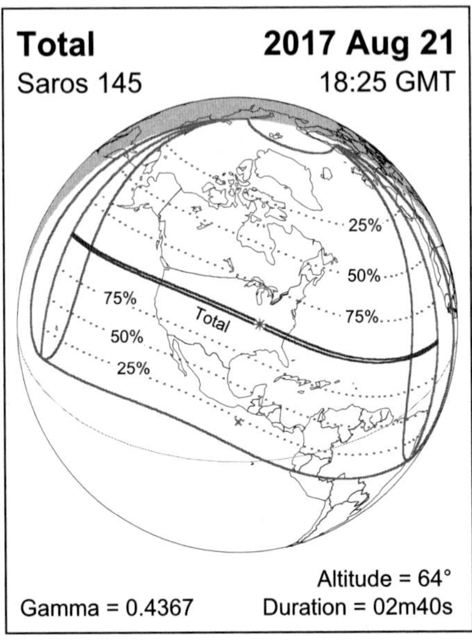
Total
2017 Aug 21
Saros 145
18:25 GMT
25%
50%
75%
Total
75%
50%
25%
Altitude = 64°
Gamma = 0.4367
Duration = 02m40s

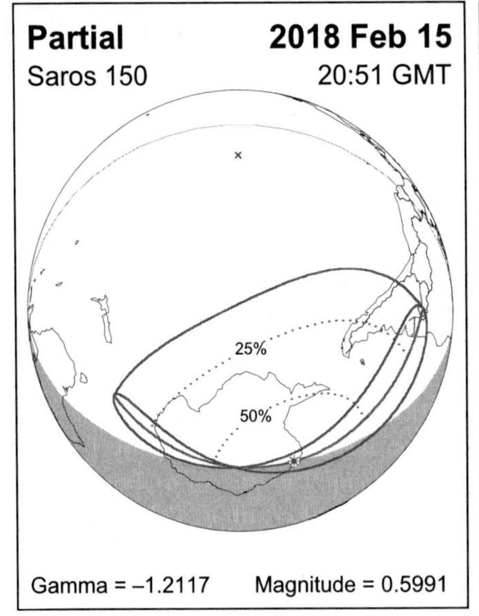
Partial
2018 Feb 15
Saros 150
20:51 GMT
25%
50%
Gamma = –1.2117
Magnitude = 0.5991

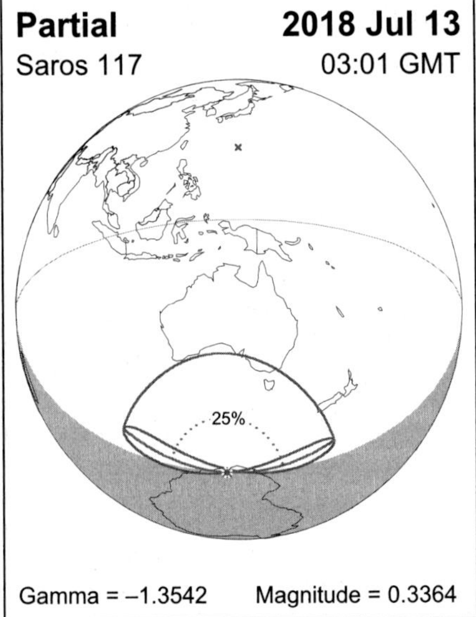
Partial
2018 Jul 13
Saros 117
03:01 GMT
25%
Gamma = –1.3542
Magnitude = 0.3364

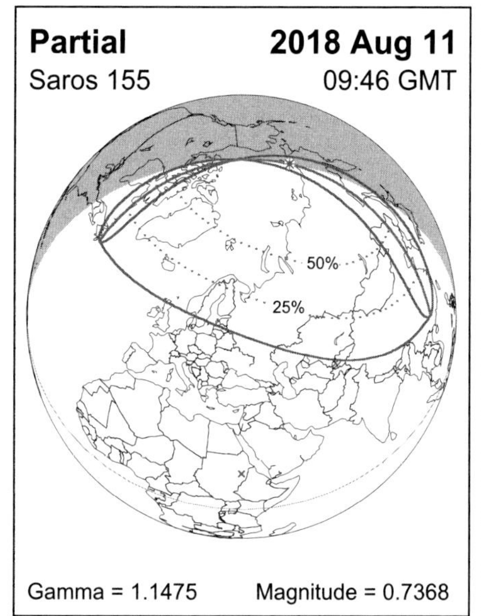

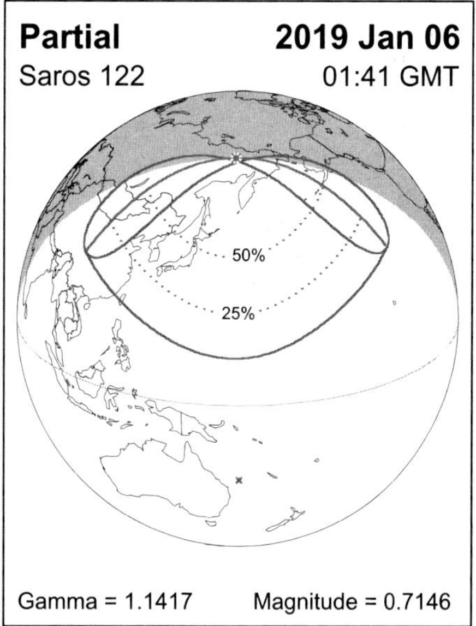

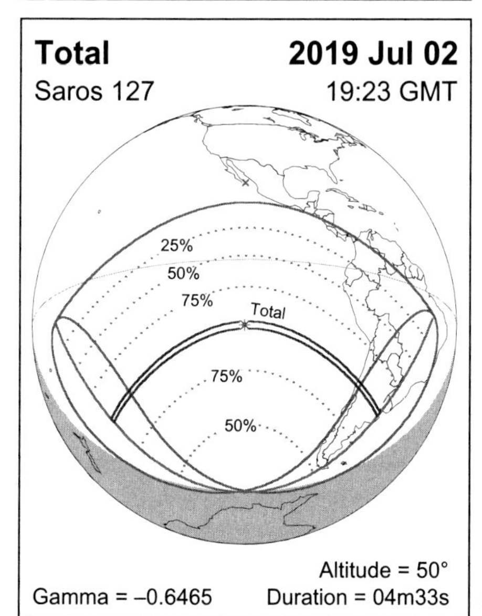

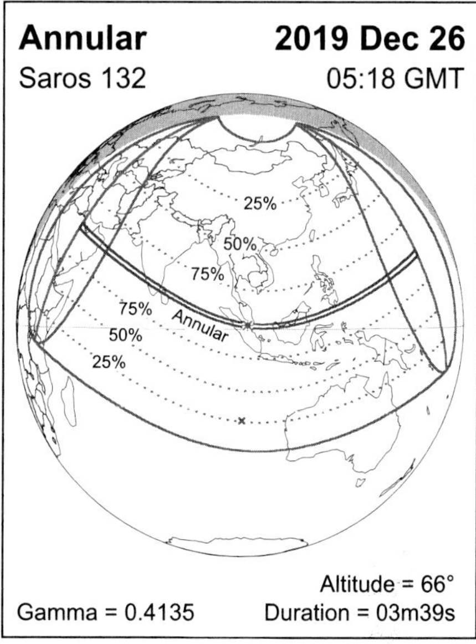

©2007 F. Espenak

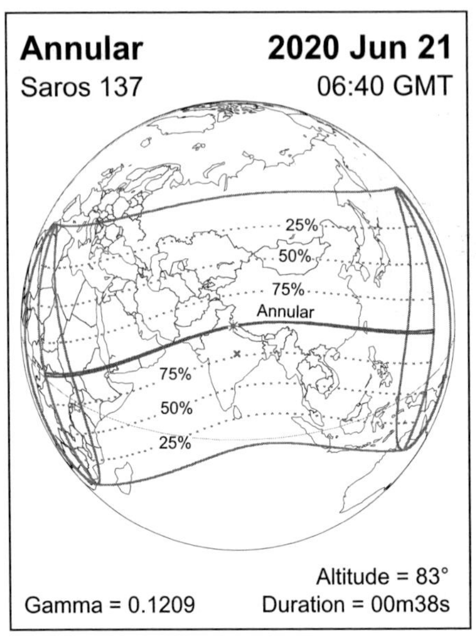

Annular
2020 Jun 21
Saros 137
06:40 GMT
25%
50%
75%
Annular
75%
50%
25%
Altitude = 83°
Gamma = 0.1209
Duration = 00m38s

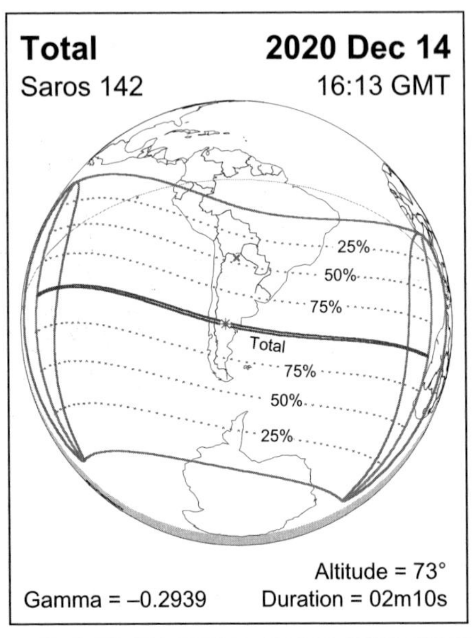

Total
2020 Dec 14
Saros 142
16:13 GMT
25%
50%
75%
Total
75%
50%
25%
Altitude = 73°
Gamma = –0.2939
Duration = 02m10s

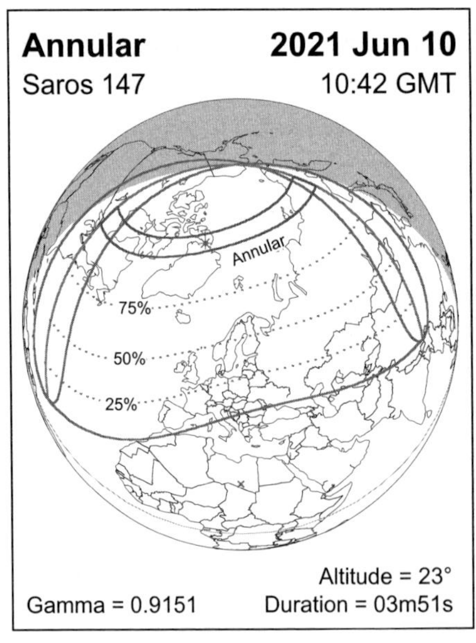

Annular
2021 Jun 10
Saros 147
10:42 GMT
Annular
75%
50%
25%
Altitude = 23°
Gamma = 0.9151
Duration = 03m51s

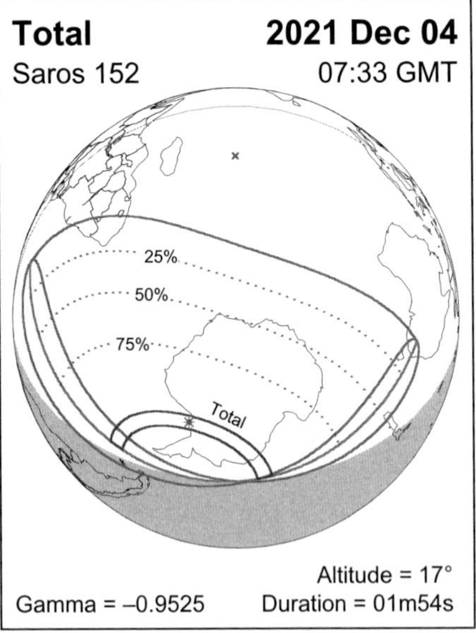

Total
2021 Dec 04
Saros 152
07:33 GMT
25%
50%
75%
Total
Altitude = 17°
Gamma = –0.9525
Duration = 01m54s

Partial **2022 Apr 30**
Saros 119 20:41 GMT

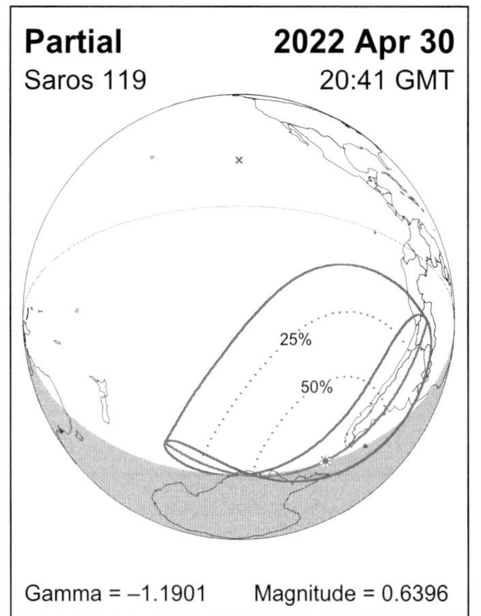

Gamma = –1.1901 Magnitude = 0.6396

Partial **2022 Oct 25**
Saros 124 11:00 GMT

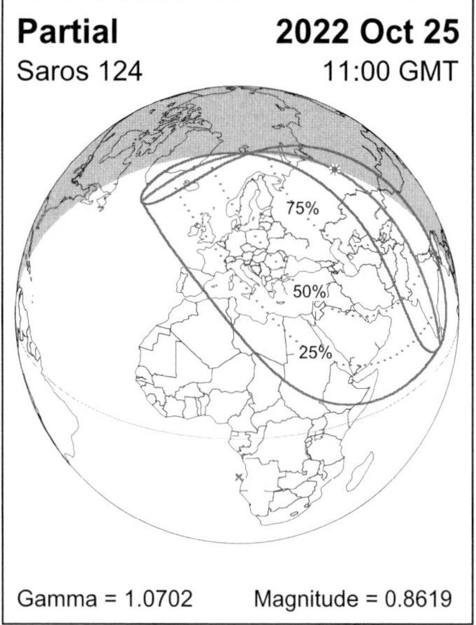

Gamma = 1.0702 Magnitude = 0.8619

Hybrid **2023 Apr 20**
Saros 129 04:17 GMT

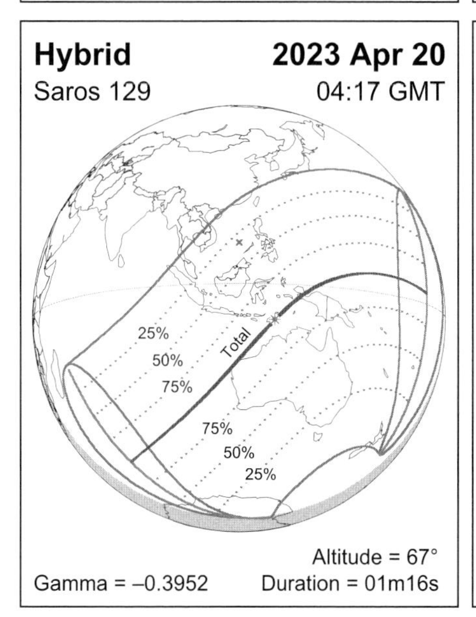

Altitude = 67°
Gamma = –0.3952 Duration = 01m16s

Annular **2023 Oct 14**
Saros 134 17:59 GMT

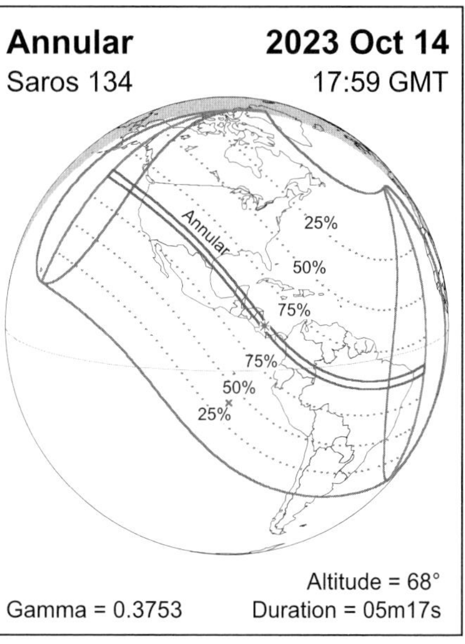

Altitude = 68°
Gamma = 0.3753 Duration = 05m17s

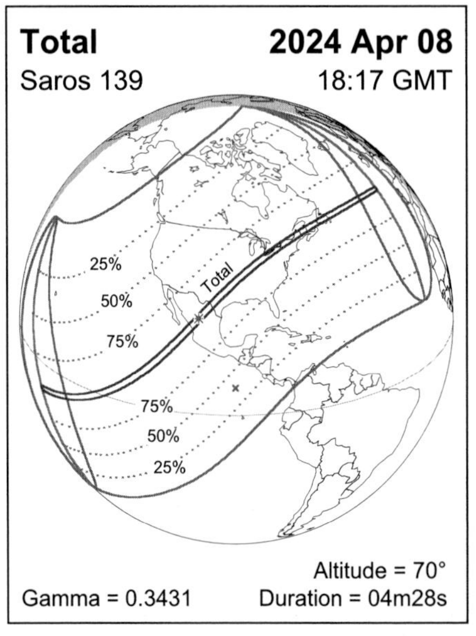
Total
2024 Apr 08
Saros 139
18:17 GMT
25%
50%
75%
Total
75%
50%
25%
Altitude = 70°
Gamma = 0.3431
Duration = 04m28s

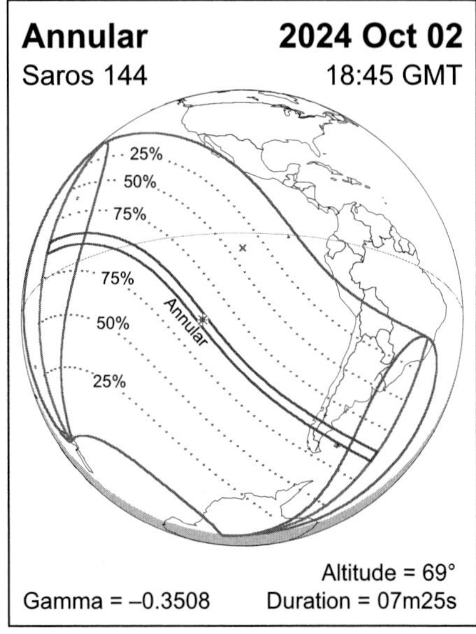
Annular
2024 Oct 02
Saros 144
18:45 GMT
25%
50%
75%
75%
50%
Annular
25%
Altitude = 69°
Gamma = –0.3508
Duration = 07m25s

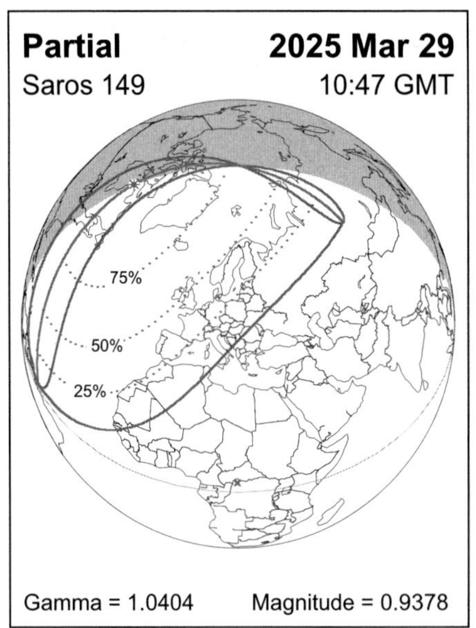
Partial
2025 Mar 29
Saros 149
10:47 GMT
75%
50%
25%
Gamma = 1.0404
Magnitude = 0.9378

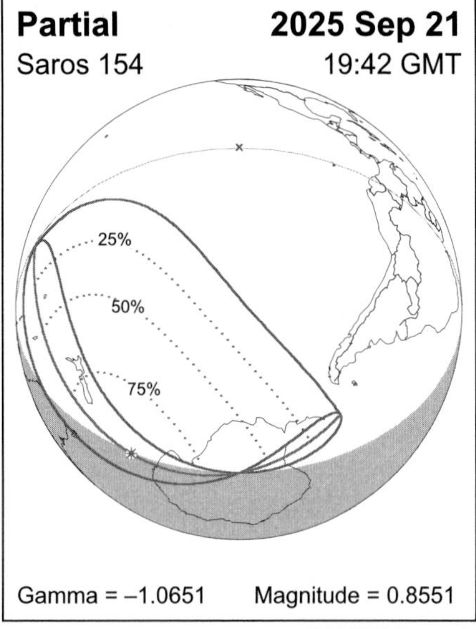
Partial
2025 Sep 21
Saros 154
19:42 GMT
25%
50%
75%
Gamma = –1.0651
Magnitude = 0.8551

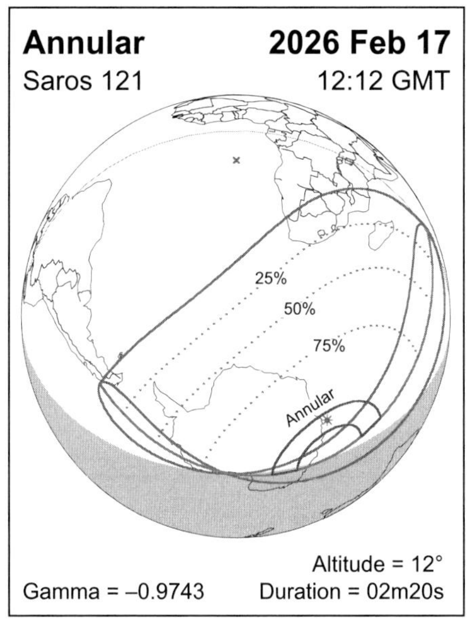
Annular
2026 Feb 17
Saros 121
12:12 GMT
25%
50%
75%
Annular
Altitude = 12°
Gamma = –0.9743
Duration = 02m20s

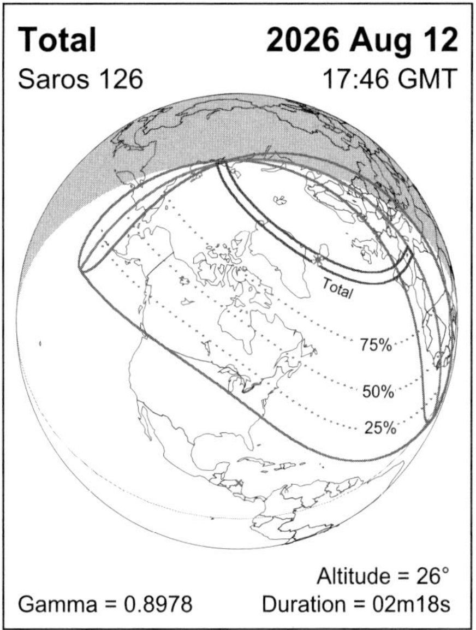
Total
2026 Aug 12
Saros 126
17:46 GMT
Total
75%
50%
25%
Altitude = 26°
Gamma = 0.8978
Duration = 02m18s

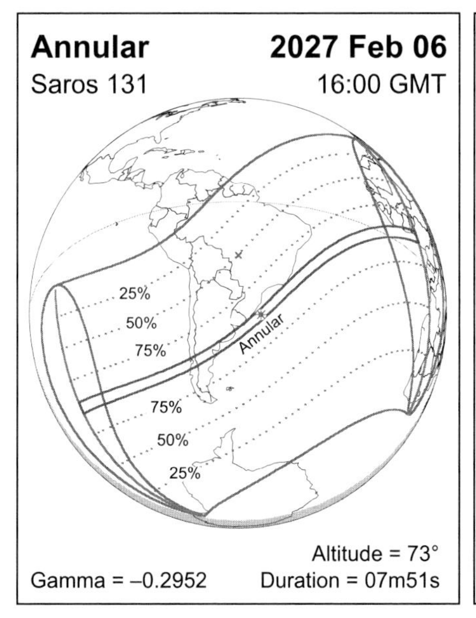
Annular
2027 Feb 06
Saros 131
16:00 GMT
25%
50%
75%
Annular
75%
50%
25%
Altitude = 73°
Gamma = –0.2952
Duration = 07m51s

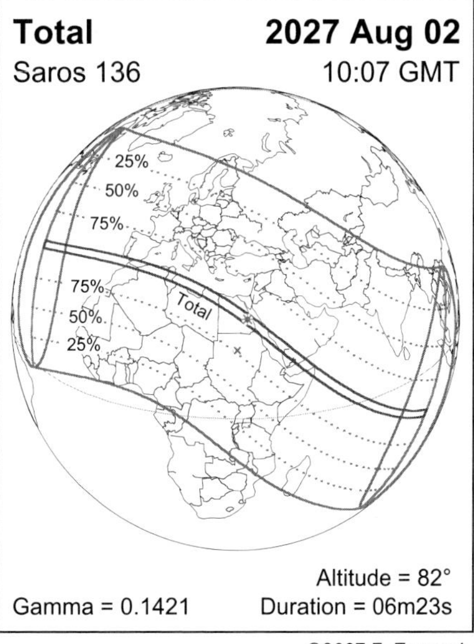
Total
2027 Aug 02
Saros 136
10:07 GMT
25%
50%
75%
75%
50%
25%
Total
Altitude = 82°
Gamma = 0.1421
Duration = 06m23s

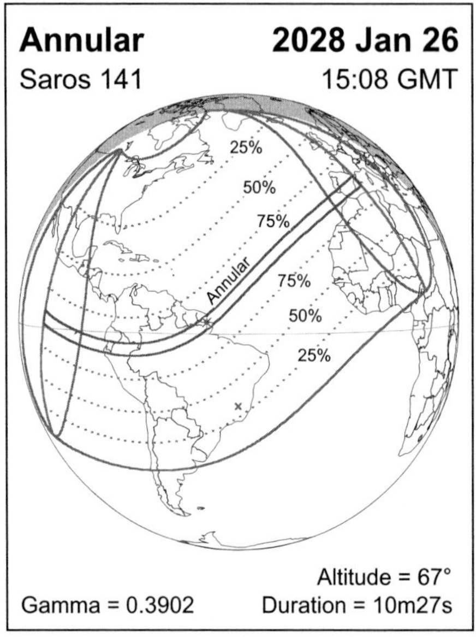
Annular
2028 Jan 26
Saros 141
15:08 GMT
25%
50%
75%
Annular
75%
50%
25%
Altitude = 67°
Gamma = 0.3902
Duration = 10m27s

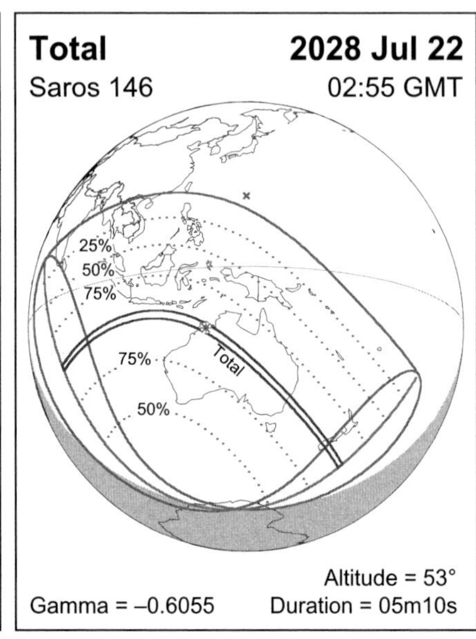
Total
2028 Jul 22
Saros 146
02:55 GMT
25%
50%
75%
75%
Total
50%
Altitude = 53°
Gamma = –0.6055
Duration = 05m10s

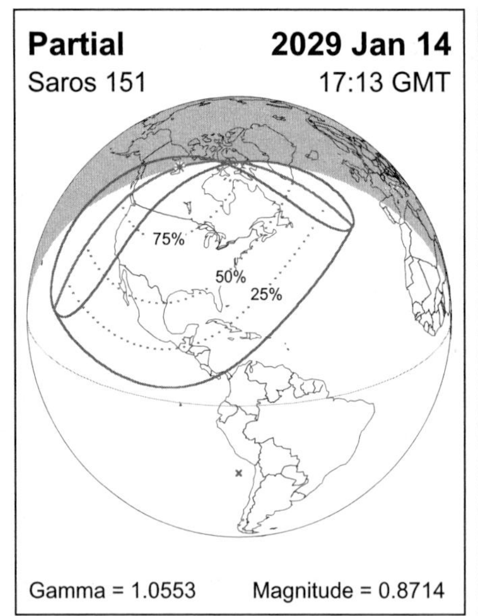
Partial
2029 Jan 14
Saros 151
17:13 GMT
75%
50%
25%
Gamma = 1.0553
Magnitude = 0.8714

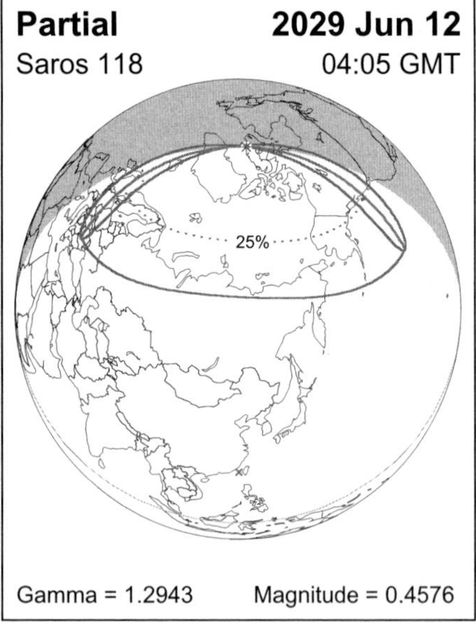
Partial
2029 Jun 12
Saros 118
04:05 GMT
25%
Gamma = 1.2943
Magnitude = 0.4576

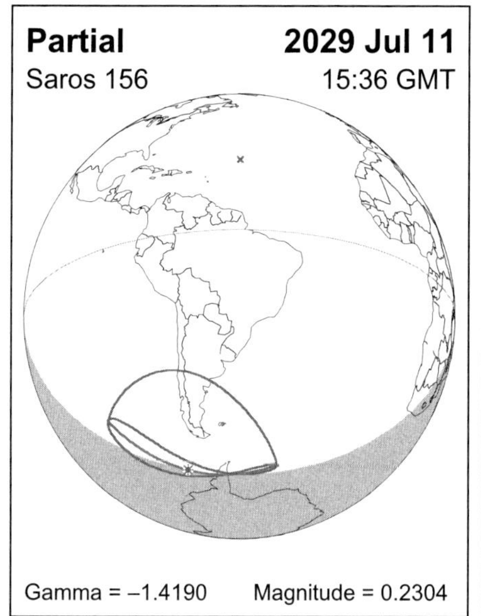
Partial
2029 Jul 11
Saros 156
15:36 GMT
Gamma = –1.4190
Magnitude = 0.2304

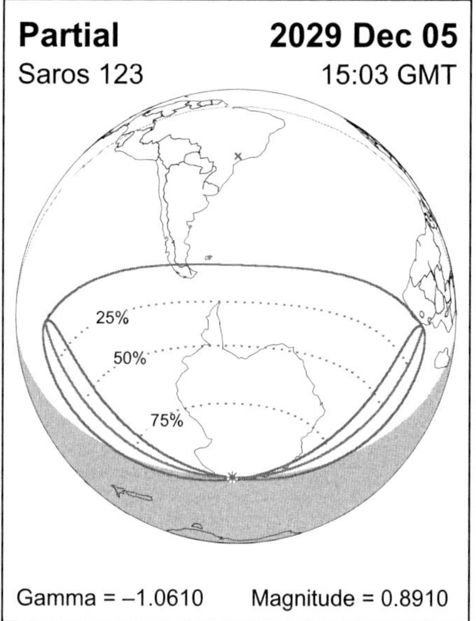
Partial
2029 Dec 05
Saros 123
15:03 GMT
25%
50%
75%
Gamma = –1.0610
Magnitude = 0.8910

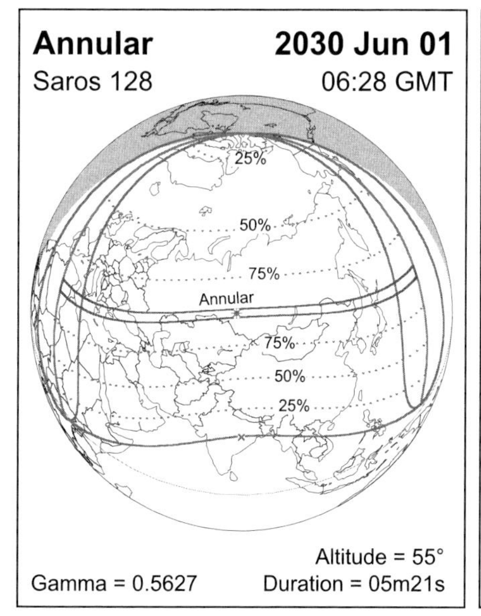
Annular
2030 Jun 01
Saros 128
06:28 GMT
25%
50%
75%
Annular
75%
50%
25%
Altitude = 55°
Gamma = 0.5627
Duration = 05m21s

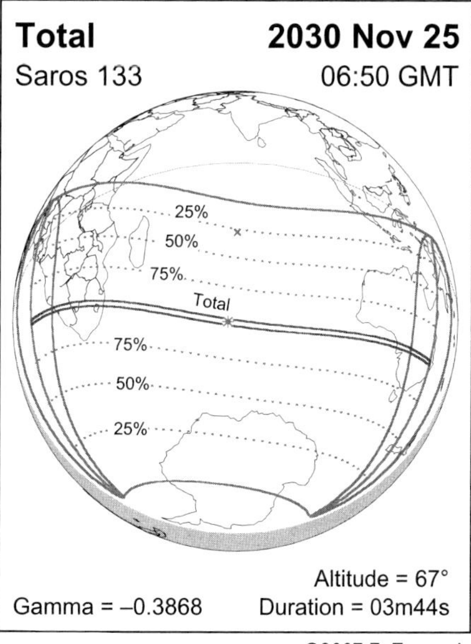
Total
2030 Nov 25
Saros 133
06:50 GMT
25%
50%
75%
Total
75%
50%
25%
Altitude = 67°
Gamma = –0.3868
Duration = 03m44s

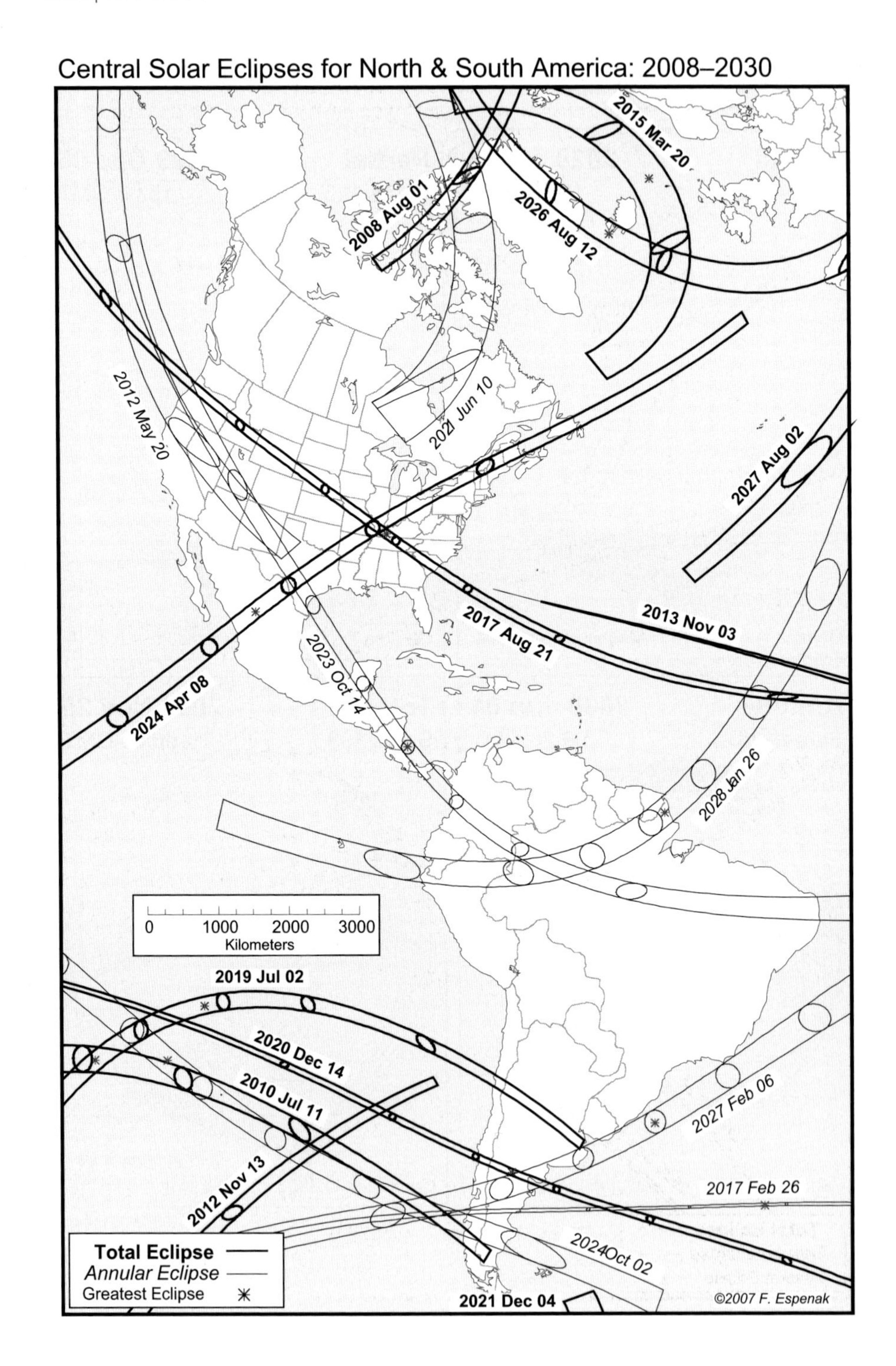
Central Solar Eclipses for North & South America: 2008–2030
2015 Mar 20
2008 Aug 01
2026 Aug 12
2012 May 20
2021 Jun 10
2027 Aug 02
2013 Nov 03
2017 Aug 21
2023 Oct 14
2024 Apr 08
2028 Jan 26
0
1000
2000
3000
Kilometers
2019 Jul 02
2020 Dec 14
2010 Jul 11
2027 Feb 06
2012 Nov 13
2017 Feb 26
2024Oct 02
Total Eclipse
Annular Eclipse
Greatest Eclipse
2021 Dec 04
©2007 F. Espenak

Central Solar Eclipses for Europe & Africa: 2008–2030

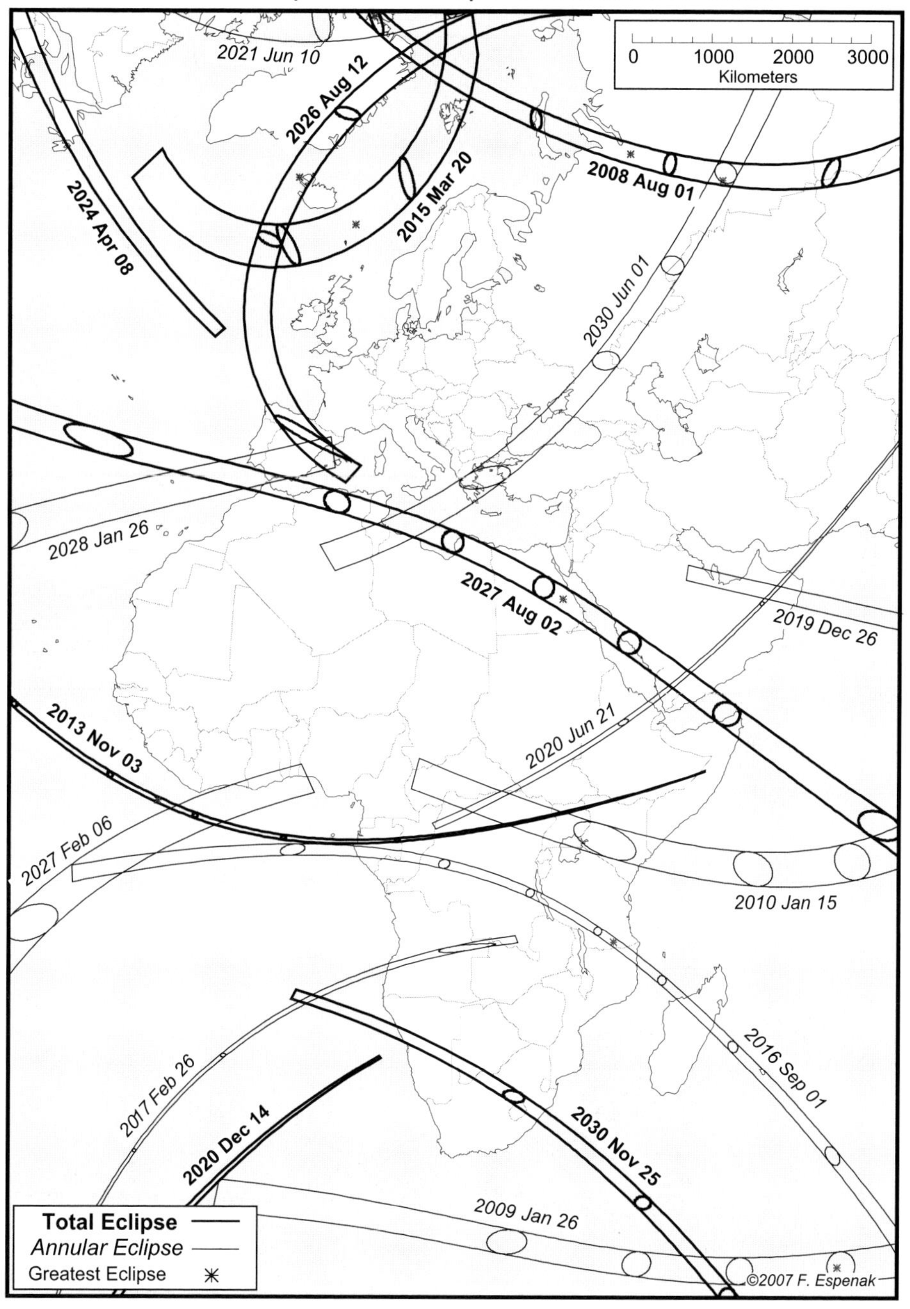

Central Solar Eclipses for Asia & Australia: 2008–2030

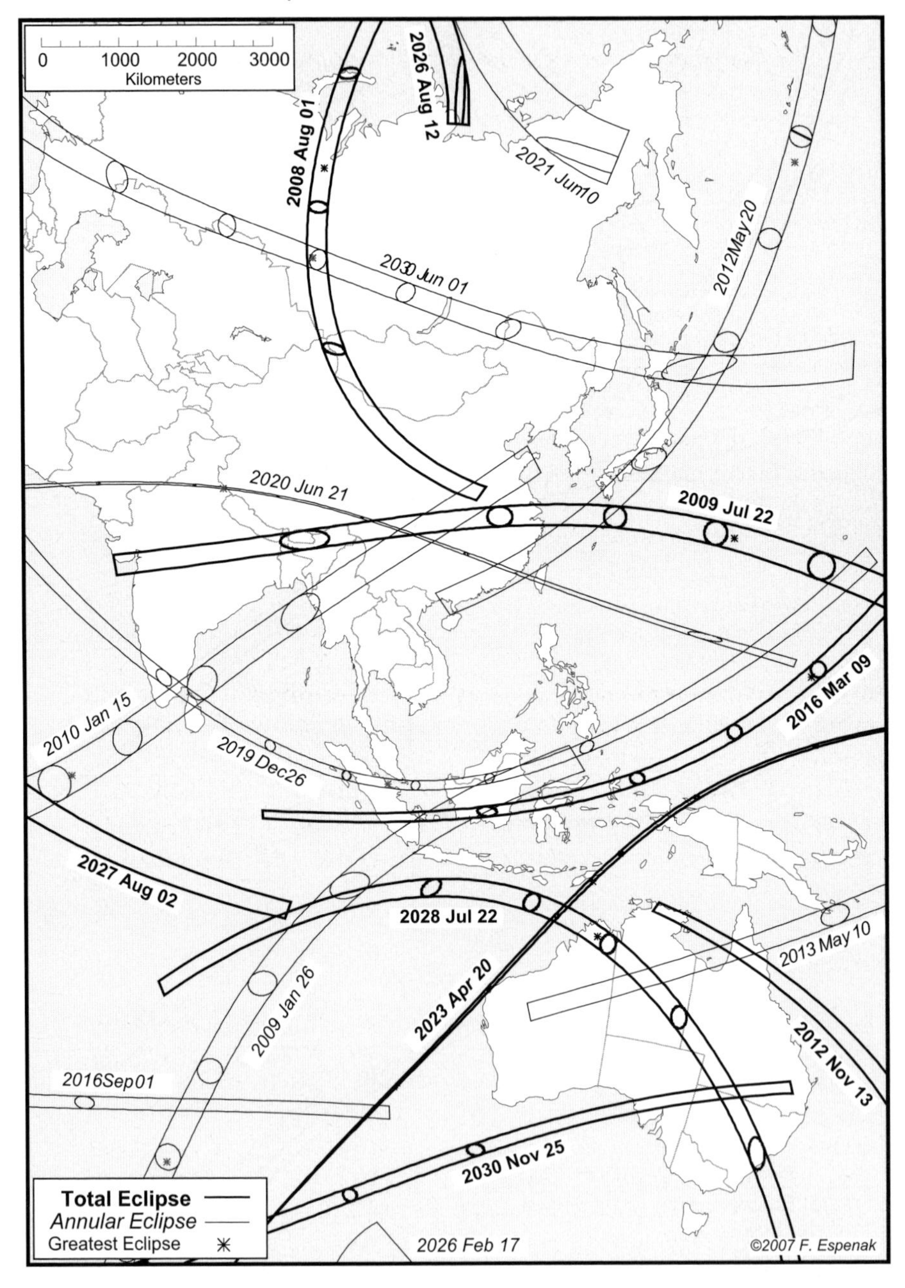

Epilogue
Eclipses—Cosmic Perspective, Human Perspective

No one should pass through life without seeing a total solar eclipse.
Leif Robinson (1999)

The Moon's shadow has darkened space for 4½ billion years, since the Sun formed and began to shine and the planets and their moons formed and could not shine. From the newly formed Sun, light streamed outward in all directions. Here and there it illuminated a body of rock or gas or ice. The Sun's dominant light identified that world as one of its own children.

A little over eight minutes of light-travel outbound from the Sun, a portion of the light encountered two small dark bodies and bathed the sunward half of each in brightness. The surrounding flood of unimpeded light sped on, leaving a long cone of darkness—a shadow—behind the planet and its one large moon.

The two worlds lay the same average distance from the Sun. Not long before—measured on a cosmic timescale—Earth and Moon had been a single body, but they were separate and utterly different now.

A Cosmic Birth

The Sun, planets, moons, comets, and asteroids had all begun within a cloud of gas and dust, a cloud so large that there was material enough to make dozens or hundreds of solar systems. Where the density was great enough, fragments of the cloud began to condense by gravity. At the heart of each fragment, a star was coalescing. Near those stars-to-be, other bodies, too low in mass ever to reach starhood, also began to form. These small bodies were planetesimals, the beginnings of the planets. At first they grew by gentle collisions and adhesions, gathering up a grain of ice, a fleck of dust along their paths around the Sun until icy or rocky planetesimals had taken shape. And still they gathered dust and small debris until they were so massive that gravity became their prime means of growth, gathering to them still more materials—and other planetesimals. The number of small bodies declined. The size of a few large bodies increased. The planets had been formed by convergence.

Convergence brought near disaster as well. Another planet-size body (the size of Mars today) wandered across the path of a body that would become the Earth. No living thing witnessed the collision. No living thing could have survived the collision that obliterated the smaller world and nearly shattered the Earth. The Earth recoiled from the impact by spewing molten fragments of its crust and mantle outward. Some escaped from Earth; others rained down from the skies, pelting the surface in a rock storm of unimaginable proportions. But many of the fragments, caught in the Earth's gravity, stayed aloft, orbiting the Earth as the Earth orbited the Sun. Quickly, in 10,000 years or less, the fragments joined together by collisions and accretion to form a new world circling the first. That new world was the Moon.[1]

From convergence had come divergence. The two worlds, sprung from one, continued to diverge. Both were the same distance from the Sun, but the Earth was 81 times more massive than the Moon. That mass allowed the Earth to hold an atmosphere by gravity, while the Moon could not.

The eons passed. Life arose on Earth and covered the planet. Plants and animals responded to the tides raised by the Moon. The lunar tides slowed the Earth and caused the Moon to spiral slowly outward, diverging ever farther from the Earth in distance and ever further from the Earth in environment as well.

The lifeless Moon withdrew until today its shadow can just barely reach the Earth. As shadows in the universe go, this one is of no great size: a cone of darkness extending at most only 236,000 miles (379,900 kilometers) in length before dwindling to a point. It is long enough to touch the Earth only occasionally and very briefly and with a single narrow stroke.

The black insubstantial cone reaches out, but for most of the time there is nothing to touch. The shadow sweeps on through space unseen, unnoticed.

Contact

Yet now ahead lies the Earth. It is a special day. Suddenly the Moon's shadow becomes visible as a point of darkness collides with a world of rock and water and air. The shadow swoops in from the heavens, silently darkening the sky as it alights upon the Earth and begins its ceaseless rush across the planet's surface.

Awesome totality—June 21, 2001 from Zambia. [Nikon 8008, MF-21 program back, Sigma 18–35 mm at 18 mm, f/5.6, autoexposure, Kodak Royal Gold 200 negatives. © 2001 Fred Espenak]

On these occasions, the people of Earth have gathered and still gather scientific knowledge from the Sun. And more. For long before we drew information from eclipses, we stared in wonder at them. And long after all knowledge from eclipses has been gleaned, people will still travel to the ends of the Earth to treasure their majesty and beauty.

So compelling are they that when at last the Moon has drifted too far to touch the Earth with darkness, the beings of that era may use solar power or other means to gently, gradually halt and reverse the Moon's retreat so that they too will see the sight that their books and visual records can only hint of.

Notes and Reference

Epigraph: Leif Robinson: Foreword to *Totality: Eclipses of the Sun*, 2nd edition, 1999.

[1] The leading theory of the Moon's formation involves a collision between a large planetesimal and the proto Earth.

Appendix A
Total, Annular, and Hybrid Eclipses: 2008–2060[1]

During the 52-year period 2008–2060, there are 119 eclipses of the Sun—total, annular, hybrid (annular–total), and partial. Here's how they break down:

Partial	40 = 33.6%
Annular	39 = 32.8%
Total	35 = 29.4%
Hybrid	5 = 4.2%

The table below gives the basic details for the 79 total, annular, and hybrid eclipses occurring during this period. When an eclipse path crosses the International Dateline, it can be seen on one of two dates depending on the observer's geographic position.

Date	*Maximum Duration*[a]	*Eclipse Type*	*Geographic Region of Visibility*
2008 Feb 07	*2m12s*	*Annular*	*South Pacific Ocean, Antarctica*
2008 Aug 01	**2m27s**	**Total**	**Northern Canada, Greenland, Arctic Ocean, Russia, Mongolia, China**
2009 Jan 26	*7m54s*	*Annular*	*South Atlantic Ocean, Indian Ocean, Indonesia (Sumatra, Java, Borneo)*
2009 Jul 22	**6m39s**	**Total**	**India, Nepal, Bhutan, Myanmar, China, Pacific Ocean**
2010 Jan 15	*11m08s*	*Annular*	*Central Africa Republic, Congo, Uganda, Kenya, Somalia, Indian Ocean, Maldives, southernmost India, Sri Lanka, Bangladesh, Myanmar, China*
2010 Jul 10/11	**5m20s**	**Total**	**South Pacific Ocean, Easter Island, Chile, Argentina**

Date	*Maximum Duration*[a]	*Eclipse Type*	*Geographic Region of Visibility*
2012 May 20/21	*5m46s*	*Annular*	*China, Taiwan, Japan, Pacific Ocean, USA (Oregon, California, Nevada, Utah, Arizona, Colorado, New Mexico, Texas)*
2012 Nov 13/14	**4m02s**	**Total**	**Northern Australia, south Pacific Ocean**
2013 May 9/10	*6m03s*	*Annular*	*Australia, Solomon Islands, Nauru, Pacific Ocean*
2013 Nov 03	***1m40s***	***Hybrid***	***Atlantic Ocean, Gabon, Congo, Democratic Republic of the Congo, Uganda, Kenya, Ethiopia (annular only at beginning of path)***
2014 Apr 29	—	*Annular*	*Antarctica (extension of shadow cone only grazes Earth)*
2015 Mar 20	**2m47s**	**Total**	**North Atlantic Ocean, Faroe Islands (Denmark), Arctic Ocean, Svalbard (Norway)**
2016 Mar 09	**4m09s**	**Total**	**Indonesia (Sumatra, Borneo, Sulawesi, Halmahera), Pacific Ocean**
2016 Sep 01	*3m06s*	*Annular*	*Atlantic Ocean, Gabon, Congo, Tanzania, Mozambique, Madagascar, Indian Ocean*
2017 Feb 26	*0m44s*	*Annular*	*Chile, Argentina, south Atlantic Ocean, Angola, Zambia, Congo*
2017 Aug 21	**2m40s**	**Total**	**Pacific Ocean, USA (Oregon, Idaho, Wyoming, Nebraska, Missouri, Illinois, Kentucky, Tennessee, North Carolina, Georgia, South Carolina), Atlantic Ocean**
2019 Jul 02	**4m33s**	**Total**	**South Pacific Ocean, Chile, Argentina**
2019 Dec 26	*3m39s*	*Annular*	*Saudi Arabia, Bahrain, United Arab Emirates, Oman, southern India, Indonesia, Malaysia*
2020 Jun 21	*0m39s*	*Annular*	*Congo, Sudan, Ethiopia, Yemen, Saudi Arabia, Oman, Pakistan, northern India, China, Taiwan*
2020 Dec 14	**2m10s**	**Total**	**Pacific Ocean, Chile, Argentina, south Atlantic Ocean**
2021 Jun 10	*3m51s*	*Annular*	*Central Canada, north pole, Russia (Siberia)*
2021 Dec 04	**1m54s**	**Total**	**Antarctica**
2023 Apr 20	***1m16s***	***Hybrid***	***South Indian Ocean, western Australia, Indonesia, Pacific Ocean (total except at beginning and end of path)***

Date	*Maximum Duration*[a]	*Eclipse Type*	*Geographic Region of Visibility*
2023 Oct 14	*5m17s*	*Annular*	*USA (Oregon through Texas), Mexico (Yucatán peninsula), Central America, Colombia, Brazil*
2024 Apr 8	**4m28s**	**Total**	**Pacific Ocean, Mexico, USA (Texas, Oklahoma, Arkansas, Missouri, Kentucky, Illinois, Indiana, Ohio, Pennsylvania, New York, Vermont, New Hampshire, Maine), southeastern Canada, Atlantic Ocean**
2024 Oct 02	*7m25s*	*Annular*	*South Pacific Ocean, southern Chile and Argentina*
2026 Feb 17	*2m20s*	*Annular*	*Antarctica*
2026 Aug 12	**2m18s**	**Total**	**Greenland, Iceland, Spain**
2027 Feb 06	*7m51s*	*Annular*	*Southern Chile, Argentina, south Atlantic Ocean, Ivory Coast, Ghana, Togo, Benin, Nigeria*
2027 Aug 02	**6m23s**	**Total**	**Atlantic Ocean, Morocco, Spain, Algeria, Libya, Egypt, Saudi Arabia, Yemen, Somalia**
2028 Jan 26	*10m27s*	*Annular*	*Ecuador, Peru, Brazil, French Guiana, Portugal, Spain*
2028 Jul 22	**5m10s**	**Total**	**South Indian Ocean, Australia, New Zealand**
2030 Jun 01	*5m21s*	*Annular*	*Algeria, Tunisia, Greece, Turkey, Kazakhstan, Russia, northern China, Japan*
2030 Nov 25	**3m44s**	**Total**	**South West Africa, Botswana, South Africa, south Indian Ocean, southeastern Australia**
2031 May 21	*5m26s*	*Annular*	*Angola, Zambia, Congo, Tanzania, southern India, Sri Lanka, Malaysia, Indonesia*
2031 Nov 14/15	***1m08s***	***Hybrid***	***Pacific Ocean (total), Panama (annular)***
2032 May 09	*0m22s*	*Annular*	*South Atlantic Ocean*
2033 Mar 30	**2m37s**	**Total**	**Alaska, Arctic Ocean**
2034 Mar 20	**4m09s**	**Total**	**Atlantic, Nigeria, Cameroon, Chad, Sudan, Egypt, Saudi Arabia, Iran, Afghanistan, Pakistan, India, China**
2034 Sep 12	*2m58s*	*Annular*	*Pacific, Chile, Bolivia, Argentina, Paraguay, Brazil, Atlantic*
2035 Mar 09/10	*0m48s*	*Annular*	*South Pacific Ocean*

Date	*Maximum Duration*[a]	*Eclipse Type*	*Geographic Region of Visibility*
2035 Sep 02	**2m54s**	**Total**	**China, Korea, Japan, Pacific Ocean**
2037 Jul 13	**3m58s**	**Total**	**Australia, New Zealand, south Pacific Ocean**
2038 Jan 05	*3m19s*	*Annular*	*Cuba, Haiti, Dominican Republic, southern Caribbean islands, Atlantic Ocean, Liberia, Ivory Coast, Ghana, Togo, Benin, Niger, Libya, Chad, Egypt*
2038 Jul 02	*1m00s*	*Annular*	*Colombia, Venezuela, Atlantic Ocean, Western Sahara, Morocco, Mauritania, Mali, Algeria, Niger, Chad, Sudan, Ethiopia, Kenya*
2038 Dec 25/26	**2m18s**	**Total**	**Indian Ocean, western and southern Australia, New Zealand, Pacific Ocean**
2039 Jun 21	*4m05s*	*Annular*	*USA (Alaska), Canada, Greenland, Norway, Sweden, Finland, Estonia, Russia*
2039 Dec 15	**1m51s**	**Total**	**Antarctica**
2041 Apr 30	**1m51s**	**Total**	**South Atlantic, Angola, Congo, Uganda, Kenya, Somalia**
2041 Oct 24/25	*6m07s*	*Annular*	*Mongolia, China, Korea, Japan, Pacific Ocean*
2042 Apr 19/20	**4m51s**	**Total**	**Indonesia, Malaysia, Philippines, Pacific Ocean**
2042 Oct 14	*7m44s*	*Annular*	*Thailand, Malaysia, Indonesia, Australia, New Zealand*
2043 Apr 09	**—**	**Total**	**Northeastern Russia (shadow cone only grazes Earth)**
2043 Oct 03	*—*	*Annular*	*South Indian Ocean (shadow cone only grazes Earth)*
2044 Feb 28	*2m27s*	*Annular*	*South Atlantic Ocean*
2044 Aug 23	**2m04s**	**Total**	**Greenland, Canada, USA (Montana, North Dakota)**
2045 Feb 16/17	*7m47s*	*Annular*	*New Zealand, Pacific Ocean*
2045 Aug 12	**6m06s**	**Total**	**USA (California, Nevada, Utah, Colorado, Kansas, Oklahoma, Arkansas, Mississippi, Alabama, Georgia, Florida), Haiti, Dominican Republic, Venezuela, Guyana, Suriname, French Guiana, Brazil**
2046 Feb 05/06	*9m42s*	*Annular*	*Indonesia, Papua New Guinea, Pacific Ocean, USA (California, Oregon, Nevada, Idaho)*

Date	*Maximum Duration*[a]	*Eclipse Type*	*Geographic Region of Visibility*
2046 Aug 02	**4m51s**	**Total**	**Eastern Brazil, Atlantic Ocean, Angola, Namibia, Botswana, South Africa, south Indian Ocean**
2048 Jun 11	*4m58s*	*Annular*	*USA (Kansas, Nebraska, Missouri, Iowa, Minnesota, Illinois, Wisconsin, Michigan), Canada, Greenland, Iceland, Norway, Sweden, Latvia, Estonia, Belarus, Russia, Afghanistan*
2048 Dec 05	**3m28s**	**Total**	**South Pacific Ocean, Chile, Argentina, south Atlantic Ocean, Namibia, Botswana**
2049 May 31	*4m45s*	*Annular*	*Peru, Colombia, Brazil, Venezuela, Guyana, Suriname, Atlantic Ocean, Senegal, Mali, Guinea, Burkina, Ghana, Togo, Benin, Nigeria, Cameroon, Congo*
2049 Nov 25	***0m38s***	***Hybrid***	***Total for part of Indian Ocean and parts of Indonesia; annular for Saudi Arabia, Oman, part of Indian Ocean, eastern Indonesia, Pacific Ocean***
2050 May 20/21	**0m21s**	**Total**	**South Pacific Ocean**
2052 Mar 30	**04m08s**	**Total**	**Pacific Ocean, Mexico, USA (Texas, Louisiana, Mississippi, Alabama, Georgia, Florida, South Carolina), Atlantic Ocean**
2052 Sep 22/23	*2m51s*	*Annular*	*Indonesia, Australia, south Pacific Ocean*
2053 Mar 20	*0m49s*	*Annular*	*Indian Ocean, Indonesia, Papua New Guinea*
2053 Sep 12	**3m04s**	**Total**	**Morocco, Algeria, Tunisia, Libya, Egypt, Saudi Arabia, Indonesia (Sumatra)**
2055 Jul 24	**3m17s**	**Total**	**South Atlantic Ocean, South Africa, south Indian Ocean**
2056 Jan 16/17	*2m52s*	*Annular*	*Central Pacific Ocean, Mexico, USA (Texas)*
2056 Jul 12	*1m26s*	*Annular*	*Central Pacific Ocean, Colombia, Ecuador, Peru, Brazil*
2057 Jan 05	**2m29s**	**Total**	**South Atlantic Ocean, south Indian Ocean**
2057 Jul 01/02	*4m22s*	*Annular*	*China, Mongolia, Russia, Canada, USA (Alaska, Michigan)*
2057 Dec 26	**1m50s**	**Total**	**Antarctica**
2059 May 11	**2m23s**	**Total**	**Pacific Ocean, Ecuador, Peru, Brazil**

Date	*Maximum Duration*[a]	*Eclipse Type*	*Geographic Region of Visibility*
2059 Nov 05	*7m00s*	*Annular*	*France, Libya, Egypt, Sudan, Ethiopia, Eritrea, Somalia, Indian Ocean, Indonesia (Sumatra)*
2060 Apr 30	**5m15s**	**Total**	**Ivory Coast, Ghana, Togo, Benin, Nigeria, Niger, Chad, Libya, Egypt, Cyprus, Turkey, Syria, Armenia, Azerbaijan, Turkmenistan, Kazakhstan, Uzbekistan, Kyrgyzstan, China**
2060 Oct 24	*8m06s*	*Annular*	*Guinea, Sierra Leone, Liberia, Ivory Coast, Angola, Namibia, Botswana, South Africa*

[a] Maximum duration of totality or annularity as seen from the central line in minutes and seconds.

Notes and Reference

[1] Based on Fred Espenak: *Fifty Year Canon of Solar Eclipses: 1986–2035* (Cambridge, Massachusetts: Sky Publishing, 1987; NASA Reference Publication 1178, revised) and Fred Espenak and Jean Meeus: *Five Millennium Canon of Solar Eclipses: −1999 to +3000* (Greenbelt, Maryland: Goddard Space Flight Center, 2006; NASA Technical Publication 2006–214141).

Appendix B
Recent Total, Annular, and Hybrid Eclipses: 1970–2008[1]

The table below gives the basic details for the 55 total, annular, and hybrid eclipses occurring during the 39-year period 1970–2008. When an eclipse path crosses the International Dateline, it can be seen on one of two dates depending on the observer's geographic position.

Date	*Maximum Duration*[a]	*Eclipse Type*	*Geographic Region of Visibility*
1970 Mar 07	**3m28s**	**Total**	**Pacific Ocean, Mexico, USA (Florida, Georgia, South Carolina, North Carolina, Virginia), eastern Canada, north Atlantic Ocean**
1970 Aug 31	*6m48s*	*Annular*	*Papua New Guinea, south Pacific Ocean*
1972 Jan 16	*1m53s*	*Annular*	*Antarctica*
1972 Jul 10	**2m36s**	**Total**	**Russia, USA (Alaska), Canada, Atlantic Ocean**
1973 Jan 04	*7m48s*	*Annular*	*South Pacific Ocean, Chile, Argentina, south Atlantic Ocean*
1973 Jun 30	**7m04s**	**Total**	**Guyana, Suriname, French Guiana, Atlantic Ocean, Mauritania, Mali, Algeria, Niger, Chad, Central African Republic, Sudan, Uganda, Kenya, Somalia, Indian Ocean**
1973 Dec 24	*12m03s*	*Annular*	*Costa Rica, Panama, Colombia, Venezuela, Brazil, Guyana, Atlantic Ocean, Mauritania, Mali, Algeria*
1974 Jun 20	**5m09s**	**Total**	**South Indian Ocean, southwestern Australia**
1976 Apr 29	*6m41s*	*Annular*	*Atlantic Ocean, Senegal, Mauritania, Mali, Algeria, Tunisia, Libya, Turkey, Iran, Turkmenistan, Aghanistan, India, China*

Date	*Maximum Duration*[a]	*Eclipse Type*	*Geographic Region of Visibility*
1976 Oct 23	**4m46s**	**Total**	**Tanzania, south Indian Ocean, southern Australia**
1977 Apr 18	*7m04s*	*Annular*	*South Atlantic Ocean, Namibia, Angola, Zambia, Congo, Tanzania, Indian Ocean*
1977 Oct 12	**2m37s**	**Total**	**Pacific Ocean, Colombia, Venezuela**
1979 Feb 26	**2m49s**	**Total**	**USA (Washington, Oregon, Idaho, Montana, North Dakota), Canada, Greenland**
1979 Aug 22	*6m03s*	*Annular*	*South Pacific Ocean, Antarctica*
1980 Feb 16	**4m08s**	**Total**	**Atlantic Ocean, Angola, Congo, Tanzania, Kenya, India, Bangladesh, Burma, China**
1980 Aug 10	*3m23s*	*Annular*	*Pacific Ocean, Peru, Bolivia, Paraguay, Brazil*
1981 Feb 04	*0m33s*	*Annular*	*Southern Australia, south Pacific Ocean*
1981 Jul 31	**2m02s**	**Total**	**Russia, Pacific Ocean**
1983 Jun 11	**5m11s**	**Total**	**Indian Ocean, Indonesia, Papua New Guinea, Pacific Ocean**
1983 Dec 04	*4m01s*	*Annular*	*Atlantic Ocean, Gabon, Congo, Uganda, Kenya, Ethiopia, Somalia*
1984 May 30	*0m12s*	*Annular*	*Pacific Ocean, Mexico, USA (Louisiana, Mississippi, Alabama, Georgia, South Carolina, North Carolina, Virginia, Maryland), Atlantic Ocean, Morocco, Algeria*
1984 Nov 22/23	**2m00s**	**Total**	**Indonesia, Papua New Guinea, south Pacific Ocean**
1985 Nov 12	**1m59s**	**Total**	**South Pacific Ocean**
1986 Oct 03	***0m00.2s***	***Hybrid***	***North Atlantic Ocean***
1987 Mar 29	***0m08s***	***Hybrid***	***annular in Argentina; total for part of south Atlantic Ocean, Gabon, Cameroon, and part of Central African Republic; then annular in Sudan, Ethiopia, Somalia***
1987 Sep 23	*3m49s*	*Annular*	*Kazakstan, China, Mongolia, Pacific Ocean*
1988 Mar 18	**3m47s**	**Total**	**Indonesia, Philippines, Pacific Ocean**
1988 Sep 11	*6m57s*	*Annular*	*South Pacific Ocean*
1990 Jan 26	*2m03s*	*Annular*	*South Atlantic Ocean, Antarctica*
1990 Jul 22	**2m33s**	**Total**	**Finland, Russia, north Pacific Ocean**
1991 Jan 15	*7m53s*	*Annular*	*Southern Australia, New Zealand, south Pacific Ocean*

Date	*Maximum Duration*[a]	*Eclipse Type*	*Geographic Region of Visibility*
1991 Jul 11	**6m53s**	**Total**	**USA (Hawaii), Mexico, Guatemala, El Salvador, Honduras, Nicaragua, Costa Rica, Panama, Colombia, Brazil**
1992 Jan 04	*11m41s*	*Annular*	*Pacific Ocean, USA (California)*
1992 Jun 30	**5m21s**	**Total**	**Uruguay, south Atlantic Ocean**
1994 May 10	*6m14s*	*Annular*	*Mexico, USA (diagonally from Arizona through Missouri, New York, and Maine), southeastern Canada, Atlantic Ocean, Morocco*
1994 Nov 03	**4m23s**	**Total**	**Peru, Chile, Bolivia, Paraguay, Brazil, south Atlantic Ocean**
1995 Apr 29	*6m37s*	*Annular*	*South Pacific Ocean, Peru, Ecuador, Colombia, Brazil*
1995 Oct 24	**2m10s**	**Total**	**Iran, Afghanistan, Pakistan, India, Bangladesh, Burma, Cambodia, Vietnam, Indonesia, Pacific Ocean**
1997 Mar 09	**2m51s**	**Total**	**Mongolia, Russia, China, Arctic Ocean**
1998 Feb 26	**4m09s**	**Total**	**Pacific Ocean, Colombia, Panama, Venezuela, southern Caribbean, Atlantic Ocean**
1998 Aug 22	*3m14s*	*Annular*	*Indonesia, Malaysia, south Pacific islands*
1999 Feb 16	*0m40s*	*Annular*	*South Indian Ocean, Australia*
1999 Aug 11	**2m23s**	**Total**	**Atlantic Ocean, England, France, Luxembourg, Germany, Austria, Hungary, Romania, Bulgaria, Turkey, Iraq, Iran, Pakistan, India**
2001 Jun 21	**4m57s**	**Total**	**Atlantic Ocean, Angola, Zambia, Zimbabwe, Mozambique, Madagascar**
2001 Dec 14	*3m53s*	*Annular*	*Pacific Ocean, Costa Rica, Nicaragua, Caribbean Sea*
2002 Jun 10/11	*0m23s*	*Annular*	*Pacific Ocean, touching the islands of Sangihe, Talaud, Rota, and Tinian*
2002 Dec 04	**2m04s**	**Total**	**Angola, Zambia, Botswana, Zimbabwe, South Africa, Mozambique, south Indian Ocean, southern Australia**
2003 May 31	*3m37s*	*Annular*	*Scotland, Iceland, Greenland (the Moon's shadow passes over the north pole of the sunward-inclined northern hemisphere, so the path of the eclipse in this unusual case moves east to west)*
2003 Nov 23	**1m57s**	**Total**	**Antarctica**

Date	*Maximum Duration*[a]	*Eclipse Type*	*Geographic Region of Visibility*
2005 Apr 08	***0m42s***	***Hybrid***	***starts annular in western Pacific Ocean, becomes total in eastern Pacific, then becomes annular again for Costa Rica, Panama, Colombia, and Venezuela***
2005 Oct 03	*4m31s*	*Annular*	*Atlantic Ocean, Portugal, Spain, Algeria, Tunisia, Libya, Chad, Sudan, Ethiopia, Kenya, Somalia, Indian Ocean*
2006 Mar 29	**4m07s**	**Total**	**Eastern Brazil, Atlantic Ocean, Ghana, Togo, Benin, Nigeria, Niger, Chad, Libya, Egypt, Turkey, Russia**
2006 Sep 22	*7m09s*	*Annular*	*Guyana, Suriname, French Guiana, Brazil, south Atlantic Ocean*
2008 Feb 07	*2m12s*	*Annular*	*South Pacific Ocean, Antarctica*
2008 Aug 01	**2m27s**	**Total**	**Northern Canada, Greenland, Arctic Ocean, Russia, Mongolia, China**

[a] Maximum duration of totality or annularity as seen from the central line in minutes and seconds.

Notes and Reference

[1] Based on Fred Espenak: *Fifty Year Canon of Solar Eclipses: 1986–2035* (Cambridge, Massachusetts: Sky Publishing, 1987; NASA Reference Publication 1178, revised) and Fred Espenak and Jean Meeus: *Five Millennium Canon of Solar Eclipses: –1999 to +3000* (Greenbelt, Maryland: Goddard Space Flight Center, 2006; NASA Technical Publication 2006–214141).

Appendix C
Chronology of Discoveries about the Sun

Year	Discoveries about the Sun Made during Solar Eclipses/ *Other Discoveries about the Sun*
2159 to 1948 B.C.	Legendary dates from China in the *Shu Ching* of the first recorded solar eclipse. In this myth, Chinese astronomers Hsi and Ho fail to prevent or predict or properly react to an eclipse and are ordered to be executed by an angry emperor
1307 B.C.	First recorded observation of the corona (or prominences?) during a solar eclipse — in China on oracle bones: "three flames ate up the Sun, and a great star was visible"
1223 B.C., March 5	Oldest record of a verifiable solar eclipse — on a clay tablet in the ruins of Ugarit (Syria)
1217 B.C.	Oldest Chinese record of a verifiable solar eclipse — inscriptions on an oracle bone by the Shang people
585 B.C., May 28	A total eclipse in the midst of a battle between the Lydians and Medes scares both sides; hostilities are suspended, according to the Greek historian Herodotus (several other dates are possible)
6th century B.C.	Babylonians (Chaldeans) are said to be able to predict eclipses of the Sun and Moon, supposedly based on cycles such as the saros, but more likely from the positions of the Moon's nodes
450 B.C.?	Anaxagoras (Greece) is probably the first to realize that the Moon is illuminated by the Sun, thus providing a scientific explanation of eclipses
	Greek philosopher Anaxagoras proposes that the Sun is a giant glowing ball of rock; he is banished from Athens for blasphemy
431 B.C., August 3	Oldest European record of a verifiable solar eclipse (annular) — by the Greek historian Thucydides
330 B.C.?	*Aristotle (Greece) proposes that the Sun is a sphere of pure fire, unchanging and without imperfections*
130 B.C.	Greek astronomer Hipparchus uses the position of the Moon's shadow during a solar eclipse to estimate the distance to the Moon (accurate to about 13%)

With advice from Charles Lindsey, Joe Hollweg, Terry Forbes, Hugh Hudson, Jay M. Pasachoff, and Alan Clark

Year	Discoveries about the Sun Made during Solar Eclipses/ *Other Discoveries about the Sun*
20 B.C.?	Liu Hsiang first in China to explain that the Moon's motion hides the Sun to cause solar eclipses
150 A.D.?	Ptolemy (Alexandria) demonstrates the computation of solar and lunar eclipses based on their apparent motions rather than the periodic repetition of eclipses
334, July 17	Firmicus (Sicily) is first to report solar prominences, seen during an annular eclipse
418, July 19	First report of a comet discovered during a solar eclipse, seen by the historian Philostorgius in Asia Minor
9th century	Shadow bands during a total eclipse are described for the first time — in the *Völuspá*, part of the old German Poetic Edda
968, December 22	First clear description of the corona seen during a total eclipse — by a chronicler in Constantinople
1605	Johannes Kepler (Germany) is the first to comment scientifically on the solar corona, suggesting that it is light reflected from matter around the Sun (based on reports of eclipses; he never saw a total eclipse)
1609	*Galileo Galilei (Italy) and Johannes Fabricius (Holland) independently observe sunspots to be part of the Sun, disproving Aristotle's contention that the Sun is unchanging and free of "imperfections"*
1683	*Gian Domenico Cassini (Italy/France) explains zodiacal light as sunlight reflected from small solid particles in the plane of the solar system*
1687	*Isaac Newton (England) publishes his* Principia, *including the law of universal gravitation, which makes precise long-range eclipse prediction possible*
1695	Edmond Halley (England) is first to notice that the reported times and places of ancient eclipses do not correlate with calculations backward from his era; he concludes correctly that the Moon's orbit has changed slightly (secular acceleration)
1706, May 12	An English ship captain named Stannyan, on vacation in Switzerland, reports a reddish streak (chromosphere? prominence?) along the rim of the Sun as the eclipse becomes total
1715, May 3	Edmond Halley (England), during an eclipse in England, is the first to report the phenomenon later known as Baily's Beads; also notes bright red prominences and the east–west asymmetry in the corona, which he attributes to an atmosphere on the Moon or Sun
1715	Gian Domenico Cassini (Italy/France) proposes that the light responsible for the corona also causes the zodiacal light

Year	**Discoveries about the Sun Made during Solar Eclipses/** ***Other Discoveries about the Sun***
1724, May 22	Giacomo Filippo Maraldi (Italy/France) concludes that the corona is part of the Sun because the Moon traverses the corona during an eclipse
1733, May 13	Birger Wassenius (Sweden), observing an eclipse near Göteborg, is the first to report prominences visible to the unaided eye; he attributes them to the Moon
1749	Richard Dunthorne (United Kingdom) calculates the approximate secular acceleration of the Moon based on Halley's 1693 findings
1774	*Alexander Wilson (Scotland) shows that sunspots are depressions in the Sun's photosphere, not clouds above it or mountains or volcanic deposits on it*
1795	*William Herschel (United Kingdom) proposes that sunspots are holes in the Sun's hot clouds through which the dark, cool, solid surface of the Sun can be seen; he also suggests that this surface may be inhabited*
1800	*William Herschel (United Kingdom) founds science of solar physics by measuring the temperature of various colors in the Sun's spectrum; he detects infrared radiation beyond the visible spectrum*
1801	*Johann Wilhelm Ritter (Germany), following the lead of Herschel, uses the spectrum of the Sun to establish the existence of ultraviolet radiation*
1802	*Dark (absorption) lines are discovered in the Sun's spectrum by William Wollaston (United Kingdom)*
1806, June 16	José Joaquin de Ferrer (Spain), observing at Kinderhook, New York, gives the name *corona* to the glow of the faint outer atmosphere of the Sun seen during a total eclipse; he proposes that the corona must belong to the Sun, not the Moon, because of its great size
1817	*Joseph Fraunhofer (Germany) independently discovers and catalogs many hundreds of dark lines in the Sun's spectrum*
1820	Carl Wolfgang Benjamin Goldschmidt (Germany) calls attention to the shadow bands visible just before and after totality at some eclipses (based on the eclipse of November 19, 1816?)
1824	*Friedrich Bessel (Prussia) introduces an easier way (Besselian elements) to make eclipse predictions*
1833	*David Brewster (United Kingdom) shows that some of the dark lines in the Sun's spectrum are due to absorption in the Earth's atmosphere, but most are intrinsic to the Sun*

Year	Discoveries about the Sun Made during Solar Eclipses/ *Other Discoveries about the Sun*
1836, May 15	Francis Baily (United Kingdom), during an annular eclipse in Scotland, calls attention to the brief bright beads of light that appear close to totality as the Sun's disk is blocked except for sunlight streaming through lunar valleys along the limb. This phenomenon becomes known as *Baily's Beads*
1837–38	*John F. W. Herschel (United Kingdom) and Claude Servais Mathias Pouillet (France) independently make the first quantitative measurements of the heat emitted by the Sun (about half the actual value). James David Forbes (United Kingdom) obtains a more accurate value in 1842, but it is considered less reliable*
1842, July 8	Francis Baily (United Kingdom), at an eclipse in Italy, focuses attention on the corona and prominences and identifies them as part of the Sun's atmosphere
1843	*Heinrich Schwabe (Germany), after a 17-year study, discovers the sunspot cycle of 10 years (now known to average 11 years)*
1845	*First clear photograph of the Sun: a daguerreotype by Hippolyte Fizeau and Léon Foucault (France)*
1848	*Julius Robert Mayer (Germany) shows through calculations that the Sun cannot shine a significant length of time by ordinary chemical burning; he incorrectly attributes the energy of the Sun to the heat released by the impact of meteoroids on the Sun*
1851, July 28	First astronomical photograph of a total eclipse: a daguerreotype by Berkowski at Königsberg, Prussia
1851, July 28	Robert Grant and William Swan (United Kingdom) and Karl Ludwig von Littrow (Austria) determine that prominences are part of the Sun because the Moon is seen to cover and uncover them as it moves in front of the Sun
1851, July 28	George B. Airy (United Kingdom) is the first to describe the Sun's chromosphere: he calls it the *sierra*, thinking that he is seeing mountains on the Sun, but he is actually seeing small prominences (spicules) that give the chromosphere a jagged appearance. Because of its reddish color, J. Norman Lockyer names this layer of the Sun's atmosphere the *chromosphere* in 1868
1852	*Johann von Lamont (Germany), after 15 years of observations, discovers a 10.3-year activity cycle in the Earth's magnetic field, but does not correlate it with the sunspot cycle*
1852	*Edward Sabine (Ireland/United Kingdom), Johann Rudolf Wolf (Germany), and Alfrede Gautier (France) independently link the sunspot cycle to magnetic fluctuations on Earth; the study of solar–terrestrial relationships begins*

Year	Discoveries about the Sun Made during Solar Eclipses/ *Other Discoveries about the Sun*
1854	*Hermann von Helmholtz (Germany) attributes the energy of the Sun to heat from gravitational contraction*
1858	*Richard C. Carrington (United Kingdom), studying sunspots, identifies the Sun's axis of rotation and discovers that the Sun's rotational period varies with latitude. He also discovers that the latitude of sunspots migrates toward the equator during the course of a sunspot cycle*
1859, September 1	*Richard C. Carrington and R. Hodgson (United Kingdom) are the first to observe a flare on the Sun. They both also note that a magnetic storm in progress on Earth intensifies soon afterwards, but they refrain from connecting the two events*
1859	*Gustav Kirchhoff (Germany) uses spectroscopy to show that the surface of the Sun cannot be solid and that the Sun's atmosphere (which he identifies with the corona) is responsible for the dark lines in the Sun's spectrum*
1860, July 18	First wet plate photographs of an eclipse; they require 1/30 of the exposure time of a daguerreotype
1860, July 18	Warren De La Rue (United Kingdom) and Angelo Secchi (Italy) use photography during a solar eclipse in Spain to demonstrate that prominences (and hence at least that region of the corona) are part of the Sun, not light scattered by the Earth's atmosphere or the edge of the Moon, because the corona looks the same from sites 250 miles apart
1859–61	*Gustav Kirchhoff and (in part) Robert Bunsen (Germany) identify 12 elements in the Sun based on lines in its spectrum*
1861–62	*Gustav Kirchhoff (Germany) maps the solar spectrum*
1868, August 18	During an eclipse seen from the Red Sea through India to Malaysia and New Guinea, prominences are first studied with spectroscopes and shown to be composed primarily of hydrogen by James Francis Tennant (United Kingdom), John Herschel (United Kingdom — son of John F. W. Herschel, grandson of William), Jules Janssen (France), Georges Rayet (France) and Norman Pogson (United Kingdom/India)
1868	Pierre Jules César Janssen (France) and J. Norman Lockyer (United Kingdom) independently demonstrate that prominences are part of the Sun (not Moon) by observing them in days after the eclipse of August 18
1868	J. Norman Lockyer (United Kingdom) identifies a yellow spectral line in the Sun's corona as the signature of a chemical element as yet unknown on Earth. He later names it helium, after the Greek word *helios*, the Sun. Helium is first identified on Earth by William Ramsay in 1895

Year	**Discoveries about the Sun Made during Solar Eclipses/** ***Other Discoveries about the Sun***
1869, August 7	Charles Augustus Young and William Harkness (United States) independently discover a new bright (emission) line in the spectrum of the Sun's corona, never before observed on Earth; they ascribe it to a new element and it is named coronium. In 1941, this green line is identified by Bengt Edlén (Sweden) as iron that has lost 13 electrons
1869	*Anders Jonas Angström (Sweden) produces an improved map of the solar spectrum (using a diffraction grating rather than a prism) and introduces angstrom unit for measuring wavelength*
1869	*Thomas Andrews (Ireland) shows more conclusively than those before him that the Sun must be made essentially of hot gas*
1870, December 2	Jules Janssen (France) uses a balloon to escape the German siege of Paris in order to study the December 22 eclipse in Algeria. He reaches Algeria, but the eclipse is clouded out
1870, December 22	Charles A. Young (United States), observing an eclipse in Spain, discovers that the chromosphere is the layer in the solar atmosphere that produces the dark lines in the Sun's spectrum
1870	*Samuel P. Langley (United States) uses the Doppler Effect of the Sun's rotation to show that it displaces the dark (absorption) lines in the solar spectrum*
1871, December 12	Jules Janssen (France) uses spectroscopy from an eclipse in India to propose that the corona consists of both hot gases and cooler particles and hence is part of the Sun
1872, August 3	*Charles A. Young (United States) observes a flare on the Sun with a spectroscope; he calls attention to its coincidence with a magnetic storm on Earth*
1874	*Samuel P. Langley (United States) proposes that the bright granules in the Sun's photosphere are columns of hot gas rising from the interior and the dark interstices are cooler gases descending; he also proposes that the bright granules are responsible for almost all of the Sun's light*
1871/1878	Jules Janssen (France) notices that the shape of the corona changes with the sunspot cycle. At sunspot maximum, the corona is rounder (1871); at sunspot minimum, the corona is more equatorial (1878). This discovery is the most convincing evidence that the corona is part of the Sun
1878, July 29	Height of search for intra-Mercurial planet Vulcan using eclipses to block the Sun. Several observers claim sightings, but they were never confirmed. The problem is finally resolved by Einstein in his general theory of relativity in 1916

Year	Discoveries about the Sun Made during Solar Eclipses/ *Other Discoveries about the Sun*
1878, July 29	Samuel P. Langley and Cleveland Abbe (United States), observing from Pike's Peak in Colorado, and Simon Newcomb (United States), observing from Wyoming, notice coronal streamers extending more than 6° from the Sun along the ecliptic and suggest that this glow is the origin of the zodiacal light
1882, May 17	A comet is discovered and photographed by Arthur Schuster (Germany/United Kingdom) during an eclipse in Egypt: the first time a comet discovered in this way has been photographed
1884	*Marie Alfred Cornu (France) applies Langley's work using the Doppler Effect of the Sun's rotation to distinguish between dark (absorption) lines created in the Sun's atmosphere and those created in the Earth's atmosphere*
1887	Theodor von Oppolzer's (Czechoslavakia) monumental *Canon of Eclipses* published, giving details of almost all solar and lunar eclipses from 1207 B.C. to 2161 A.D.
1887, August 19	Dmitry Ivanovich Mendeleev (Russia) uses a balloon to ascend above the cloud cover to an altitude of 11,500 feet (3.5 kilometers) to observe an eclipse in Russia
1889	*Henry Augustus Rowland (United States) produces an essentially modern map of the solar spectrum, identifying a total of 36 elements present in the Sun*
1893 & 1894	*George Ellery Hale (United States) and Henri Deslandres independently develop spectroheliographs to photograph the Sun's chromosphere, prominences, and flares without waiting for eclipses*
1894	*William E. Wilson and P. L. Gray (Ireland) are the first to measure with reasonable accuracy the effective temperature of the Sun's photosphere: 11,200 °F (6,200 °C), about 800 °F (400 °C) too high*
1899	Friedrich Ginzel (Austria), Oppolzer's co-worker, uses data from the *Canon of Eclipses* for his *Special Canon of Solar and Lunar Eclipses*, which lists references in classical literature to eclipses between 900 B.C. and 600 A.D.
1904	*George Ellery Hale (United States) establishes on Mt. Wilson in California the first large solar observatory*
1908	*George Ellery Hale (United States) shows that sunspots are regions with strong magnetic fields*
1911	*Albert Einstein (Germany), working on his general theory of relativity, proposes that gravity bends light and that this phenomenon might be observed during a solar eclipse*
1916	Einstein publishes his complete general theory of relativity with a revised prediction for the gravitational deflection of starlight

Year	Discoveries about the Sun Made during Solar Eclipses/ *Other Discoveries about the Sun*
1919, May 29	Arthur S. Eddington (United Kingdom) and co-workers, observing a total solar eclipse from Principe and Brazil, confirm the bending of starlight by gravity as predicted by Einstein in his general theory of relativity
1922, September 21	William Wallace Campbell and Robert J. Trumpler (United States) reconfirm Einstein's relativistic bending of starlight during an eclipse in Wallal, Australia
1926	*Arthur Eddington (United Kingdom) proposes that the Sun and stars derive their energy from nuclear reactions at their core*
1930	*Bernard Lyot (France) invents the coronagraph, which creates an artificial eclipse inside a telescope so that the corona can be studied outside of eclipses*
1932, August 31	G. G. Cillié (United Kingdom) and Donald H. Menzel (United States) use eclipse spectra to show that the Sun's corona has a higher temperature (faster atomic motion) than the photosphere. Confirmed, with much higher temperatures, by R. O. Redman during an eclipse in South Africa on October 1, 1940
1942	*Hannes Alfvén (Sweden) predicts the existence of electromagnetic-hydromagnetic waves (Alfvén waves). They will prove vital in understanding the Sun's corona and the solar wind*
1951	*Ludwig F. Biermann (Germany) discovers the solar wind, a stream of charged gas particles continuously ejected from the Sun in all directions, based on observations of charged gases in comet tails*
1952	*Discovery of strong absorption lines of the molecule carbon monoxide in the solar infrared spectrum by Leo Goldberg and colleagues (United States)*
1955	*Max Waldmeier (Switzerland) finds that sunspot pairs are slightly tilted, with the leading spot closer to the equator. The farther from the equator, the greater the tilt, but the tilt decreases as the pair develops. He also finds that there is an overlap of approximately two years as one sunspot cycle ends and another begins. His studies of the sunspot cycle lays a foundation for forecasting solar activity*
1958	*Eugene Parker (United States) provides a theoretical model for the solar wind by showing that the corona must be expanding and by demonstrating how to calculate flow speeds and densities*
1961	*Marcia Neugebauer and her coworkers (United States) use NASA's Mariner 2 spacecraft to confirm major features of Eugene Parker's 1958 model of the solar wind*

Year	Discoveries about the Sun Made during Solar Eclipses/ *Other Discoveries about the Sun*
1962	*Robert Leighton, Robert Noyes, and George Simon (United States) and, independently, John Evans (United States) and Raymond Michard (France) discover 5-minute oscillations of the surface of the Sun, leading to the birth of helioseismology and the "acoustic" exploration of the Sun's interior*
1962–1975	*NASA launches 8 Orbiting Solar Observatories (OSOs)*
1970	*Roger Ulrich (United States) explains how 5-minute oscillations on the Sun's surface — surface ripples caused by waves propagating in the solar interior — can be used to look deep inside the Sun, thus opening the field of helioseismology*
1971	*Robert Stein and John Leibacher (United States) explain that the 5-minute oscillations are acoustic noise caused by convective turbulence near the Sun's surface*
1971	*John Belcher and Leverett Davis (with earlier help from Ed Smith) (United States) demonstrate that the solar wind is full of Alfvén waves, stimulating the study of coronal heating and solar wind acceleration by waves*
1972	*Eugene Parker (United States) shows that all but the simplest magnetic fields must undergo "topological dissipation" and field-line merging — simplifying themselves — with important astrophysical consequences*
1972	*Using data from an Orbiting Solar Observatory (OSO 4), Richard H. Munro and George L. Withbroe (United States) discover* coronal holes, *later shown by Werner M. Neupert and Victor Pizzo (United States) to be the source of solar wind because they correlate with magnetic storms on Earth*
1972	*Robert Noyes (United States) and Donald Hall (Australia) recognize that the carbon monoxide molecules on the Sun (discovered 1952) must be colder than the gases around them, with a temperature significantly below the minimum temperature measured by other techniques*
1973, June 30	John Beckman (United Kingdom) and other scientists use a Concorde supersonic passenger jet flying at 1,250 miles per hour (2,000 kilometers per hour) over Africa to extend the duration of solar eclipse totality to 74 minutes — 10 times longer than can ever be observed from the ground
1973	*Jack Eddy (United States) demonstrates that the sunspot cycle turned off in the seventeenth century, showing that the Sun changes in more ways than the 11-year sunspot cycle*

Year	Discoveries about the Sun Made during Solar Eclipses/ *Other Discoveries about the Sun*
1973–1974	*Astronauts on NASA's Skylab orbiting laboratory study the corona over a nine-month period using the Apollo Telescope Mount: (1) X-ray images show coronal holes and the relationship between these "open-field" regions with the mysterious "M-regions" postulated by Sydney Chapman and Julius Bartels (Germany) in 1940 but never seen; (2) the high-speed solar wind escaping from coronal holes produces geomagnetic disturbances when it reaches the Earth about two days after the coronal hole passes through the central meridian of the Sun; (3) the active corona is composed of loops of plasma following strong magnetic field lines, indicating that magnetism heats the corona*
1973–1978	*Michael Schulz (United States) begins a shift from thinking about the Sun as being magnetically sectored north–south to thinking of the sectors essentially following the solar equatorial plane. Ed Smith and co-workers confirm this discovery by using the NASA spaceprobe* Pioneer *11 (1978) to observe the Sun at high solar latitudes. The magnetic sectors form a "ballerina skirt" fluttering around the Sun's equatorial plane*
1974	*Hannes Alfvén (Sweden) and, independently, Leif Svalgaard and John Wilcox (United States) describe the basic magnetic structure of the solar wind as a warped neutral current sheet that divides the solar wind into hemispheres of opposite magnetic polarity*
1975	*Raymond Davis Jr. (United States) detects neutrinos (subatomic particles released by hydrogen fusion reactions at the core of the Sun) by placing a huge tank of dry-cleaning fluid nearly one mile (1.5 km) underground in a South Dakota gold mine. Almost all neutrinos pass through the Earth without interacting with any matter, but a very few strike chlorine atoms in the dry-cleaning fluid and convert them to argon 40. Davis bubbles helium through the tank to collect the few atoms of argon 40 to determine how many neutrinos have been captured. He finds only one-third the number expected — a mystery that remains unsolved for more than 3 decades*
1975	*Franz-Ludwig Deubner (Germany) measures fringes in the solar power spectra predicted by Ulrich (1970) that show how the temperature of the Sun increases with depth*
1975	*Edward Rhodes and Roger Ulrich (United States) use improved measurements of the fringes in the solar power spectra to show the profile of differential solar rotation beneath the photosphere.*

Year	Discoveries about the Sun Made during Solar Eclipses/ *Other Discoveries about the Sun*
1978, October 25	*NASA & NOAA launch Nimbus-7 weather satellite, with a detector to measure fluctuations in solar energy reaching Earth. It detects that the Sun is continuously and significantly varying its output of energy, an observation to be investigated further by subsequent solar observatories*
1979, February 26	Alan Clark (United Kingdom/Canada) and Rita Boreiko (Canada) begin airborne infrared observations during eclipses (through 1988)
1979	*Roger Ulrich and Edward Rhodes (United States) and Franz-Ludwig Deubner (Germany) (1975) use helioseismology (solar seismic waves) to probe solar rotation beneath the photosphere. They discover that the differential rotation seen at the Sun's surface — the equator revolving in a shorter period than the poles — extends to the base of the convection zone, deep in the Sun's interior, crucial to understanding the solar dynamo*
1979–1980	*By observing the Sun continuously for a week from sites around the world, George Isaaks and colleagues (United Kingdom) determine the Sun's fundamental resonant frequencies. Eric Fossat and colleagues (France) independently get similar results by observing the Sun continuously for a week from the South Pole*
1980, February 14	*NASA's Solar Maximum Mission satellite is launched; proves that the Sun varies its energy output (craft crippled by blown fuses after 9 months; repaired in orbit by Space Shuttle astronauts in 1984 and operates until 1989)*
1981	*Richard Willson, Hugh Hudson, and colleagues (United States) show that sunspots and faculae have measurable effects on total solar irradiance (the solar constant)*
1981–1988	Airborne far infrared observations during eclipses by Eric Becklin, Charles Lindsey, and colleagues (United States) show an unexpected extension of the solar limb at wavelengths between 20 and 800 micrometers. They suggest this extension is a manifestation of the mechanism that heats the solar chromosphere to anomalously high temperatures, producing most of the Sun's excess ultraviolet radiation
1982	*Barry LaBonte and Robert Howard (United States) discover torsional oscillations on the Sun. The fast-moving lower-latitude gases on the Sun flow by the slower-moving higher-latitude gases, but this differential rotation varies as the sunspot cycle progresses. Later work in helioseismology eventually shows that the torsional oscillations extend below the Sun's surface deep into the convection zone*

Year	Discoveries about the Sun Made during Solar Eclipses/ *Other Discoveries about the Sun*
1982–1988	*Joseph Hollweg and Alphonse Sterling (United States) introduce the "rebound shock model" for solar spicules, in which spicules are driven by large amplitude acoustic-gravity waves in the solar chromosphere*
1983	*Eugene Parker (United States) proposes that the coronal magnetic field must be continuously reconnecting and releasing energy and that this mechanism can heat the corona*
1983	*Martin Woodard and Hugh Hudson (United States), using the active-cavity radiometer on board the Solar Maximum Mission, find variations in the Sun's total irradiance over periods of about 5 minutes*
1986	*Hans Bethe (United States) suggests that "neutrino oscillations" proposed by Stanislav Mikheyev and Alexei Smirnov (Russia) explain Raymond Davis Jr.'s "missing neutrinos problem." On their way from the Sun to the Earth, electron neutrinos, which are produced in the Sun and to which Davis's detector in South Dakota is sensitive, change back and forth to and from muon and tau neutrinos to which it is insensitive. This oscillation would explain why Davis's detector sees only one-third of the solar neutrinos expected*
1987	*The first comprehensive observation of the solar infrared spectrum between 2 and 16 micrometers from space by the ATMOS Space Shuttle experiment by Crofton Farmer (United Kingdom/United States) and Robert Norton (United States)*
1988	*Joseph Hollweg and Walter Johnson (United States) present a model of coronal holes in which high-frequency ion-cyclotron waves heat the corona and the resulting fast solar wind is driven by hot protons. Although data at the time indicate that the protons are not hot, later SOHO data (John Kohl, 1995) show that the protons are hot, as predicted*
1988	*Martin Woodard and Robert Noyes (United States), based on work by Woodard and Hudson, discover that the resonant frequencies of the Sun's seismic oscillations vary by about one part in ten thousand between sunspot minimum and maximum, leading to further advances in helioseismology*
1988	*Douglas Braun, Thomas Duvall Jr. and Barry LaBonte (United States) discover that acoustic waves in the solar interior react strongly to sunspots and other magnetic regions on the Sun's surface. Their discovery opens the field of "local helioseismology," enabling scientists to monitor small regions of the Sun*

Year	Discoveries about the Sun Made during Solar Eclipses/ *Other Discoveries about the Sun*
1990	*Observations of solar irradiance from the Solar Maximum Mission and Nimbus 7 satellites show the Sun's energy reaching Earth is about 0.1% greater at sunspot maximum than at sunspot minimum, despite the irradiance-blocking effect by sunspots, which are dark. The Sun's ultraviolet irradiance undergoes a much greater variation. This increased UV expands the Earth's upper atmosphere, increasing the drag on satellites in low Earth orbit*
1990, October 6	*ESA/NASA Ulysses spacecraft launched from the Space Shuttle into polar orbit around the Sun (via gravitational assist from Jupiter) to explore the polar regions of the Sun and its corona, which are poorly seen from Earth; confirms that the polar regions of the Sun are dominated by a steady, fast solar wind flow, especially around the time of solar minimum*
1991, July 11	A total solar eclipse passes directly over the world's largest observatory complex at the summit of Mauna Kea in Hawaii. Observations of the eclipse are made in the radio, infrared, and visible spectrum from the United Kingdom's James Clerk Maxwell Telescope, the Caltech Submillimeter Observatory, NASA's Infrared Telescope Facility, the Canada-France-Hawaii Telescope, and the University of Hawaii's 2.2-meter telescope
1991	*Jørgen Christensen-Dalsgaard (Denmark), Douglas Gough, and Michael Thompson (United Kingdom), and others use helioseismology to measure the temperature of the Sun deep into its interior. They locate the base of the Sun's convection zone, where heat is conveyed to the Sun's surface by rising gases (rather than radiation), at about 30% of the way from the surface to the core — 124,000 miles; 200,000 kilometers). Their profiles show enhanced differential rotation at the base of the convection zone, which may create the magnetic anomalies that rise to become sunspots on the Sun's surface*
1991, August 31	*Japan's Yohkoh spacecraft launched. A primary objective is the study of solar flares*
1994, September 13–15	*NASA's Spartan 201 satellite, designed to examine the Sun's outer corona and solar wind, is deployed and retrieved for a second time on a Space Shuttle mission; shows that the fast solar wind accelerates to very high speeds (~2 million mph; ~3 million km/hr — about twice previous estimates) very close to the coronal base, thus challenging conventionally held views of the origin of the solar wind*

Year	Discoveries about the Sun Made during Solar Eclipses/ *Other Discoveries about the Sun*
1995, December 2	*ESA/NASA satellite SOHO (Solar and Heliospheric Observatory) launched into solar orbit 1.5 million kilometers on the sunward side of Earth*
1995	*Günter Brückner (United States) and colleagues use SOHO to discover that coronal mass ejections occur daily*
1997	*SOHO scientists propose that loops in the "magnetic carpet" structure of the surface of the Sun are the source of the multi-million-degree temperatures of the corona*
1997	*Dan Moses (United States) and colleagues, using the Extreme Ultraviolet Imaging Telescope (EIT) on SOHO, discover EIT waves (coronal waves), which later prove to be associated with coronal mass ejections*
1998, April 1	*NASA TRACE (Transition Region and Coronal Explorer) satellite launched. Its purpose is to provide observations of the Sun with far better resolution than previously possible.*
1998	*Alexander Kosovichev (United States) and Valentina Zharkova (United Kingdom) use SOHO to discover that some flares on the Sun drive strong acoustic waves into the solar interior, causing an expanding ring of ripples when they refract back to the surface. These "sunquakes" are now known to occur frequently in flares and offer a way of exploring the interior of the Sun beneath flare sites*
1998	*Jesper Schou (United States) and colleagues use the Michelson Doppler Imager aboard SOHO to discover that differential rotation of the Sun takes place not only at its surface (known since 1858: equatorial gases revolve in about 25 days while polar gases take about 35) but differential rotation also occurs throughout the convective zone beneath the Sun's surface and continues deeper into the Sun's radiative zone, still closer to the core*
1998–1999	*Barbara Thompson, Meredith Wills-Davey, and others use SOHO and TRACE to track EIT waves in the corona and demonstrate their association with coronal mass ejections*
1999	*David McKenzie and Hugh Hudson (United States), using instruments on the Yohkoh spacecraft, make the first observation of a reconnection outflow in a solar flare, confirming the essential role of reconnection in flares*
1999	*John Kohl (United States) and colleagues use ESA/ NASA SOHO and NASA's Spartan 201 to discover that the fast solar wind is accelerated away from the Sun by rapidly vibrating magnetic waves that preferentially heat protons and heavy ions; oxygen ions are found to have temperatures of hundreds of millions of degrees*

Year	Discoveries about the Sun Made during Solar Eclipses/ *Other Discoveries about the Sun*
1999	*Markus Aschwanden (United States) and colleagues use TRACE to discover that coronal loops oscillate ("twang") rapidly in the coronal waves produced by flares and coronal mass ejections, providing a tool ("coronal seismology") to explore how the corona is heated*
1999	*Valery Nakariakov (United Kingdom) and colleagues, using TRACE, find that the oscillations of coronal loops damp down rapidly after a solar flare, suggesting that the plasma of the corona provides far more resistance than expected and that this damping contributes in a major way to the extraordinarily high temperatures of the corona*
2000	*Deborah Haber, Bradley Hindman (United States), and others use helioseismology to discover a flow of gases in the Sun's upper convection zone from the equator toward the poles at ~10–20 m/sec. This "meridional flow" is thought to require a return flow at or near the base of the convection zone of ~1–2 m/sec, the speed at which the active region bands at the surface (photosphere) drift toward the equator as the sunspot cycle progresses*
2000	*SOHO and TRACE observations of waves and oscillations in the solar corona give birth to "coronal seismology" — using the waves to infer coronal properties*
2001	*Charles Lindsey and Douglas Braun (United States) use "helioseismic holography" to detect active regions on the far side of the Sun. Seismic imaging of far-side solar activity allows scientists to anticipate the appearance of large active regions more than a week ahead of their arrival on the side of the Sun facing Earth*
2001	*Philip Scherrer, Alexander Kosovichev, and Junwei Zhao (United States) use the Michelson Doppler Imager (MDI) on SOHO to infer the magnetic structure and fluid flows beneath sunspots and active regions*
2001–2002	*Scientists at the Sudbury Neutrino Observatory in Canada confirm the existence of neutrino oscillations, solving the "missing" solar neutrinos problem. Although the Sun generates only electron neutrinos, they constantly oscillate form to muon and tau neutrinos en route to Earth, so Raymond Davis Jr.'s 1970–1984 South Dakota mine experiment, sensitive only to electron neutrinos, detected only one-third of the neutrinos the Sun was emitting. It also proves that neutrinos, long thought to have no mass, actually have a significant but minuscule mass. In the case of the electron neutrino, this mass is only a few millionths the mass of an electron*

Year	Discoveries about the Sun Made during Solar Eclipses/ *Other Discoveries about the Sun*
2002, February 5	*NASA's Reuven Ramaty High Energy Solar Spectroscopic Imager is launched by a rocket fired from a Lockheed 1011 passenger/cargo jet into low Earth orbit to study solar flares*
2002, May 21	*The Swedish 1m Solar Telescope (LaPalma, Spain) is opened to full aperture and the adaptive optics system is switched on for the first time, delivering diffraction-limited images despite atmospheric turbulence that would blur the view of ordinary telescopes*
2002, October 8	*Raymond Davis Jr. (United States) and Masatoshi Koshiba (Japan) receive the Nobel Prize in Physics "for pioneering contributions to astrophysics, in particular for the detection of cosmic neutrinos." Davis was the first to detect solar neutrinos (1975), ultimately leading to a new understanding of neutrinos and to the validation of models of the Sun's generation of energy by nuclear fusion in its core*
2003, January 25	*NASA launches SORCE (SOlar Radiation and Climate Experiment) spacecraft to continue measuring variations in the Sun's energy reaching Earth (irradiance) and its effect on Earth's climate*
2003, June 18	*Advanced adaptive optics installed on the 30-inch (76-cm) Dunn Solar Telescope at the National Solar Observatory in New Mexico produces sharp images of the Sun despite atmospheric turbulence that would blur the view of ordinary telescopes. This adaptive optics system may be installed on other solar telescopes*
2004	*Bart De Pontieu (United States) and colleagues (United Kingdom) show by numerical simulations that observed photospheric p-mode motions can produce the observed occurrence and properties of individual spicules*
2004	*R. M. Close (Scotland) and colleagues report that the timescale for magnetic flux in the quiet-Sun corona is, surprisingly, only 1.4 hr., implying that the quiet-Sun corona is far more dynamic than previously thought*
2005, August 5	*SOHO images its 1,000th Sun-grazing comet*
2005	*Greg Kopp (Canada) and colleagues use the Total Irradiance Monitor on SORCE to make the first measurement of the total radiant energy output of a flare*
2006	*A new computer model of Sun's corona accurately predicts its appearance at the March 29, 2006 total solar eclipse — pointing toward improved predictions of the solar wind*

Year	Discoveries about the Sun Made during Solar Eclipses/ *Other Discoveries about the Sun*
2006	*Mausumi Dikpati, Peter Gilman, and Giuliana de Toma (United States) propose that the return of the meridional flow shown by helioseismic observations at and below the solar surface explains the equatorward drift of solar activity bands. They use this concept to develop a technique for forecasting the strength and phase of future sunspot cycles*
2006, September 23	*Hinode ("sunrise" —formerly Solar B), a Japanese X-ray solar satellite with U.S. and U.K. scientific instruments, is launched into an Earth orbit that allows a continuous view of the Sun. A key objective is to study solar outbursts — to watch coronal mass ejections all the way from the Sun to the Earth*
2006, October 25	*NASA Solar-Terrestrial Relations Observatory (STEREO), a pair of probes, is launched into orbit around the Sun ahead of and behind the Earth to provide three-dimensional images of the Sun. A key objective is to study solar outbursts — to watch coronal mass ejections all the way from the Sun to the Earth*
2007	*SOHO observations are interpreted as showing the presence of g-mode standing waves trapped in the deep interior of the Sun. This technique opens the possibility of probing the energy-generating core regions of stars via asteroseismology*
2007	*NASA's STEREO, a pair of solar observatories leading and trailing the Earth around the Sun, provides the first 3-D images of the Sun*

Appendix D
NASA Solar Eclipse Bulletins

NASA astronomer Fred Espenak and Canadian meteorologist Jay Anderson publish special NASA bulletins for each major eclipse of the Sun. Produced through NASA's Technical Publication (TP) series, the eclipse bulletins are prepared in cooperation with the Working Group on Eclipses of the International Astronomical Union and are provided as a public service to both the professional and lay communities, including educators and the mass media. Each eclipse bulletin is a complete reference for a specific eclipse and contains detailed predictions, tables, maps, and weather prospects.

To date, eleven eclipse bulletins have been published and cover eclipses from 1994 through 2009. Future bulletins currently planned include the eclipses of 2010 July 11; 2012 May 20 (annular); and 2012 November 13. To allow readers a reasonable lead time for planning purposes, bulletins are published one to two years before each eclipse.

Single copies of the bulletins are available at no cost by sending a 9-by-12-inch self-addressed envelope stamped with sufficient postage for 11 ounces (310 grams) and with the eclipse date printed in the lower left corner. Complete details for ordering past and future NASA eclipse bulletins can be found at:

<http://eclipse.gsfc.nasa.gov/SEpubs/bulletin.html>

The NASA eclipse bulletins are also available on-line as web pages or as downloadable PDF documents. They can be found at the link above.

The NASA Eclipse Website is NASA's official source for solar and lunar eclipse predictions. It contains maps and tables for 5,000 years of eclipses and includes information on eclipse photography, observing tips, and eye safety. The NASA Eclipse Website is located at:

<http://gsfc.nasa.gov/eclipse.html>

Glossary

annular eclipse A central eclipse of the Sun in which the angular diameter of the Moon is too small to completely cover the disk of the Sun and a thin ring (*annulus*) of the Sun's bright apparent surface surrounds the dark disk of the Moon. Thus an annular eclipse is actually a special kind of partial eclipse of the Sun. There are more annular eclipses than total eclipses.

annular–total eclipse A solar eclipse that begins as an annular eclipse, changes to a total eclipse along its path, and then returns to annular before the end of the eclipse path. Rare annular-total eclipses start as total eclipses and end as annulars or start as annulars and end as totals. Also called a **hybrid eclipse**.

anomalistic month The time it takes (27.55 days) for the Moon to orbit the Earth as measured from its closest point to Earth (perigee) to its farthest point (apogee) and back to its closest point again.

anomalistic year The time it takes (365.26 days) for the Earth to orbit the Sun as measured from its closest point to the Sun (perihelion) to its farthest point (aphelion) and back to its closest point again.

antumbra The extension of the Moon's shadow cone so that it forms a mirror image of itself. A region experiencing an annular eclipse lies in the antumbra of the Moon.

aphelion The point for any object orbiting the Sun where it is farthest from the Sun.

apogee The point for any object orbiting the Earth where it is farthest from the Earth.

arc minute An angular measurement: 1 minute of arc is 60 seconds of arc and 1/60 of a degree of arc.

arc second An angular measurement: 1 second of arc is 1/60 of a minute of arc and 1/3600 of a degree of arc.

ascending node (of the Moon) The point on the Moon's orbit where it crosses the ecliptic (orbit of the Earth) going north.

Baily's beads An effect seen just before and after the total phase of a solar eclipse in which the Moon hides all the light from the Sun's disk except for a few bright points of sunlight passing through valleys at the rim of the Moon.

central eclipse (of the Sun) An eclipse in which the axis of the Moon's shadow touches the Earth. A central eclipse can be either total or annular.

chromosphere The reddish lower atmosphere of the Sun just above the photosphere. The chromosphere is only about 600 miles (1,000 kilometers) thick and the temperature is about 7,100 °F (4,200 °C).

contact (in a solar eclipse) Special numbered stages of a solar eclipse. For all eclipses (total, annular, and partial), **first contact** occurs when the leading edge of the Moon appears tangent to the western rim of the Sun, iniating the eclipse.

In a *total eclipse*, **second contact** occurs when the Moon's leading edge appears tangent to the eastern rim of the Sun, iniating the total phase of the eclipse. In a *total eclipse*, **third contact** occurs when the Moon's trailing edge appears tangent to the western rim of the Sun, concluding the total phase of the eclipse. In an *annular eclipse*, **second contact** occurs when the Moon's trailing edge appears tangent to the Sun's western rim, initiating the ring of sunlight completely around the Moon. In an *annular eclipse*, **third contact** occurs when the Moon's leading edge appears tangent to the Sun's eastern rim, ending the ring of sunlight. **Fourth contact** (for all eclipses) occurs when the trailing edge of the Moon appears tangent to the eastern rim of the Sun, concluding the eclipse. Note that a partial eclipse has only first and fourth contacts (no second or third contacts). These same four contact points are used to describe lunar eclipses, planet transits across the face of the Sun, satellite transits across the face of a planet, and binary star transits across the face of one another.

core (of the Sun) The central regions of the Sun where it produces its energy in a nuclear fusion reaction by converting hydrogen into helium at a temperature of 27 million °F (15 million °C).

corona The rarefied upper atmosphere of the Sun that appears as a white halo around the totally eclipsed Sun. Speeds of atomic particles in the corona give it a temperature sometimes exceeding 2 million °F (1.1 million °C).

coronagraph A special telescope that produces an artificial solar eclipse by masking the Sun's apparent surface with an opaque disk; invented by Bernard Lyot in 1930.

coronal hole A region of the corona low in brightness and density. It is from coronal holes that solar particles escape most easily into space to become the solar wind.

coronal mass ejections Vast bubbles of gas ejected from the corona into space. The bubbles can expand to a size larger than the Sun. Coronal mass ejections seem to be the principal cause of aurorae and a significant hazard to electric power grids and spacecraft electronics when these high-velocity particles and ropes of magnetic fields hit the Earth.

degree of obscuration The fraction of the area of the Sun's disk obscured by the Moon at eclipse maximum, usually expressed as a percentage. (Degree of obscuration is not the same as the magnitude of an eclipse, which is the fraction of the Sun's diameter that is covered.)

descending node (of the Moon) The point on the Moon's orbit where it crosses the ecliptic (orbit of the Earth) going south.

diamond ring effect The stage of a solar eclipse when only a tiny sliver of the Sun's photosphere shines along the edge of the Moon as the corona appears. The diamond ring effect occurs in the seconds before and after totality.

draconic month The time it takes (27.21 days) for the Moon to orbit the Earth as measured from ascending node through descending node and back to ascending node again. Eclipses of the Sun and Moon can only take place near a node.

eclipse limit (for the Sun) The maximum angular distance that the Sun can be from a node of the Moon and still be involved in an eclipse seen from Earth. For partial eclipses, the limit ranges from 15°21' to 18°31' according to the varying angular sizes of the Moon and Sun due to the elliptical orbits of the Moon and Earth. For a central eclipse, the maximum and minimum limits are 11°50' and 9°55'.

eclipse maximum The moment and position in a solar eclipse when the cone of the Moon's shadow is aimed most nearly at the center of the Earth.

eclipse season The period of time in which the apparent motion of the Sun places it close enough to a node of the Moon so that an eclipse is possible. The Sun crosses the ascending and descending nodes of the Moon in a period of 346.62 days, so eclipse seasons occur about 173.3 days apart. Depending on where the Sun is on the ecliptic (how fast the Earth is moving), a solar eclipse season may last from 31 to 37 days.

eclipse year The time (346.62 days) it takes for the apparent motion of the Sun to carry it from ascending node to the descending node and back to the ascending node of the Moon.

ecliptic The apparent annual path of the Sun around the star field as seen from the Earth as the Earth orbits the Sun in the course of a year. (Thus the ecliptic is the plane of the Earth's orbit around the Sun.) The Sun's apparent path is called the ecliptic because all eclipses of the Sun and Moon occur on or very close to this track in the sky.

exeligmos An eclipse repetition cycle of 54 years 34 days, equal to 3 saros cycles and often called the triple saros. After one exeligmos cycle, a solar eclipse returns to almost the same longitude, but occurs about 600 miles (1,000 kilometers) north or south of its predecessor.

filament A dark threadlike feature seen against the face of the Sun. A filament is a **prominence** seen from the top rather than at the edge of the Sun.

flare Intense brightening in the upper atmosphere of the Sun which erupts vast amounts of charged particles into space. Flares can reach temperatures of 36 million °F (20 million °C).

greatest eclipse The moment and position in a solar eclipse when the Moon is apparently largest in comparison to the Sun, so that the duration of the total phase of a total eclipse is *longest* and the duration of the annulus (ring of sunlight around the Moon) in an annular eclipse is *shortest*.

hybrid eclipse A solar eclipse that begins as an annular eclipse, changes to a total eclipse along its path, and then returns to annular before the end of the eclipse path. Rare hybrids start as total eclipses and end as annulars or start as annulars and end as totals. Also called an **annular–total eclipse**.

inex A period of 10,571.95 days (29 years less 20.1 days) after which another eclipse of the Sun or Moon will occur (although not of the same type, such as total). This period equals 358 synodic months and 388.5 draconic months.

lunation The time it takes (29.53 days) for the Moon to complete a phasing cycle (also called a synodic period).

magnitude (of a solar eclipse) The fraction of the apparent diameter of the solar disk covered by the Moon at eclipse maximum. Eclipse magnitude is usually expressed as a decimal fraction: below 1.000 is a partial eclipse; 1.000 or above is a total eclipse. (The magnitude of a solar eclipse is not the same as the degree of obscuration, which is the percentage of the area of the Sun's disk that is covered.)

maximum eclipse The moment in a solar eclipse when the shadow of the Moon passes closest to the center of the Earth. This is also the instant when the greatest fraction of the Sun's disk is obscured.

mid-eclipse The instant in a central solar eclipse halfway between second and third contacts.

new moon The phase of the Moon when it is most nearly in conjunction with the Sun (also called dark-of-the-Moon). Solar eclipses can occur only at new moon. (In ancient times, new moon had a different meaning: the crescent Moon when it became visible after dark-of-the-Moon.)

nodes The two points at which the orbit of a celestial body crosses a reference plane. The Moon crosses the orbital plane of the Earth (the ecliptic) going northward at the ascending node and going southward at the descending node.

partial eclipse (of the Sun) An eclipse in which only part of the Sun's disk is not covered by the Moon.

penumbra The portion of a shadow from which only part of the light source is occulted by an opaque body. Seen from outside the shadow region, the penumbra is a fuzzy fringe to the dark umbra that declines in darkness outward from the umbra. A region experiencing a partial solar eclipse lies in the penumbra of the Moon.

perigee The point for any object orbiting the Earth where it is closest to the Earth.

perihelion The point for any object orbiting the Sun where it is closest to the Sun.

photosphere The apparent "surface" of the Sun. It is actually a layer of hot gases only about 300 miles (500 kilometers) thick where the Sun's atmosphere changes from opaque to transparent, and visible light escapes from the Sun. The temperature of the photosphere is about 10,000 °F (5,500 °C).

prominence An arch or filament of denser gas in the Sun's corona, shaped by the magnetic field of the Sun. Some prominences rise but most are descending, as if raining.

regression of the nodes The westward shift of the Moon's nodes along the ecliptic due to tidal forces on the Moon's orbit exerted by the Sun and Earth. The regression of the nodes is responsible for the eclipse year being 18.62 days shorter than the seasonal (tropical) year. The nodes complete a westward regression entirely around the ecliptic in 18.6 years.

saros An eclipse cycle of 6,585.32 days (18 years 111/3 days or 18 years 101/3 days if five leap years occur in the interval) in which an eclipse will occur that is very similar to the one that preceded it. The saros results from the near equivalence of 223 synodic months, 19 eclipse years, and 239 anomalistic months.

shadow bands Faint flickers or ripples of light sometimes seen on the ground or buildings shortly before or after the total phase of a solar eclipse. Shadow bands are caused by light from the thin crescent of the Sun passing through parcels of rising and falling air that have different densities and hence act as lenses to bend the light continuously in varying amounts.

solar constant The amount of power from the Sun falling on an average square meter of the Earth's surface (1.35 kilowatts).

solar wind A stream of charged particles (mostly protons, electrons, and helium nuclei) ejected from the Sun which flows by the Earth at 720,000 to 1.8 million miles per hour (320 to 800 kilometers per second). When enhanced by flares, the particles collide with molecules in the Earth's upper atmosphere so intensely that they cause the upper atmosphere to glow by fluorescence in displays of the aurora (the northern and southern lights).

spectroheliosope A solar spectroscope that blocks unwanted colors so that an observer can view the Sun in the light of one spectral line at a time; invented by Jules Janssen in 1868.

spectroscope A device (usually employing a prism or a diffraction grating) to spread out a beam of light into its component wavelengths for study. Spectroscopy can reveal the composition, temperature, radial velocity, rotation, magnetic fields, and other features of a light source.

spicule A jetlike spike of upward-moving gas in the chromosphere of the Sun. Viewed near the edge of the Sun, spicules resemble a forest. Each spicule lasts 10 minutes or so and ejects material into the corona at speeds of 12–19 miles per second (20–30 kilometers per second).

sunspot A darker area in the Sun's photosphere where magnetic fields are very strong. The temperatures of sunspots are 2,500 to 3,600 °F (1,400 to 2,400 °C) cooler than their surroundings, making them appear darker. Sunspots can last for a day up to several months.

sunspot cycle A period averaging 11.1 years in which the number of sunspots increases, decreases, and then begins to increase again.

synodic month The period of time (29.53 days) required for the Moon to orbit the Earth and catch up with the Sun again. Because the Moon's position with respect to the Sun determines the phase of the Moon, the synodic period is the time required for a complete set of phases by the Moon.

transition region The thin, irregular layer that separates the chromosphere from the corona. In this layer the temperature rises suddenly from about 7,200 °F (4,000 °C) to about 1.8 million °F (1 million °C). The transition region is of variable thickness, sometimes no more than tens of miles.

tritos A period of 3,986.6295 days (11 years less 31 days) after which another eclipse will occur (although not the same type, such as total). This period, less accurate than the saros or inex, equals 135 synodic months, 146.5 draconic months, and roughly 144.5 anomalistic months.

total eclipse (of the Sun) An eclipse in which the angular size of the Moon is sufficient to totally cover the disk of the Sun. In a total eclipse, the umbral shadow of the Moon touches the surface of the Earth.

umbra The central, completely dark portion of a shadow from which all of the light source is occulted by an opaque body. A region experiencing a total solar eclipse lies in the umbra of the Moon.

Selected Bibliography

Allen, David; and Carol Allen. *Eclipse*. Sydney; Boston: Allen & Unwin, 1987.

Arago, François. *Popular Astronomy*. 2 volumes. Translated by W. H. Smyth and Robert Grant. London: Longman, Brown, Green, Longmans, and Roberts, 1858.

Ashbrook, Joseph. *The Astronomical Scrapbook*. Edited by Leif J. Robinson. Cambridge: Cambridge University Press; Cambridge, Massachusetts: Sky Publishing, 1984.

Astrophotography Basics. (Kodak Publication P-150.) Rochester, N.Y.: Eastman Kodak, 1988.

Aveni, Anthony F. *Skywatchers of Ancient Mexico*. Austin: University of Texas Press, 1980.

Baily, Francis. "On a Remarkable Phenomenon that Occurs in Total and Annular Eclipses of the Sun." *Memoirs of the Royal Astronomical Society*. Volume 10, 1838, pages 1–40.

Baily, Francis. "Some Remarks on the Total Eclipse of the Sun, on July 8th, 1842." *Memoirs of the Royal Astronomical Society*. Volume 15, 1846, pages 1–8.

Brewer, Bryan. *Eclipse*. Seattle: Earth View, 1978; 2nd edition 1991.

Bruce, Ian. *Eclipse: An Introduction to Total and Partial Eclipses of the Sun and Moon*. Harrogate, England: Take That, 1999.

Brunier, Serge; and Jean-Pierre Luminet. *Glorious Eclipses: Their Past, Present, and Future*. Translated by Storm Dunlop. Cambridge: Cambridge University Press, 2000.

Chambers, George F. *The Story of Eclipses*. Library of Valuable Knowledge. New York: D. Appleton, 1912.

Clerke, Agnes M. *A Popular History of Astronomy During the Nineteenth Century*. 4th edition. London: A. and C. Black, 1902.

Couderc, Paul. *Les éclipses*. (Que sais-je? series no. 940.) Paris: Presses Universitaires de France, 1961.

Covington, Michael. *Astrophotography for the Amateur*. Cambridge: Cambridge University Press, 1988.

Douglas, Allie Vibert. *The Life of Arthur Stanley Eddington*. London: T. Nelson, 1956.

Dyson, Frank; and Richard v.d. R. Woolley. *Eclipses of the Sun and Moon*. Oxford: At the Clarendon Press, 1937.

Espenak, Fred. *Fifty Year Canon of Solar Eclipses: 1986–2035*. Washington, D.C.: NASA; Cambridge, Massachusetts: Sky Publishing, 1987. NASA Reference Publication 1178 Revised.

Espenak, Fred; and Jay Anderson. "Predictions for Total Solar Eclipses of 2008, 2009 and 2010." *Proceedings of IAU Symposium 233: Solar Activity and Its Magnetic Origin*. Volker Bothmer and Ahmed Abdel Hady, editors. Pages 34–41. Cambridge: Cambridge University Press, 2006.

Espenak, Fred; and Jay Anderson. *Total Solar Eclipse of 1999 August 11*. NASA Reference Publication 1398. Greenbelt, Maryland: NASA Goddard Space Flight Center, 1997.

Espenak, Fred; and Jay Anderson. *Total Solar Eclipse of 2008, August 1*. NASA Technical Publication 2007–214149. Greenbelt, Maryland: NASA Goddard Space Flight Center, 2007.

Espenak, Fred; and Jay Anderson. *Total Solar Eclipse of 2009, July 22*. NASA Technical Publication 214169. Greenbelt, Maryland: NASA Goddard Space Flight Center, 2008.

Espenak, Fred; and Jean Meeus. *Five Millennium Canon of Solar Eclipses: –1999 to +3000 (2000 BCE to 3000 CE)*. NASA Technical Publication 2006–214141. Greenbelt, Maryland: NASA Goddard Space Flight Center, 2006.

Fiala, Alan D.; James A. DeYoung; and Marie R. Lukac. *Solar Eclipses, 1991–2000*. (U.S. Naval Observatory circular 170.) Washington, D.C.: U.S. Naval Observatory, 1986.

Flammarion, Camille. *The Flammarion Book of Astronomy*. Edited by Gabrielle Camille Flammarion and André Danjon. Translated by Annabel and Bernard Pagel. New York: Simon and Schuster, 1964.

Francillon, Gérard; and Patrick Menget, editors. *Soleil est mort: l'éclipse totale de soleil du 30 Juin 1973*. Nanterre: Laboratoire d'ethnologie et de sociologie comparative (Récherches thématiques, 1), 1979.

Friedman, Herbert. *Sun and Earth*. New York: W. H. Freeman (Scientific American Library), 1986.

Guillermier, Pierre; and Serge Koutchmy. *Total Eclipses: Science, Observations, Myths and Legends*. Translated by Bob Mizon. Berlin: Springer-Verlag and Chichester, United Kingdom: Praxis, 1999.

Golub, Leon; and Jay M. Pasachoff. *The Solar Corona*. Cambridge: Cambridge University Press, 1997.

Harrington, Philip S. *Eclipse! The What, Where, When, Why, and How Guide to Watching Solar and Lunar Eclipses*. New York: John Wiley & Sons, 1997.

Harris, Joel; and Richard Talcott. *Chasing the Shadow*. Waukesha, Wisconsin: Kalmbach, 1994.

Held, Wolfgang. *Eclipses: 2005–2017.* Translated by Christian von Arnim. Edinburgh: Floris, 2005.

Johnson, Samuel J. *Eclipses, Past and Future; with General Hints for Observing the Heavens.* Oxford: J. Parker, 1874.

Joslin, Rebecca R. *Chasing Eclipses: The Total Solar Eclipses of 1905, 1914, 1925.* Boston: Walton Advertising and Printing, 1929.

Kippenhahn, Rudolph. *Discovering the Secrets of the Sun.* New York: John Wiley & Sons,1994.

Kudlek, Manfred; and Erich H. Mickler. *Solar and Lunar Eclipses of the Ancient Near East from 3000 B.C. to 0 with Maps.* Neukirchen-Vluyn, Germany: Butzon & Bercker Kevelaer, 1971.

Lang, Kenneth R. *Sun, Earth, and Sky.* New York: Springer, 1995.

Lang, Kenneth R.; and Owen Gingerich, editors. *A Source Book in Astronomy and Astrophysics, 1900–1975.* Cambridge, Massachusetts: Harvard University Press, 1979.

Lewis, Isabel M. *A Handbook of Solar Eclipses.* New York: Duffield, 1924.

Lewis, Isabel M. "The Maximum Duration of a Total Solar Eclipse." *Publications of the American Astronomical Society.* Volume 6, 1931, pages 265–266.

Little, Robert T. *Astrophotography: A Step-by-Step Approach.* New York: Macmillan, 1986.

Littmann, Mark; Fred Espenak, and Ken Willcox. *Totality: Eclipses of the Sun.* 3rd edition. New York: Oxford University Press, 2008.

Littmann, Mark; Ken Willcox, and Fred Espenak. *Totality: Eclipses of the Sun.* 2nd edition. New York: Oxford University Press, 1999.

Littmann, Mark; and Ken Willcox. *Totality: Eclipses of the Sun.* Honolulu: University of Hawaii Press, 1991.

Lovell, Bernard, editor. *Astronomy.* 2 volumes. The Royal Institution Library of Science. Barking, Essex; New York: Elsevier Publishing, 1970.

Marschall, Laurence A. "A Tale of Two Eclipses." *Sky & Telescope.* Volume 57, February 1979, pages 116–118.

Maunder, Michael. "Eclipse Chasing." (On eclipse photography.) In Patrick Moore, editor. *Yearbook of Astronomy.* New York: W. W. Norton, 1989, Pages 139–157.

Maunder, Michael; and Patrick Moore. *The Sun in Eclipse.* (Patrick Moore's Practical Astronomy Series). London: Springer-Verlag London, 1997.

Meadows, A. J. *Early Solar Physics.* Oxford: Pergamon Press, 1970.

Meeus, Jean. *Astronomical Algorithms.* Richmond, Virginia: Willmann-Bell, 1991.

Meeus, Jean. *Elements of Solar Eclipses: 1951–2200.* Richmond, Virginia: Willmann-Bell, 1989.

Meeus, Jean. *Mathematical Astronomy Morsels.* Richmond, Virginia: Willmann-Bell, 1997.

Meeus, Jean; Carl C. Grosjean; and Willy Vanderleen. *Canon of Solar Eclipses.* (Solar eclipses from 1898 to 2510 A.D.) Oxford: Pergamon Press, 1966.

Menzel, Donald H.; and Jay M. Pasachoff. *A Field Guide to the Stars and Planets.* 2nd edition. Boston: Houghton Mifflin, 1983.

Mitchell, Samuel A. *Eclipses of the Sun.* 5th edition. New York: Columbia University Press, 1951.

Mobberley, Martin. *Total Solar Eclipses and How to Observe Them.* Astronomers' Observing Guides. New York: Springer-Verlag New York, 2007.

Mucke, Hermann; and Jean Meeus. *Canon of Solar Eclipses: –2003 to +2526.* Vienna: Astronomisches Büro, 1983.

Needham, Joseph; and Wang Ling. *Science and Civilisation in China.* Volume 3: *Mathematics and the Sciences of the Heavens and the Earth.* Cambridge: At the University Press, 1959.

Neugebauer, Otto. *The Exact Sciences in Antiquity.* 2nd edition. Providence: Brown University Press, 1957.

Newton, Robert R. *Ancient Astronomical Observations and the Accelerations of the Earth and Moon.* Baltimore: Johns Hopkins Press, 1970.

Oppolzer, Theodor von. *Canon of Eclipses.* (Solar and lunar eclipses from 1207 B.C. to 2161 A.D.) Translated by Owen Gingerich. New York: Dover, 1962.

Osterbrock, Donald E.; John R. Gustafson; and W. J. Shiloh Unruh. *Eye on the Sky: Lick Observatory's First Century.* Berkeley: University of California Press, 1988.

Ottewell, Guy. *Astronomical Calendar.* Greenville, North Carolina: Universal Workshop, annually.

Ottewell, Guy. *The Under-Standing of Eclipses.* 3rd edition. Greenville, North Carolina: Universal Workshop, 2004.

Pasachoff, Jay M. *A Field Guide to the Stars and Planets.* 4th edition. Boston: Houghton Mifflin, 2000 (updated 2006).

Pasachoff, Jay M., Michael A. Covington. *The Cambridge Eclipse Photography Guide.* Cambridge: Cambridge University Press, 1993.

Pepin, R. O.; J. A. Eddy; and R. B. Merrill, editors. *The Ancient Sun: Fossil Record in the Earth, Moon and Meteorites.* Proceedings of the Conference on the Ancient Sun; Boulder, Colorado; October 16–19, 1979. New York: Pergamon Press, 1980.

Rao, Joe. *Your Guide to the Great Solar Eclipse of 1991.* Cambridge, Massachusetts: Sky Publishing, 1989.

Reynolds, Michael D.; and Richard A. Sweetsir. *Observe: Eclipses.* Washington, D.C.: Astronomical League, 1995.

Sands, Charles P. *Chasing the Shadow: The Dynamics of Eclipses.* Frederick, Maryland: PublishAmerica, 2005.

Sébillot, Paul Y. *Le folk-lore de France.* Volume 1: *Le ciel et la terre.* Paris: Librairie orientale & américaine, 1904.

Silverman, Sam; and Gary Mullen. "Eclipses: A Literature of Misadventures." *Natural History.* Volume 81, June–July 1972, pages 48–51, 82.

Steel, Duncan. *Eclipse: The Celestial Phenomenon Which Has Changed the Course of History.* London: Headline, 1999, and Washington, D.C.: National Academies Press, 2001.

Stegemann, Viktor. "Finsternisse." In Hanns Bächtold-Stäubli, editor. *Handwörterbuch des Deutschen Aberglaubens.* Volume 2. Berlin: W. de Gruyter, 1930. Columns 1509–1526.

Stephenson, F. Richard; and David H. Clark. *Applications of Early Astronomical Records.* Monographs on Astronomical Subjects, 4. New York: Oxford University Press, 1978.

Thompson, J. Eric S. *A Commentary on the Dresden Codex; A Maya Hieroglyphic Book.* Philadelphia: American Philosophical Society, 1972.

Todd, Mabel Loomis. *Total Eclipses of the Sun.* Revised. Boston: Little, Brown, 1900.

Wentzel, Donat G. *The Restless Sun.* Washington, D.C.: Smithsonian Institution Press, 1989.

Zirker, Jack B. *Total Eclipses of the Sun.* 2nd edition. Princeton: Princeton University Press, 1995.

Index